BUILDING 2000

Volume 2

Office Buildings, Public Buildings, Hotels and Holiday Complexes

Commission of the European Communities

BUILDING 2000

Volume 2

Office Buildings, Public Buildings, Hotels and Holiday Complexes

Edited by

C. DEN OUDEN

EGM engineering bv, Dordrecht, The Netherlands
Technical University, Delft, The Netherlands

and

T.C. STEEMERS

Commission of the European Communities,
Brussels, Belgium

SPRINGER-SCIENCE+BUSINESS MEDIA, B.V.

ISBN 978-94-010-5127-9 ISBN 978-94-011-2554-3 (eBook)
DOI 10.1007/ 978-94-011-2554-3

Publication arrangements by
Commission of the European Communities
Directorate-General Telecommunications, Information Industries and Innovation, Scientific and Technical Communications Service, Luxembourg

Originally published by Kluwer Academic Publishers in 1992

PREFACE

This is the second volume of BUILDING 2000, a pilot project of the Commission's R&D-programme 'Solar Energy Applications to Buildings' with the purpose of encouraging the adoption of solar architecture in large buildings.

In a first volume, a similar series of studies is presented for the building categories: SCHOOLS, LABORATORIES and UNIVERSITIES, and SPORTS AND EDUCATIONAL CENTRES.

In this second volume the results of the design studies illustrating passive solar architecture in buildings in the European Community are presented in particular for the building categories: OFFICE BUILDINGS, PUBLIC BUILDINGS and HOTELS AND HOLIDAY COMPLEXES.

There was an enthousiastic response from project teams responsible for the design of 32 large buildings with a total construction budget of more than 140 million ECU.

The willingness to improve their building concepts by collaborating with R&D-experts was encouraging to the Commission's action in this field.

These two books reflect the results of the exchange of information between the actual design practitioners and the European R&D-community.

Within the BUILDING 2000 programme 'Science and Technology at the Service of Architecture' became reality. This was not only realised by the various support activities initiated by BUILDING 2000, but also by the active exchange of ideas by architects and design team members with R&D-workers during the various workshops held within the BUILDING 2000 programme.

I highly recommend architects and engineers interested in passive solar architecture and modern daylighting approaches to study these final products of the BUILDING 2000 programme.

Dr. W. Palz
Directorate General XII for Science, Research and Development
Commission of the European Communities.

Contents

Volume 2

Contents

Volume 2

Commission of the European Communities

BUILDING 2000

OFFICE BUILDINGS

O1
University Faculty Building, Seville, Spain
O2
Head Office of Electricity Supply Company, Arnhem, NL
O3
Senior Citizen's Flats, Dordrecht, NL
O4
Centre for New Business, Reze, France
O5
Innercity Studios and Offices, West Berlin, Germany

PUBLIC BUILDINGS

P1
Magistrates Courthouse, Rotherham, UK
P2
Community Centre for Cultural Activities, Le Vigen, France
P3
Public Library, Newbury, UK
P4
Community and Educational Centre, Steyerberg, Germany
P5
Town Hall, Ilhavo, Portugal

HOTELS AND HOLIDAY COMPLEXES

H1
Tourist Complex, Agios Nicolas, Crete, Greece
H2
Hotel and Conference and Service Centre, Barcelona, Spain
H3
Tourist Hotel, Bethune, France
H4
Holiday Apartments, Gournes, Crete, Greece
H5
Tourist Complex, Chania, Crete, Greece
H6
Tourist Complex, Drios, Paros, Greece

Building 2000 is a series of design studies illustrating passive solar architecture in buildings in the European Community.

VOLUME 2 / **KLUWER ACADEMIC PUBLISHERS**

Commission of the European Communities

BUILDING 2000

OFFICE BUILDINGS

O1
University Faculty Building, Seville, Spain
O2
Head Office of Electricity Supply Company, Arnhem, NL
O3
Senior Citizen's Flats, Dordrecht, NL
O4
Centre for New Business, Reze, France
O5
Innercity Studios and Offices, West Berlin, Germany

Building 2000 is a series of design studies illustrating passive solar architecture in buildings in the European Community.

VOLUME 2 / **KLUWER ACADEMIC PUBLISHERS**

BUILDING 2000

Commission of the European Communities

Modular faculty building for University of Seville's new campus to be built on Expo 92 site.

Minimum energy consumption and low maintenance.

Adequate daylighting of all rooms.

Prevention of summer-time overheating and glare.

UNIVERSITY FACULTY BUILDING

SEVILLA/SPAIN

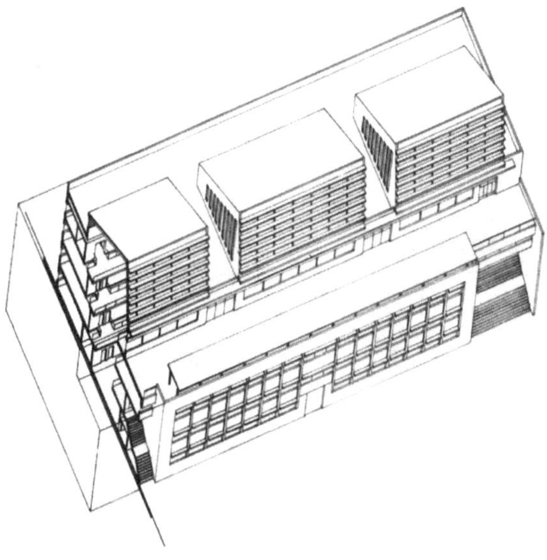

Building 2000 is a series of design studies illustrating passive solar architecture in buildings in the European Community.

ISSUE **01**

FEB 1991

PROJECT DESCRIPTION

Building Type

When the fair is over, the Expo 92 site will be transformed into a new campus for the University of Seville. This building is one of several being constructed initially as an office block for Expo 92 which will eventually be converted into accommodation for one of the university faculties. It contains teaching and administration accommodation, the usual service areas, shops and a bar and a room for use as a students' club.

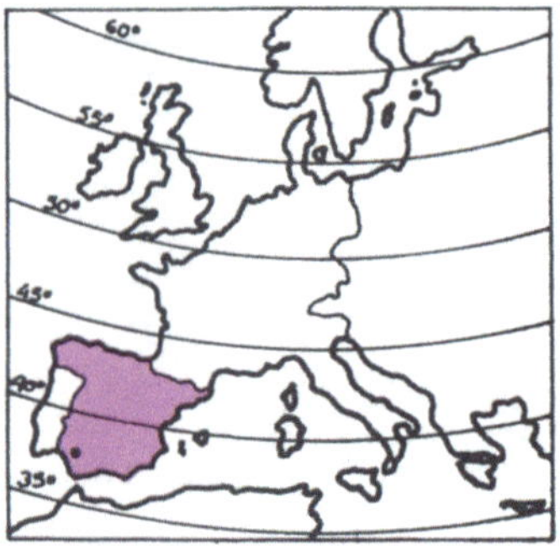

Figure 1. Site location

Location

The site (which is 7 m above sea level) is to the west of the city of Seville (latitude 37 ° 25' N, longitude 6 ° W) in the Cartuja area by the Guardalquivir river (see Figure 1).

Site Microclimate

The climate is characterized by extremes. There are cold periods and periods of overheating. The differences between daily minimum temperatures and daily maximum temperatures are considerable. Solar irradiation and degree day data are given in Figure 2 and relative humidity and mean ambient temperature data in Figure 3. These data, together with those in Table 1, were collected some 3 km north-west of the site.

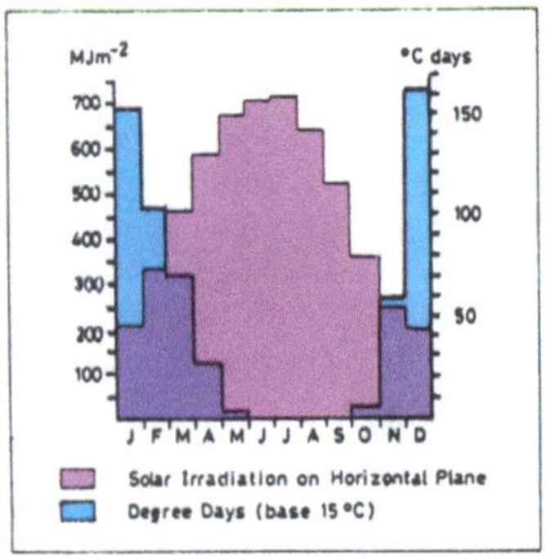

Figure 2. Solar irradiation and degree day data

Precipitation	586.8 l/m^2
Relative humidity	65%
Ambient temperature	17.3 °C
Average maximum temperature (July)	30.0 °C
Average minimum temperature (January)	4.9 °C
Prevailing wind direction	SW
Wind speed	3.2 m/s
Global irradiation on the horizontal	
in kWh/m^2	1579 kWh/m^2
in MJ/m^2	5686 MJ/m^2
Degree days (base 15 °C)	580
Sunshine hours	1475

Note: These data are annual averages unless indicated otherwise

Table 1. Some key climate data

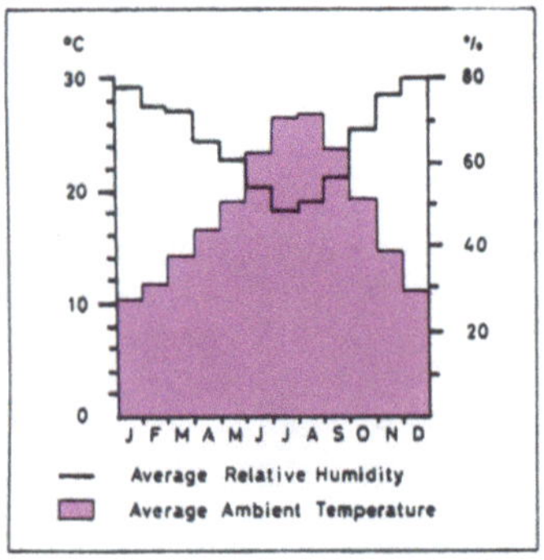

Figure 3. Relative humidity and ambient temperature data

Design and Construction Details

The building design is a modular one which can be modified to meet the specific needs of individual faculties. The arrangement of the individual faculty buildings on the site is shown in Figure 6.

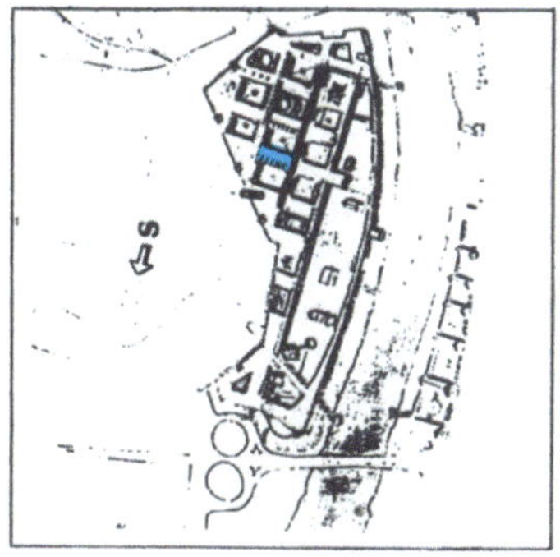

Figure 6. Arrangement of modular buildings on the site.

The main feature of each building is a central atrium which is divided into three independent units to facilitate control of heat and sound. The accommodation is arranged in two four-storey blocks on either side of the atria. The teaching block is on the north side and the administration and services block on the south side. Plans of the ground, first, second and third floors are shown in Figures 4, 5, 7 and 8. U-values of some key building elements are given in Table 2.

Roof	0.5
Wall	0.49
Glazing	4.0
Floor	0.35

Table 2. U-values (in $W/m^2 K$) of some key building elements

Because the teaching block requires better heating and lighting control than the administration block, it is arranged alongside the atria in a stepped fashion to give maximum exposure to the available solar radiation from the south (see Figure 9). The entry of the radiation into the rooms is controlled on a seasonal and daily basis by means of blinds and other shading devices. The ventilation system makes full use of the atrium. The classrooms have adequate daylighting with minimum glare.

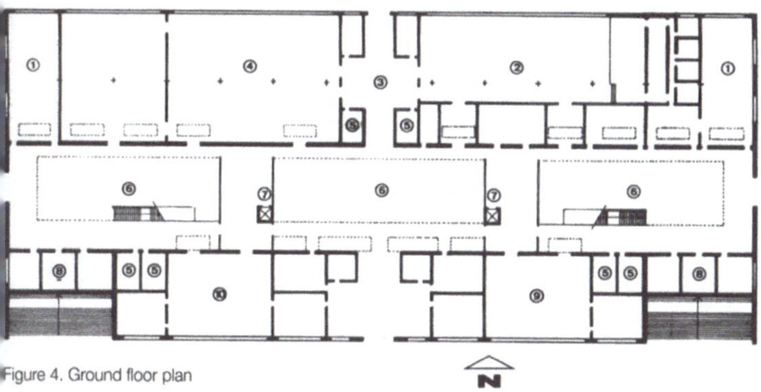

Figure 4. Ground floor plan

Key to ground floor plan:

1. Store
2. Auditorium
3. Hall
4. Library
5. Bathrooms
6. Atria
7. Elevators
8. Shops
9. Bar
10. Students' club

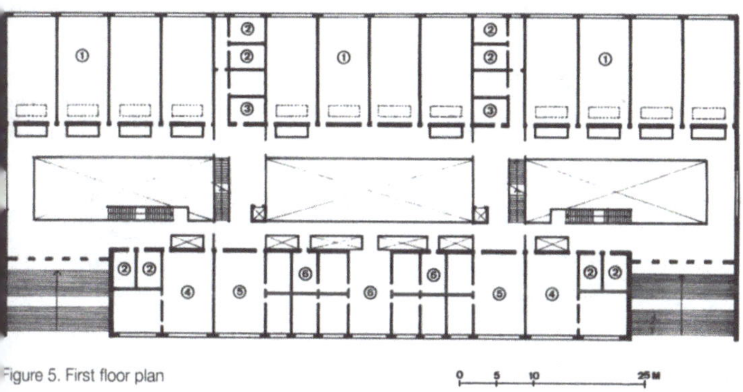

Figure 5. First floor plan

Key to first floor plan:

1. Classrooms
2. Bathrooms
3. Stores
4. Seminar rooms
5. Boardrooms
6. Manager and administration offices

Key to second floor plan:

1. Classrooms
2. Bathrooms
3. Stores
4. Seminar rooms
5. Research departments

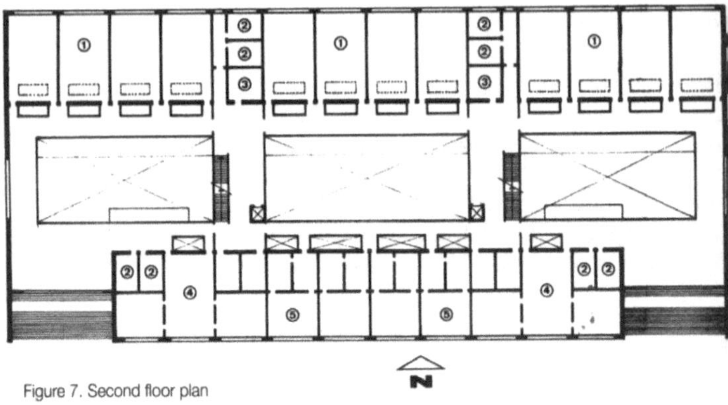

Figure 7. Second floor plan

Key to third floor plan:

1. Professors' offices and subject departments
2. Bathrooms
3. Stores
4. Terrace

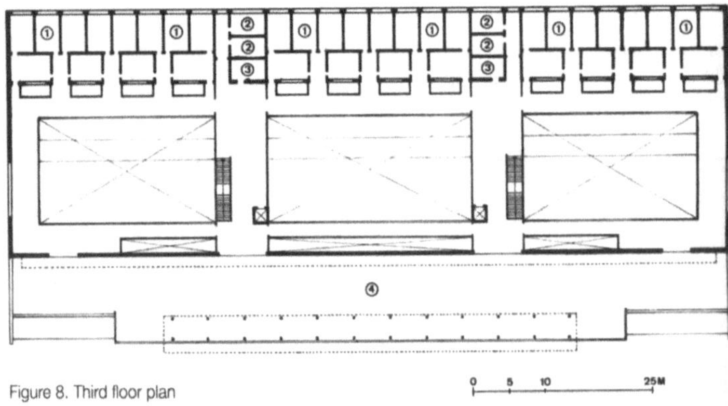

Figure 8. Third floor plan

Key to cross section:

1. Atrium
2. Professors' offices
3. Classrooms
4. Library
5. Research departments
6. Administration offices
7. Bar

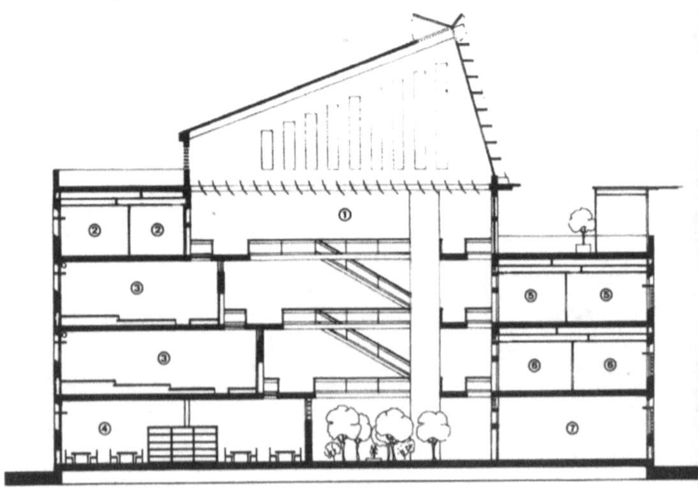

Figure 9. Cross section

DESCRIPTION OF PASSIVE SOLAR FEATURES/COMPONENTS

Introduction

The object has been to create an energy-efficient building which is thermally, acoustically and visually comfortable and conducive to study and research. Public spaces have been arranged to favour relationships between users. Passive solar features have been used with the aim of providing each room with the appropriate amount of cooling in spring and summer, heating in winter and lighting throughout the year. Care has been taken to achieve a level of ventilation in each space which is right for its occupancy. Unwanted heat gains and losses have been reduced. A good level of daylighting has been provided in work spaces and recreation areas.

Control of Thermal Environment through Atria

The three independent atria are the main means of providing environmental control. Through them, the amount of solar radiation and ventilation inside the building can be regulated. Different devices enable the thermal needs of the building to be met whatever the weather.

Mode of Operation in Summer

In Seville, the extremely high temperatures reached in summer are the major problem. In this building, summer-time cooling is achieved by shading, planting on the ground floor and controlled ventilation. The fixed blinds on the atrium glazing prevent entry of direct radiation at this time of year. Good ventilation is achieved using the Venturi effect: hot air rises in the atrium because of density differences and a hatch at the atrium ridge helps create air flow throughout the building (see Figure 10).

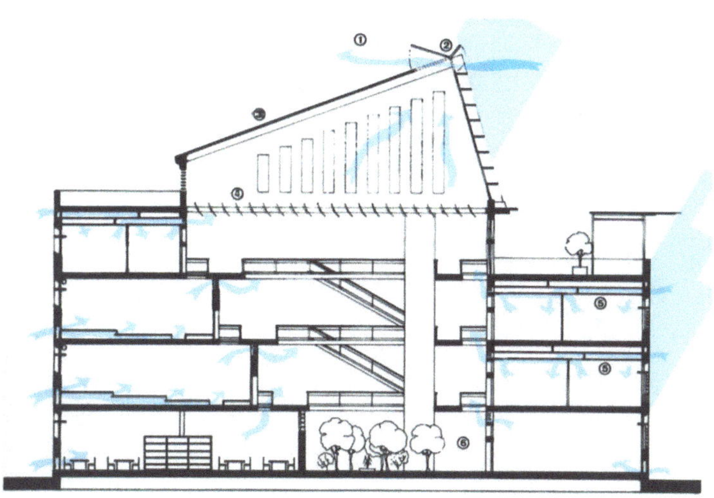

Key:

1. Prevailing wind
2. Large opening at ridge
3. Opaque roof
4. In summer, the cloth mat is removed
5. Split ceiling void vent
6. Cool zone - planting and fountains

Figure 10. Mode of operation in summer

Mode of Operation in Winter

In winter, the fixed blinds on the atrium roof permit entry of direct solar radiation. The movable mat on the bottom of the atrium roof can, according to its angle of incidence, either allow this radiation to enter the teaching block when it is needed on cold winter days or serve as a light diffuser on milder days. The mode of operation in winter is shown in Figure 11.

Daylighting

The main problem has been to provide adequate lighting of the rooms in the north block. This has been achieved by means of diffuse light from the windows on the north facade and direct sunlight through the skylight at the south end of each room, above the lecturer's desk. Entry of the direct light is controlled to prevent glare. Above the windows, light shelves reflect light towards the ceiling and help provide uniform lighting throughout the room.

Key:

1. Fixed overhangs
2. Glazing
3. Removable cloth mat to diffuse light when needed
4. Light shelves and movable louvers

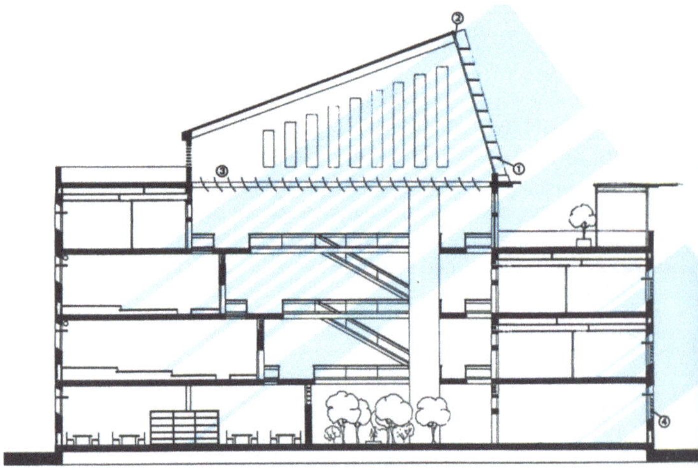

Figure 11. Mode of operation in winter

ENERGY CALCULATIONS PERFORMED AND DESIGN TOOLS USED

Design Tools

The ventilation and daylighting systems were designed with the help of various design tools. Air flow modelling and testing was carried out by Cambridge Architectural Research. Daylighting was studied using the split flux method plus direct sunlight and the results checked using a SERI protractor. Some of the results of these studies are given below.

Library

With maximum daytime temperatures in June around 31 °C, a high ventilation rate is needed in areas such as the library, lecture rooms and classrooms which are subject to direct sunlight. It was decided that the only feasible cooling strategy was to provide air flow rates of 0.5-1.0 m/s in these rooms in summer to give direct physiological cooling. The library, which is on the north side, receives a good level of daylight but does not require shading: the high ceilings (4.5 m) allow good light penetration. The solar radiation and air flows in that room are shown in Figure 12. The main seating and reading areas are arranged on the perimeter and are lit naturally. The book stocks are in the centre and are artificially lit only when and where it is necessary. The outdoor space to the north of the library can be made secure and developed as a study area. Pergolas and other landscape features can be used there to further minimize the risk of overheating in the library.

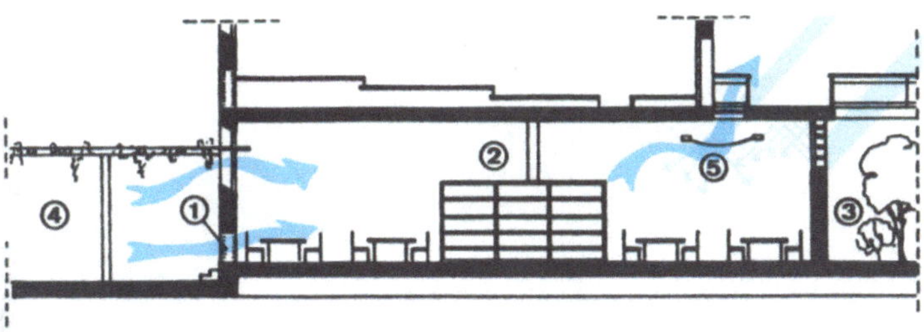

Key:

1. Louvers fixed for good ventilation
2. Bookstack
3. Cool zone
4. Secure open air reading area
5. Light diffusers

Figure 12. Ventilation and daylighting of the library

Lecture Rooms

Lecture rooms present a challenge to any architect wanting to use passive solar design - partly because they are densely occupied and therefore have high cooling loads and partly because they have to be blacked out sometimes for the showing of slides, etc., but still require ventilation. In this building, it was decided to bring fresh air in from the north wall under a raised, stepped floor and vent it into the atrium via light-proof louvers at the lecturer's end of the room (see Figure 13). In addition, there is glazing which are capable of being blacked out but which can be opened to provide extra ventilation when the blackout is not required. Bringing in air through the floor void has the advantage of providing fresh air where it is needed, close to the occupants, and creating perceptible air movement.

Calculations showed that, with an area of 4.05 m^2 at both the inlet and the outlet positions, the natural ventilation would be enough to keep the temperature in the lecture rooms down to 3 °C above ambient on the hottest June day when the ambient temperature is 31 °C but the relative humidity is 46% and the wet bulb temperature is 23 °C. This gives a corrected effective temperature of 29 °C at 0.5 m/s which is just 2 °C above the normal comfort level. As this represents the worst possible case (the building is not normally used in July and August), it is regarded as being just about acceptable.

In winter, ventilation rates can be much lower - the minimum required to provide enough fresh air.

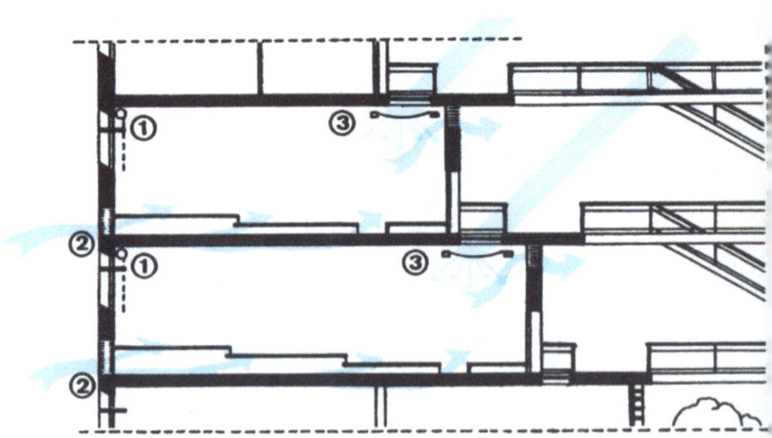

Key:

1. Blackout when needed
2. Light-proof mechanical louvers
3. Light diffusers

Figure 13. Ventilation and daylighting of the lecture rooms

Offices

The studies showed that the perimeter offices and other rooms on the south side of the building require effective shading and operable windows to create cross ventilation. It was found that a shading system which combines fixed light shelves with movable louvers was effective. The offices in this block which face the atrium needed no shading but did require openable windows. To achieve the cross ventilation, a false ceiling void (split as indicated in Figure 14) provides an exhaust for the perimeter offices and a means of supplying inlet air for those facing the atrium.

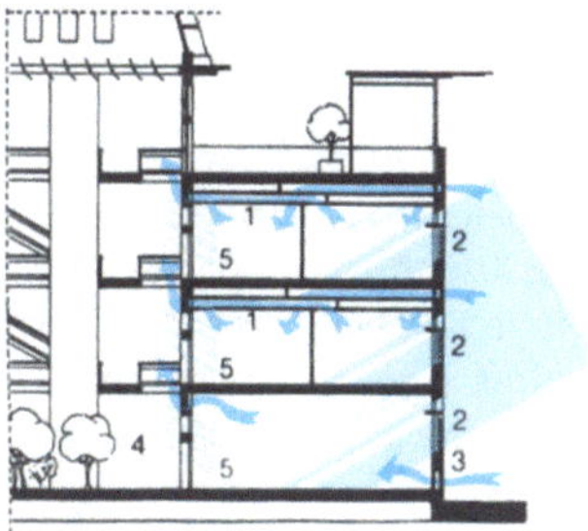

Figure 14. Ventilation and daylighting of the offices, club and bar in the south block

Key:

1. Split ceiling void vent showing supply of air to offices facing atrium
2. Light shelves and operable louvers
3. Restricted openings to maintain extraction from north rooms
4. Cool zone
5. Diffused light

Role of Atria in Ventilation and Cooling

The atria not only have an important role in daylighting but also have a part to play in driving ventilation air through the building. Each atrium will, however, only extract exhaust air from the adjacent rooms if the temperature of the air in the atrium is above ambient. Therefore the design is such that the unoccupied upper portion of the atrium is allowed to reach 5-7 °C above ambient and is vented at a relatively high rate. This enables it to serve as an exhaust for the used air from the library, lecture rooms and offices. The atrium floor is planted extensively and contains fountains to give a cooling effect. Solar gains are minimized by shading the atrium roof. This causes the ventilation rate in the bottom zone of the atrium to be relatively low. To stabilize the cool air in this zone, openings at the bottom of the south and north external walls are kept to a minimum. The vent openings to the library are at the first floor level and have been designed to direct air upwards. At the time of writing this brochure, no detailed analysis had been carried out. It seems likely, however, that it will be possible to reduce air temperatures at the bottom of the atrium to at least halfway to the wet bulb temperature and that there will be further cooling from the cold collected by massive surfaces during night-time ventilation. It is possible that the temperature will be maintained at a level at least 5 °C below maximum external daytime temperatures. The atrium is divided into three to avoid problems related to air flow stability, fire safety and noise. The provision of internal shades from cloth sails on mechanical rollers has obviated the need for movable shades on the south side of the atrium. Fixed overhangs give seasonal control.

Daylighting Analysis

The daylighting analysis was carried out for a point 2 m above floor level in the first-floor south gallery of the atrium (point P in Figure 15). This was chosen for the study because it was judged to have intermediate to bad daylighting compared with other places in the building. The daylight factor at this point due to the light entering through the south-oriented rooflight was 2.84% and that from the light from the east and west windows was 2.2%.

In December, January and February, the sky in Seville is overcast. For the winter months, therefore, the total illuminance on the vertical plane can generally be taken as 5,000 lux. In winter, the component for internal reflection is around 1%. Therefore the total lux at point P in winter would be (2.84 + 2.2 + 1)% x 5000 = 300 lux. When the transmittance of the glass and the dirt factor are taken into account, this reduces to 200 lux. In general, a figure in the range 300 to 200 lux can be said to be adequate. However, if the glass was not maintained in a clean condition a point in a classroom near P would probably be felt to have a fairly low lux. If direct sunlight enters the atrium, of course, values as high as 700 lux can be reached. Statistically, however, this is only likely to happen 56% of the working period and 200-300 lux should be taken as the "safe" winter figure. It is the result of a needed compromise because the building was designed to take advantage of sunny conditions but avoid overheating and over-illumination.

For the rest of the year, the sky in Seville is clear to very clear. The building, however, is designed so that it does not receive direct sunlight from April to September. A figure of 15,000 lux can therefore be taken on a south-oriented vertical plane and 12,000 lux on an east/west-oriented plane. The component from internal reflection at this time of year is 2%, mainly due to the ceiling. From these figures it can be calculated that the lux at point P from March to October/November is (2.84 + 2)% x 15,000 + 2.2% x 12,000 = 990 lux. This reduces to 631 lux when the transmittance of the glass and the dirt factor are taken into account. With this illuminance level, good daylighting is assured in the galleries and classrooms. In the upper galleries, indeed, there is a risk of over-illumination with glare and discomfort. Intelligent use of canvas or fabric louvers reduces this risk but care must be taken to ensure that, when these are used, enough light reaches the other parts of the building, particularly the north- facing galleries. To aid this, reflecting finishes and translucent fabrics are used within the building.

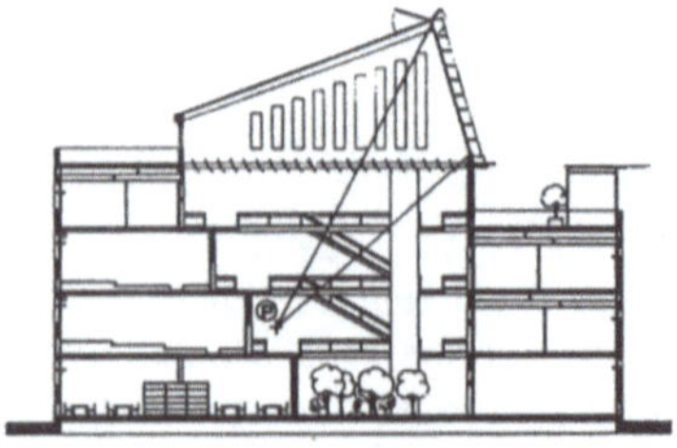

GENERAL DESIGN GUIDELINES/POINTS OF INTEREST RESULTING FROM THIS PROJECT

In warm climates the atrium is an excellent device for controlling indoor environments but it must be designed to induce ventilation by the stack effect in summer rather than heating by the greenhouse effect in winter.

The volume of the atrium must be kept small to ensure that environmental control takes place. It may sometimes be necessary to divide the atrium into several courtyards.

There must be a large volume of air above the upper walking level of the atrium which can be allowed to warm up above the comfort temperature to induce the stack effect.

In warm climates it is necessary to limit the amount of glazing in the atrium to prevent overheating in hot weather.

In the sloping lightweight roof used in this building, the aerodynamic design which generates positive and negative pressures at the ridge of the atrium and the openings to achieve a Venturi effect all improve air extraction.

To control entry of solar radiation into the building, it is better in inaccessible positions on the outside of the building to use fixed light shelves designed to operate seasonally. Indoors, in accessible places, light blinds can be used for adjustment on a daily basis.

When rooms are deep and therefore difficult to illuminate, it is a good idea to create stepped building levels. With these, daylight penetrates deeply into the rooms and the volume of the atrium is increased so that the lighting performance is better.

BUILDING 2000 Building 2000 brochures are published by Directorate General XII of the Commission of the European Communities to show how design studies can help architects and other building designers use passive solar principles to the best effect to produce attractive energy-efficient buildings. Each brochure describes studies carried out with the support of the Commission during the design phase of one of thirty-six non-domestic buildings in the EC Member States. The studies were on such topics as daylighting, heating, cooling, ventilation, comfort, control systems and urban design. They were carried out with the help of acknowledged European experts in these fields and drew heavily on lessons learned and techniques developed through the Commision's research and development programme on solar energy applications to buildings.

Commission of the European Communities/Directorate-General for Science, Research and Development

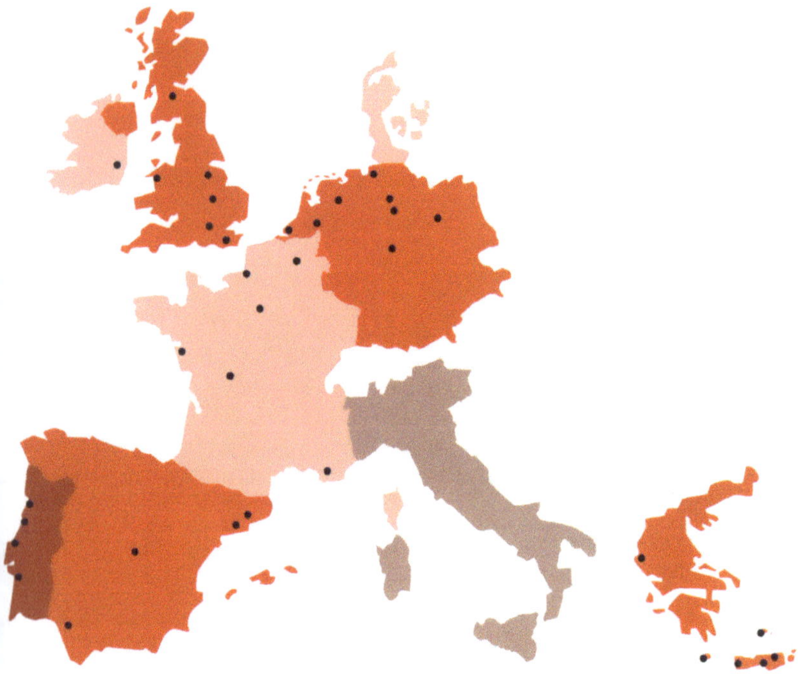

List of Design Team Participants and Advisers

Client
EXPO 92
Sociedad Estattal
Universidad de Sevilla,
Spain

Architects
Jaime Lopèz de Asiain
Alberto L Ballesteros Rodriguez
José M Cabeza Lainez
Seminario de Arquitectura
Reina Mercedes s/n
41012 Sevilla
Spain

Climate Data
Cátedra de Termodinámica y Fisicoquimica
Escuela Superior de Ingenieros Industrialles de Sevilla
Director: Valeriano Ruiz Hernández

Daylighting Data and Graphs
Cátedra de Instalaciones ETSAS
Director: Jaime Navarro Casas

Energy Consultants
Nick Baker
Cambridge Architectural Research Limited
6 Chaucher Road
Cambridge CB2 2EB
United Kingdom

This set of **Building 2000** brochures illustrates how architects and other building designers can successfully apply passive solar principles to produce energy-efficient buildings.

BUILDING 2000 Participants

Project Director
Theo C. Steemers

Coordinator
Cees den Ouden

Technical Steering Committee
Dean Hawkes
Nick Baker
Alex Lohr
Jean P. Lepoivre

Regional Liaison Agents
(D) Jörn Behnsen
(E) Vicente Sifre
(F) Michel Raoust
(GB) Alan Hildon
(GR) Matheus Santamouris
(P) Eduardo Maldonado

Further information or copies of the brochures can be obtained from prof. ir. Cees den Ouden, EGM Engineering BV, P.O. Box 1042, 3300 BA Dordrecht, The Netherlands.

BUILDING 2000

Commission of the European Communities

- office buiding and data processing centre
- atrium with glazed roof joins two parts of office building
- solar gains in atrium roof and bioclimatic walls
- shading devices prevent overheating
- heat recovery using thermal wheel and heat pump system

HEAD OFFICE OF ELECTRICITY SUPPLY COMPANY

ARNHEM/THE NETHERLANDS

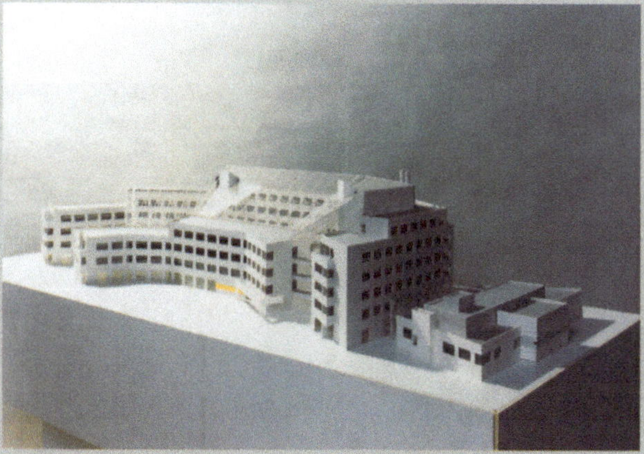

Building 2000 is a series of design studies illustrating passive solar architecture in buildings in the European Community.

ISSUE 02

FEB 1991

PROJECT DESCRIPTION

Building Type

This office complex is to be the new administrative headquarters of the Provinciale Gelderse Energie Maatschappij (PGEM), an electricity supply company. It will house the management, data processing, administrative and design departments. Work began on site in June 1988. It is expected that PGEM will move into the complex in the autumn of 1991.

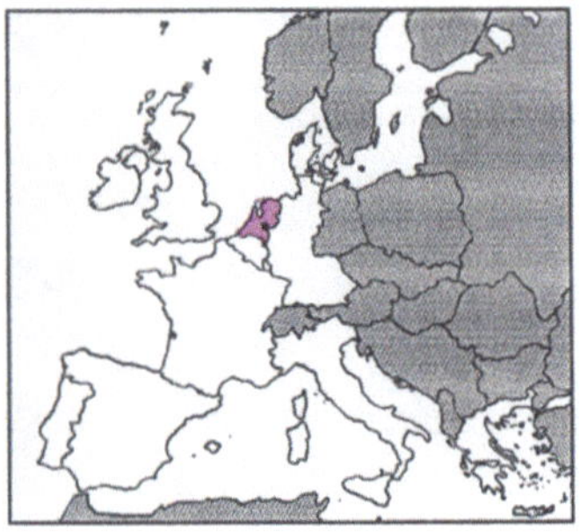

Figure 1. Location

Location

The complex is located on the outskirts of Arnhem (latitude 52 ° N) - see Figure 1. The site is 1 m above sea level in parkland on the slope overlooking the river Neder-Rijn. It is not far from the city centre and the main railway station.

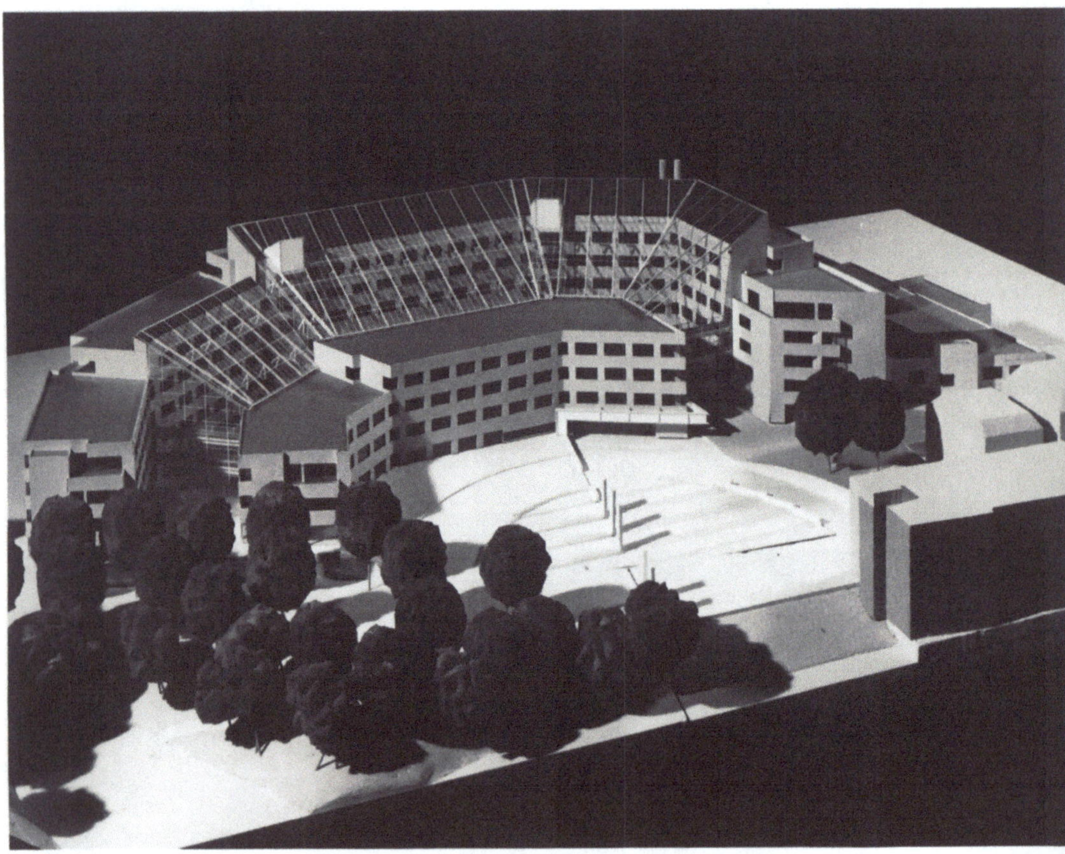

Figure 2. Model of the complex viewed from the south-west

Site Microclimate

The Netherlands are very flat with some land close to the North Sea being one or two metres below sea level. The climate is typical of the maritime type with a lot of rain throughout the country. In general, temperatures are relatively mild but inland they can become more severe in winter. Average temperature, solar irradiation, sunshine hour and degree day data for De Bilt (which is close to Arnhem) for the period 1961-1970 are given in Figures 3-7.

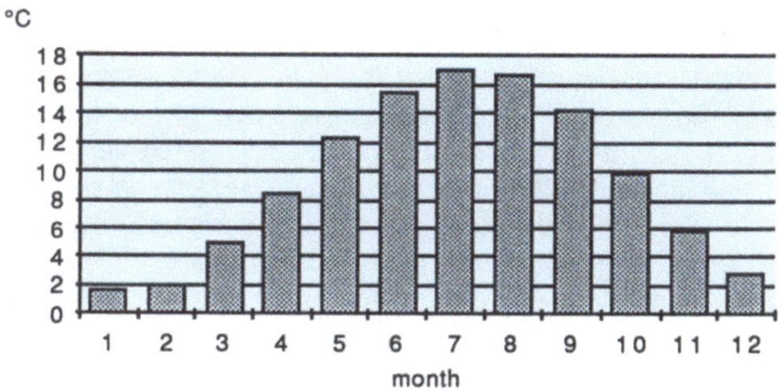

Figure 3. Average ambient temperatures. (The annual average is 9.34 ° C.)

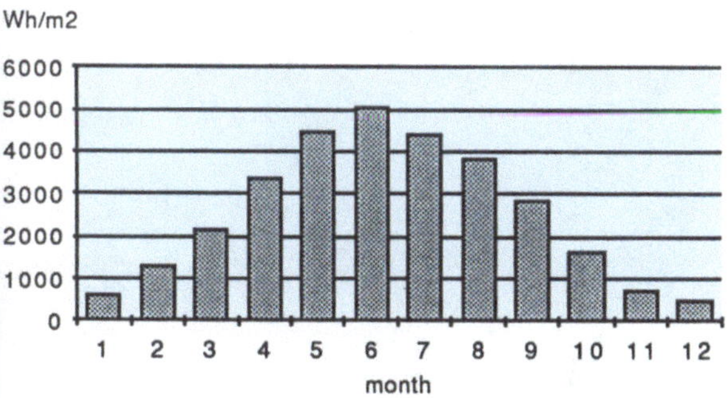

Figure 4. Mean daily solar irradiation data. (The annual average is 2.595 kWh/m^2 day (9.342 MJ/m^2 day).)

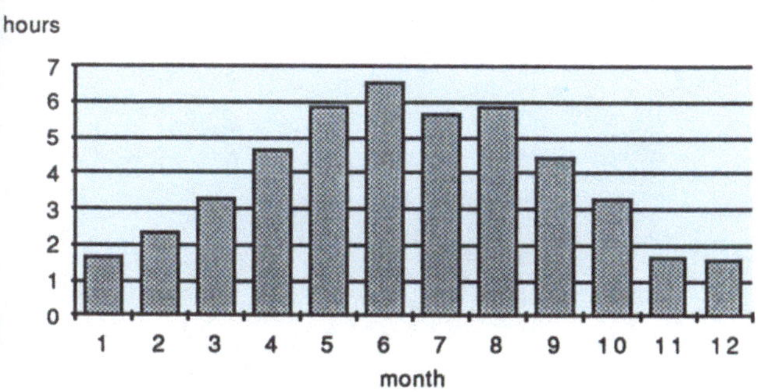

Figure 5. Mean daily sunshine hours. (The annual average is 4.0 hours a day.)

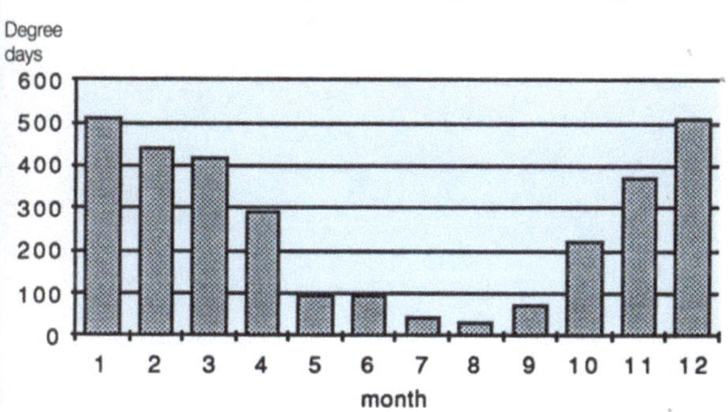

Figure 6. Degree days (base 18 ° C). (The total number of degree days (base 18 ° C) over the year is 3131.)

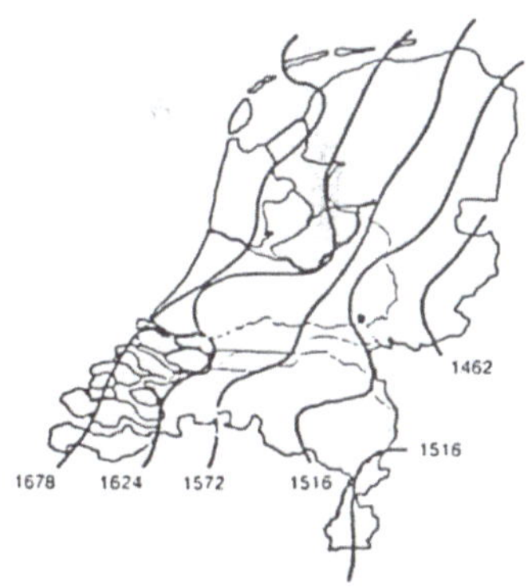

Figure 7. Average number of sunshine hours in The Netherlands

Design and Construction Details

The complex (which has a floor area of 31,000 m^2 and a volume of 125,000 m^3) consists of two south-facing office blocks of different height joined by an atrium with a transparent sloping roof plus a separate building for the data processing department. A model of the complex viewed from the south west is shown in Figure 2. The front, rear and east elevations are given in Figures 8, 9 and 10. The north-south cross section showing the atrium between the office blocks is illustrated in Figure 11. The east-west cross section is given in Figure 12. U-values of some building elements are given in Table 1 together with the solar shading factor of some of them.

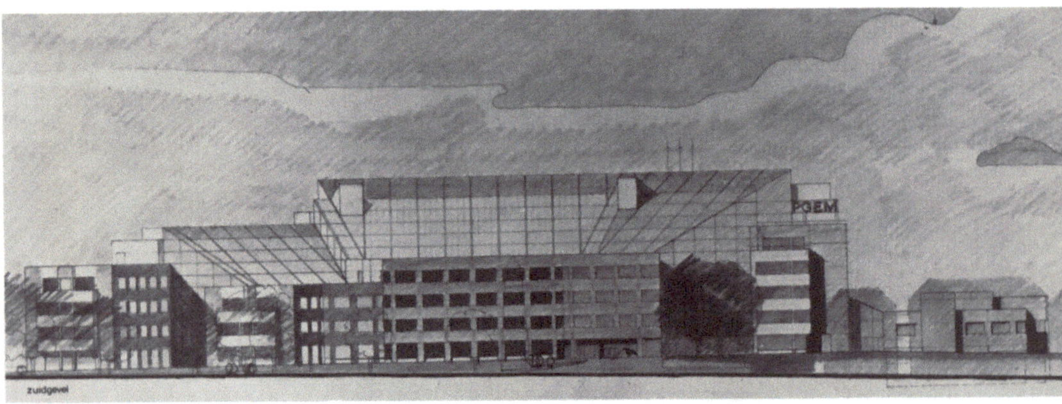

Figure 8. Front (south) elevation

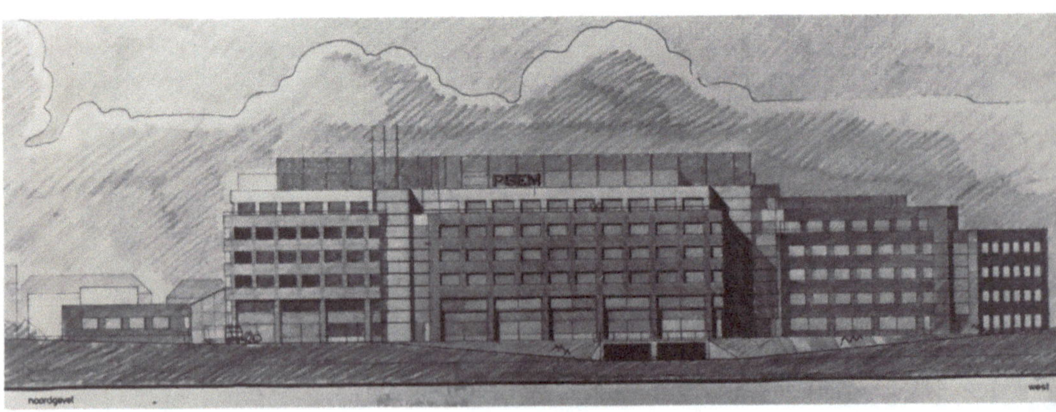

Figure 9. Rear (north) elevation

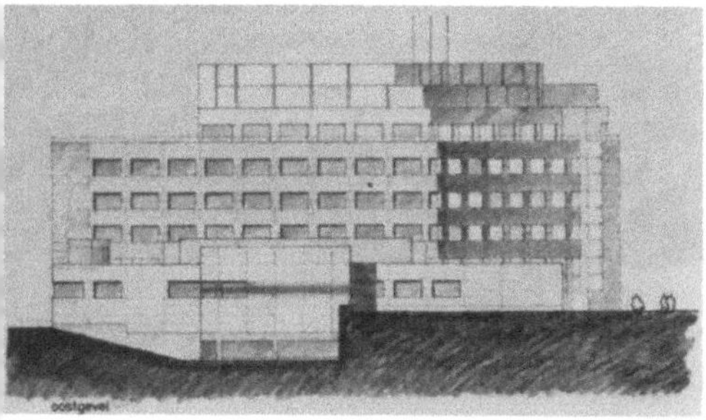

Figure 10. East elevation

	U-value ($W/m^2 K$)	Solar shading factor
Roof	0.5	-
Walls	0.6	0.2
Glazing	0.8	0.25-0.3
Floor	0.5	-

Table 1. U-values and solar shading factors of some building elements

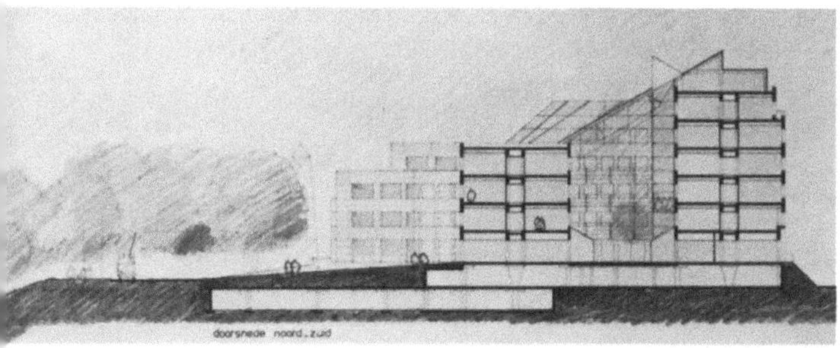

Figure 11. North-south cross section showing the atrium between the office blocks

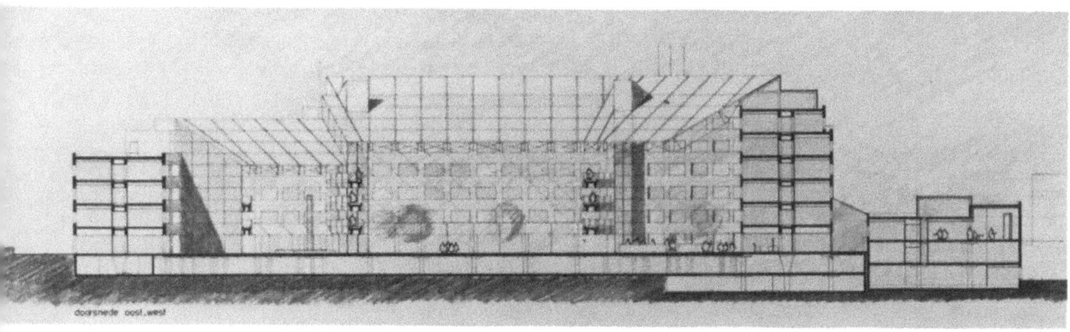

Figure 12. East-west cross section

DESCRIPTION OF PASSIVE SOLAR FEATURES/COMPONENTS

Introduction

The office blocks, which contain a number of passive solar and other energy saving features, are continuously ventilated with fresh air. The transparent atrium roof and bioclimatic walls on the south facade serve as solar collectors when needed. Both incorporate shading devices to prevent overheating. Exhaust air from the offices is circulated through this system and heat extracted using a thermal wheel (see Figure 14). In the data processing block, excess heat is recovered using a heat pump system.

Atrium Roof and Thermal Wheel

The transparent roof of the atrium is at an angle of approximately 30 and faces south. Its construction is shown in Figure 13. It can serve as a solar air collector and also as a shading device. As shown in Figure 14, exhaust air from the offices is partially extracted via the atrium and circulated through the atrium roof between an outer layer of double glazing and an inner layer of perforated shading plates. A thermal wheel extracts the useful heat.

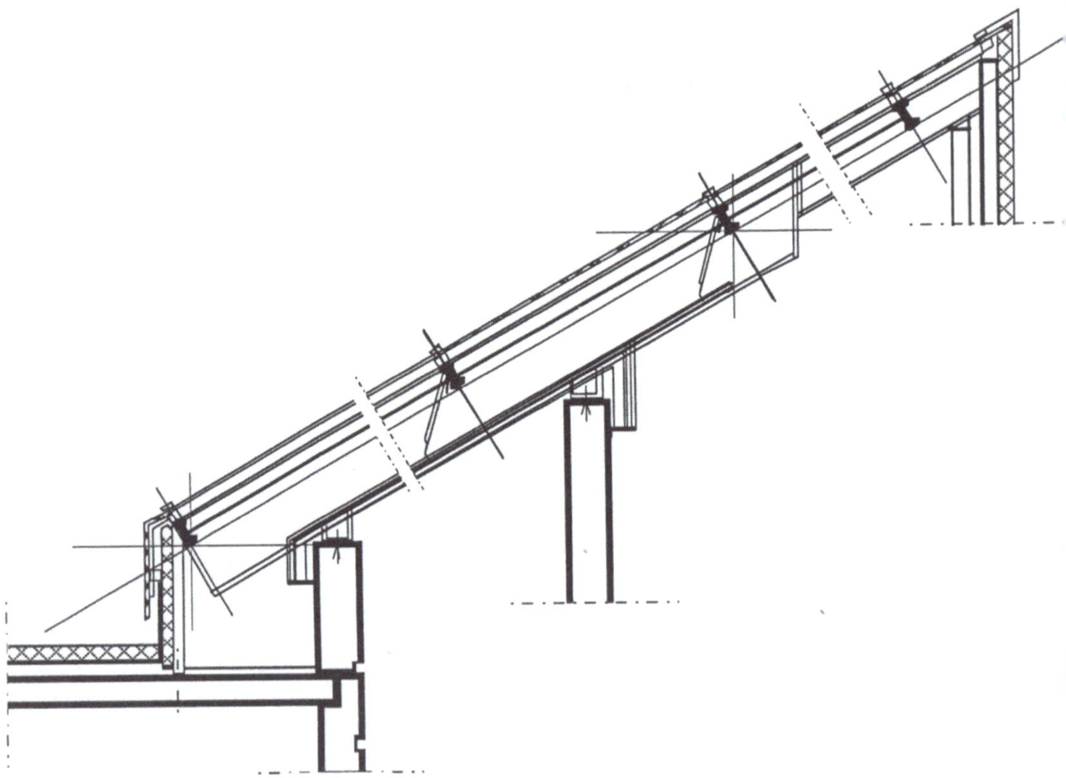

Figure 13. Atrium roof construction

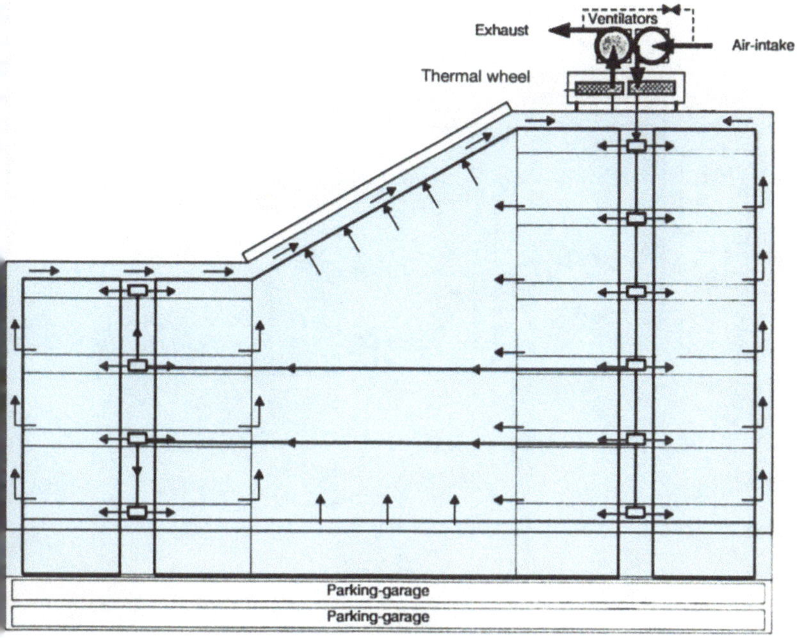

Figure 14. The offices' ventilation system

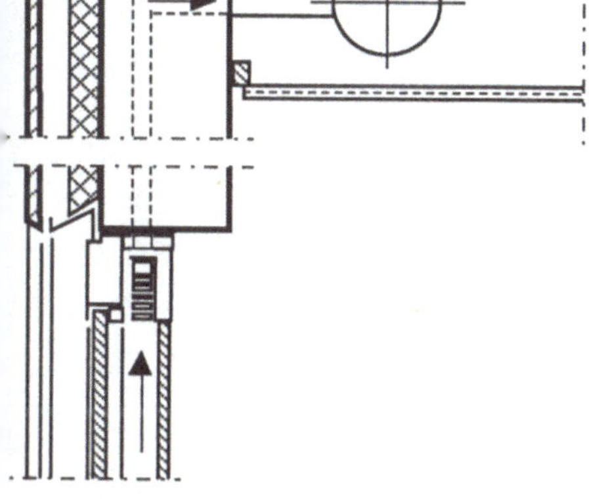

Figure 15. The bioclimatic wall

Bioclimatic Walls

The south-facing office facades consist of glazed cavity walls incorporating shading devices (see Figure 15). Along with the atrium roof, they form part of the building's exhaust air system. The air is circulated between the inner and outer skins. The walls' construction guarantees low heat losses.

Auxiliary Heating and Cooling

The complex is fully air conditioned. Details of the auxiliary heating, ventilating and cooling installations are given in Figure 16. A Frengair suspended ceiling provides radiant heating and cooling. Excess heat from the data processing centre is recovered by an electric heat pump/cold storage system and fed back into the heating system.

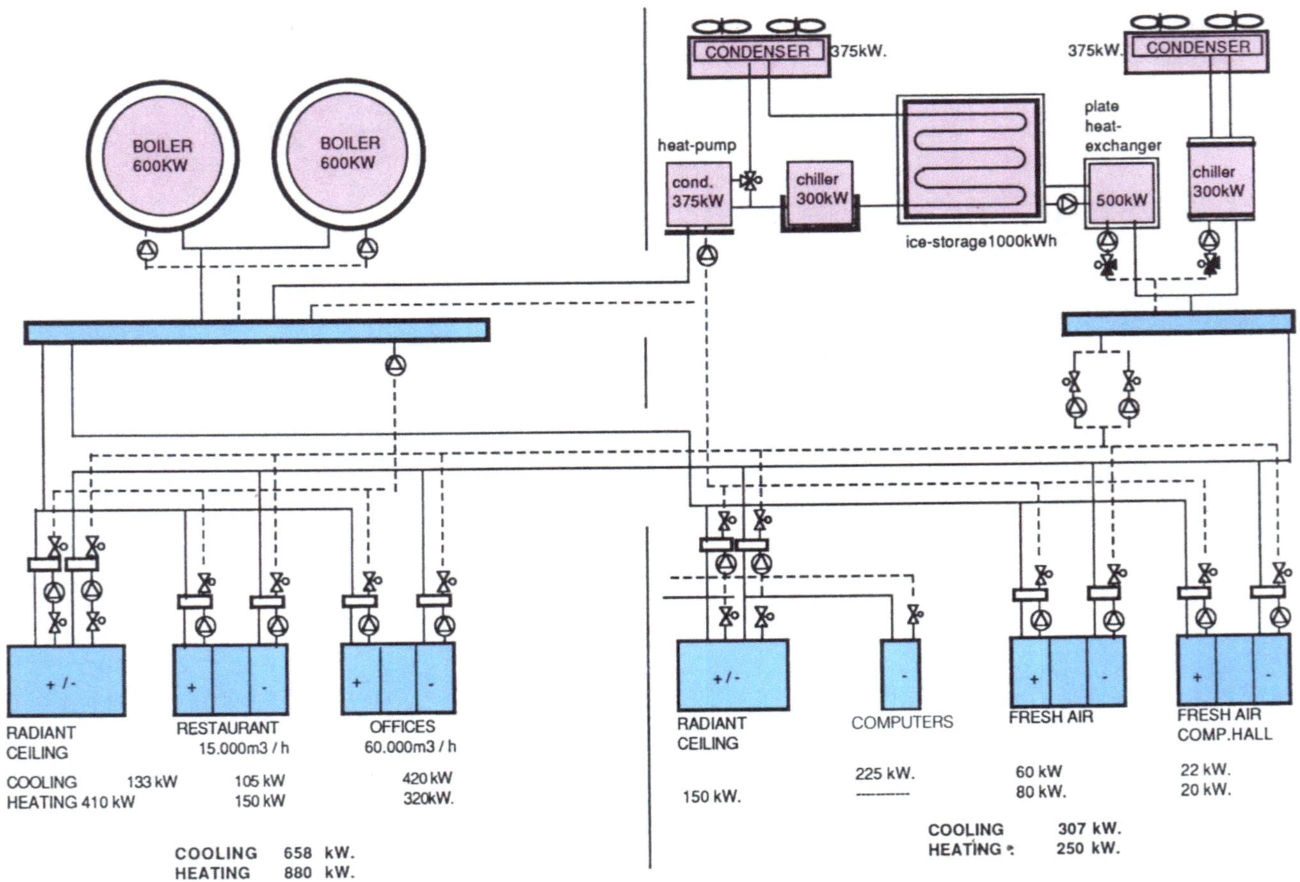

Figure 16. The complex's auxiliary heating, ventilating and cooling systems

ENERGY CALCULATIONS PERFORMED AND DESIGN TOOLS USED

Design of Atrium Roof

The atrium roof was designed with the aid of a computer model developed by TNO which allowed its solar collection and shading performance to be determined. The model was validated by testing a segment of the roof under laboratory conditions (see Figures 17-19). The results of the simulation studies (Figure 20) and laboratory tests suggested that the shading factor of the roof would not exceed 0.3 provided that:

- the amount of perforation in the steel shading plates was less than 30%;

- the coefficient of reflectance of these plates was above 0.8;

- the air circulation rate was maintained above 120 m^3/hour per m width of roof.

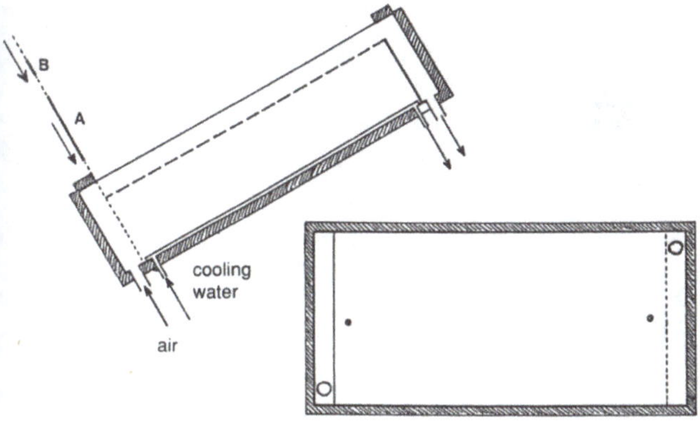

Figure 17. Sketch of the atrium roof model

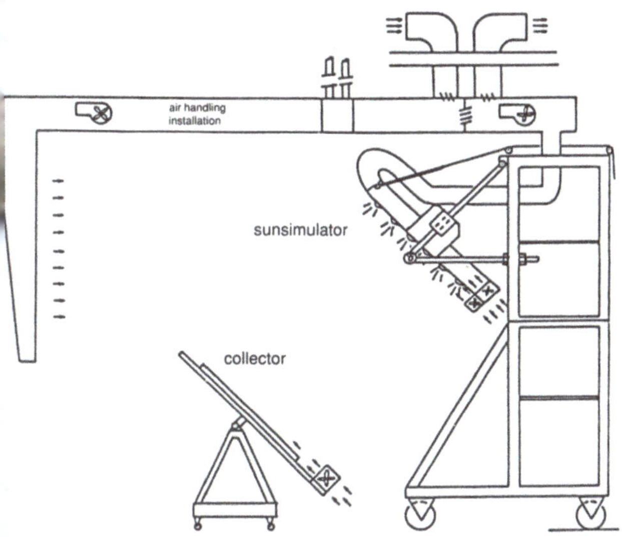

Figure 18. Sketch of the test facility

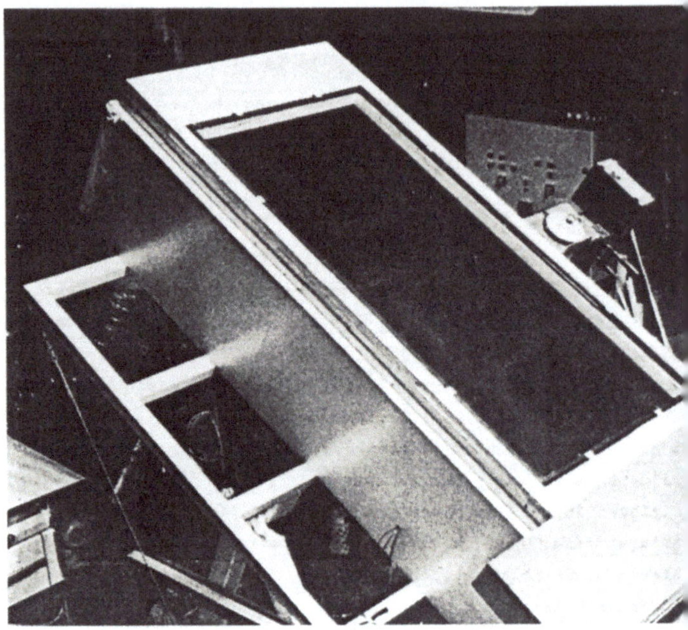

Figure 19. The model in the test facility

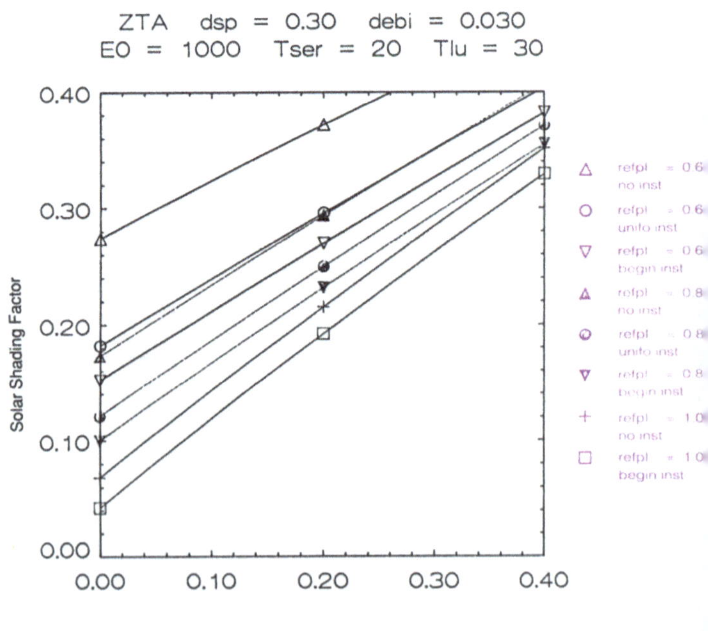

Figure 20. Results from the computer simulation studies on the atrium roof

Thermal Performance of the Complex

Thermal performance evaluations showed that the passive solar and other energy saving measures incorporated in the complex will cut the total annual energy demand from 247,000 m^3 to 100,000 m^3 gas - a reduction of some 60%. Solar and internal gains contribute just over half of this and the electric heat pump the remainder (see Figures 21 and 22).

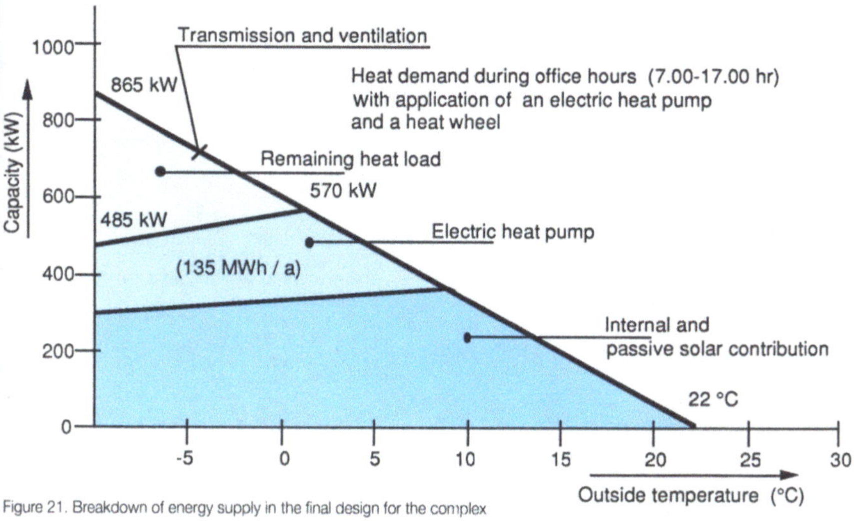

Figure 21. Breakdown of energy supply in the final design for the complex

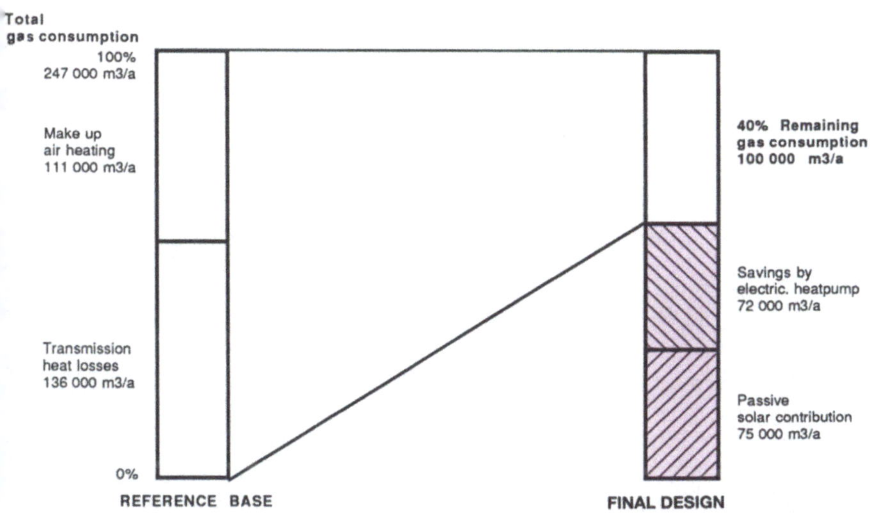

Figure 22. Comparison of energy losses in comparable reference building and energy gains in the final design

BUILDING 2000 Building 2000 brochures are published by Directorate General XII of the Commission of the European Communities to show how design studies can help architects and other building designers use passive solar principles to the best effect to produce attractive energy-efficient buildings. Each brochure describes studies carried out with the support of the Commission during the design phase of one of thirty-six non-domestic buildings in the EC Member States. The studies were on such topics as daylighting, heating, cooling, ventilation, comfort, control systems and urban design. They were carried out with the help of acknowledged European experts in these fields and drew heavily on lessons learned and techniques developed through the Commision's research and development programme on solar energy applications to buildings.

Commission of the European Communities/Directorate-General for Science, Research and Development

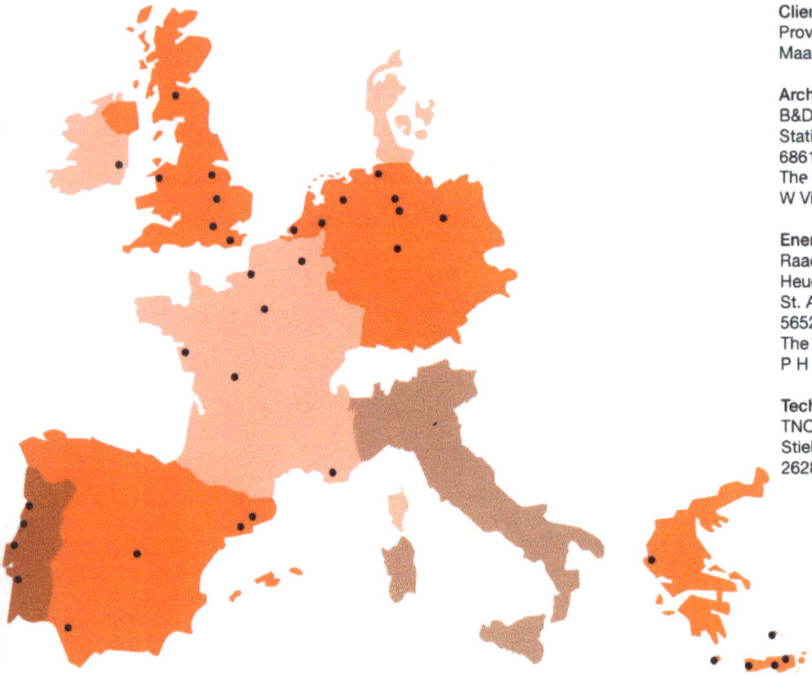

List of Design Team Participants and Advisers

Client
Provinciale Gelderse Energie Maatschappij

Architect
B&D architekten bv
Stationweg 2
6861 EG Oosterbeek
The Netherlands
W Visser and D Lafeber

Energy Consultants
Raadgevend Technies Buro Van Heugten bv
St. Annastraat 14
56524 EP Nijmegen
The Netherlands
P H H Leijendeckers

Technisch Physische Dienst
TNO-TH
Stieltjesweg 1
2628 CK Delft

This set of **Building 2000** brochures illustrates how architects and other building designers can successfully apply passive solar principles to produce energy-efficient buildings.

BUILDING 2000 Participants

Project Director
Theo C. Steemers

Coordinator
Cees den Ouden

Technical Steering Committee
Dean Hawkes
Nick Baker
Alex Lohr
Jean P. Lepoivre

Regional Liaison Agents
(D) Jörn Behnsen
(E) Vicente Sifre
(F) Michel Raoust
(GB) Alan Hildon
(GR) Matheus Santamouris
(P) Eduardo Maldonado

Further information or copies of the brochures can be obtained from prof. ir. Cees den Ouden, EGM Engineering BV, P.O. Box 1042, 3300 BA Dordrecht, The Netherlands.

BUILDING 2000

Commission of the European Communities

- Flats for the elderly, support centre and offices
- Conversion of central courtyard into atrium to reduce transmission and ventilation losses
- Preheating flats' ventilation air in atrium and recovery of heat from flats' exhaust air in winter
- High levels of insulation
- Prevention of overheating in summer by ventilation and correct glazing ratio in atrium

SENIOR CITIZEN'S FLATS

DORDRECHT/THE NETHERLANDS

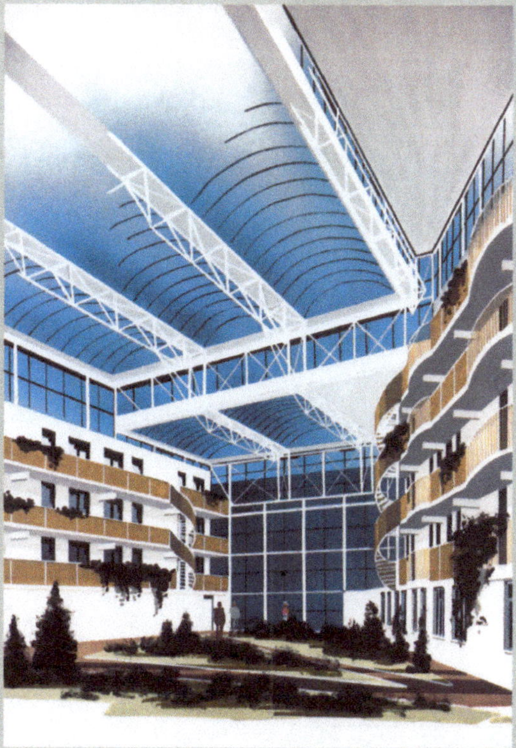

Building 2000 is a series of design studies illustrating passive solar architecture in buildings in the European Community.

ISSUE 03

FEB 1991

PROJECT DESCRIPTION

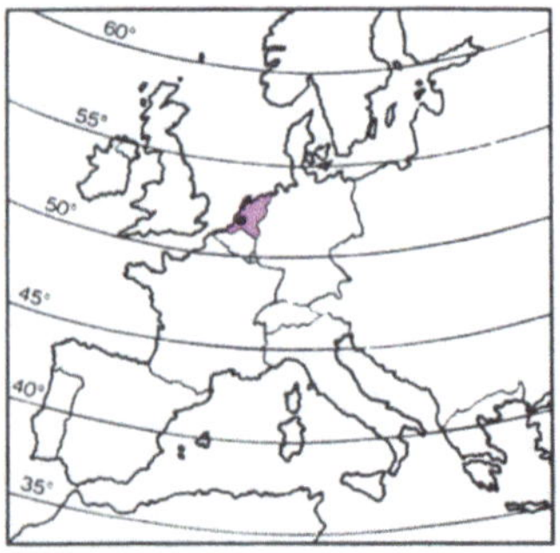

Figure 1. Location

Degree days (base 18° C)	3530
Sunshine hours	1444
Mean daily temperatures	
Jul	16.7° C
Jan	1.5° C
Global irradiation on a horizontal plane	
Jan-Dec in kWh/m²	976
(Jan-Dec in MJ/m²	3514)

Table 1. Some key climate data

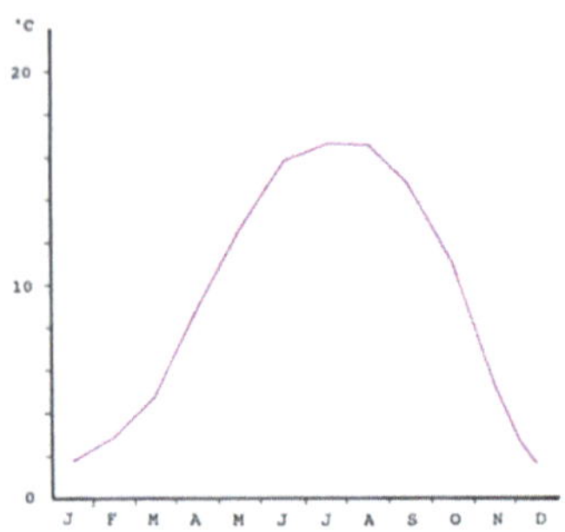

Figure 2. Average ambient temperatures

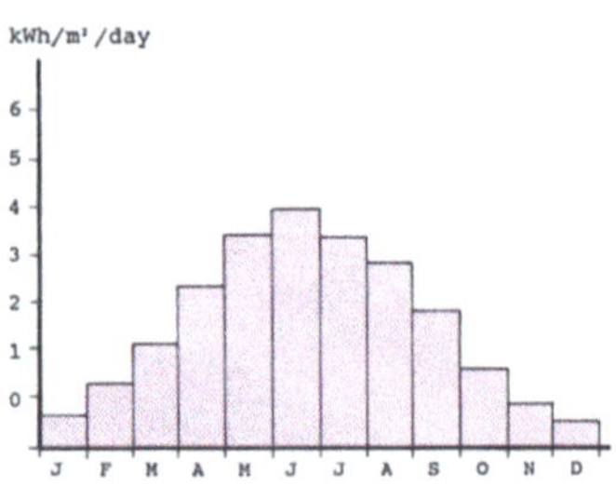

Figure 3. Daily solar irradiation on a horizontal plane

Building Type

This building contains 69 flats and a support centre for senior citizens aged 55 or over plus office accommodation for two local public services, the city harbour and ferry transport services.

The building was commissioned by the Stichting Wooncentra Ouderen Dordrecht housing association. The association's aim is to help older people live independently as long as possible. With this in mind, each flat has a modern communications system with 24-hour service from a caretaker. In addition, the support centre provides recreational, maintenance and security, etc., services for the occupants of the flats and other elderly citizens living nearby.

Location

The site is close to Dordrecht town centre (latitude 51° 45', longitude 4° 40') at sea level alongside the river Merwede near the point where the Oude Maas, Noord and Beneden Merwede merge. Thus the building overlooks a considerable stretch of water. The site is bounded by the Merwekade (along the river), the Riedijk and the Hoefijzerstraat. The site plan is shown in Figure 4. A small park will be created opposite the main entrance in 1990.

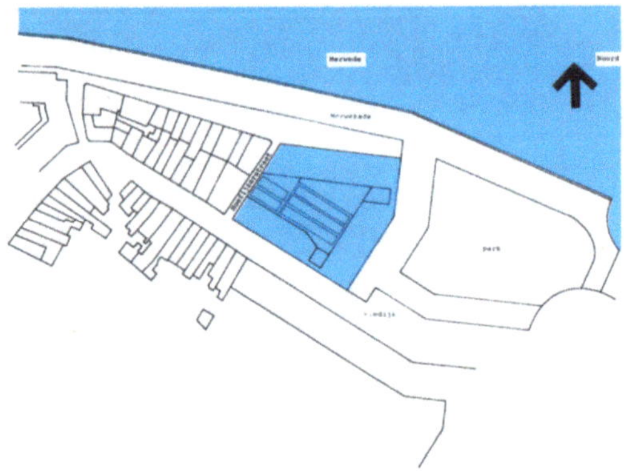

Figure 4. Site plan

Site Microclimate

The Dutch climate is diverse and exhibits both continental and maritime characteristics in what is a relatively small area. Dordrecht, which is in the west, tends to have maritime (moderate) conditions with fairly low summer temperatures, relatively high winter temperatures and rather strong west winds. Further into the interior (De Bilt) the climate is more continental in character with less wind, higher summer temperatures and greater differences between summer and winter - and between one part of the day and another. Some climate data for Dordrecht (where the project is located) are given in Table 1 and Figures 2, 3, 7 and 8.

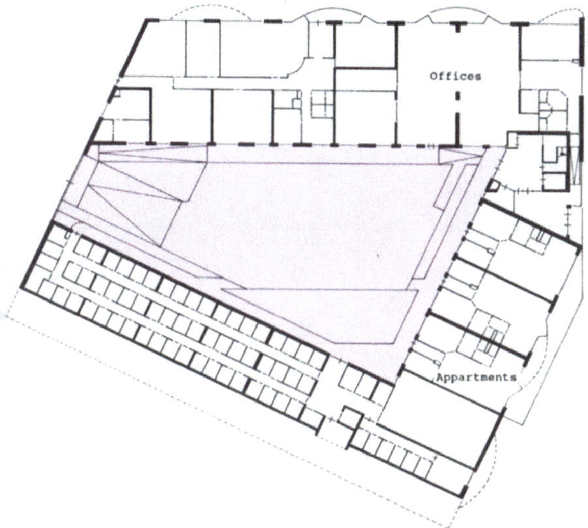

Figure 5. Plan of ground floor

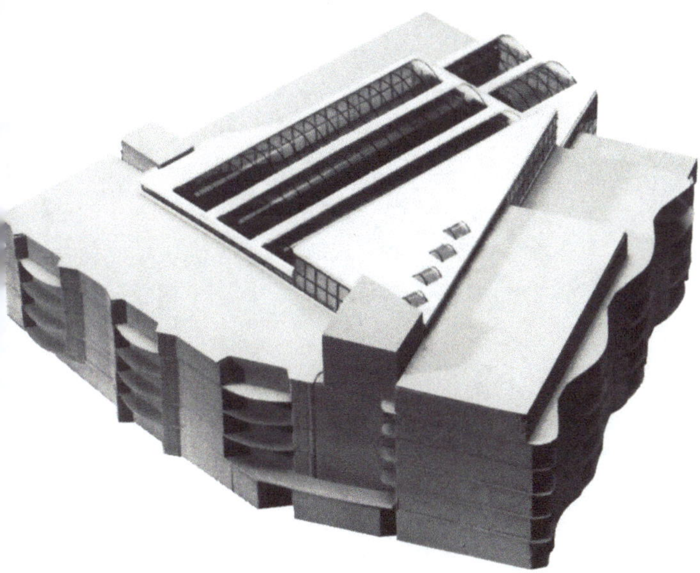

Figure 6. View from the east

Design and Construction Details

The building is U-shaped, about 55 m long and 55 m wide with 4 or 5 storeys. The total floor area is 4,855 m^2 and the volume 11,602 m^3.

On the ground floor (Figure 5) are three flats, offices, the service centre and boxrooms.

The roof of the inner courtyard and the west facade are transparent, forming an atrium. Entrance to each flat is from the atrium. Each flat is self-contained and there is no communal kitchen.

The construction is of a traditional form with concrete structural walls and floors and brick cavity external walls. Standard, commonly-used materials and components have been used to reduce maintenance costs.

In the early design stages, the inner courtyard was left open. It was felt, however, that it would be difficult to avoid wind turbulence in this area and investigations were carried out to see how the design could be modified to use passive solar features to improve comfort and reduce total energy consumption.

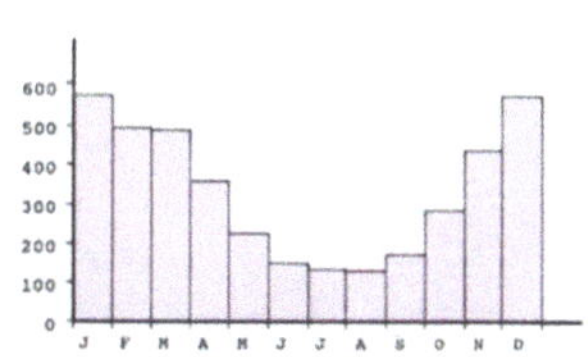

Figure 7. Degree days (base 20° C)

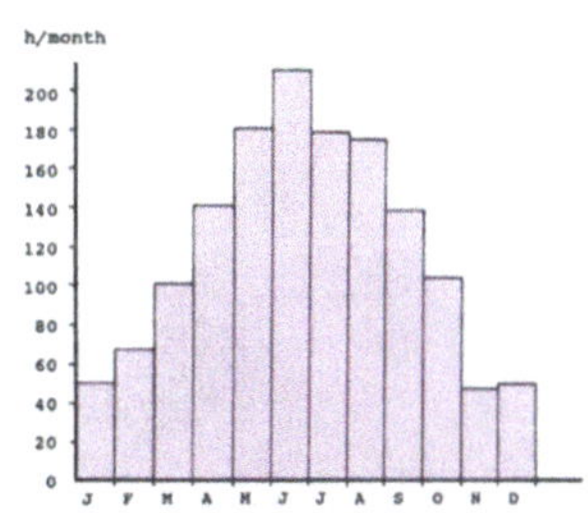

Figure 8. Sunshine hours

DESCRIPTION OF PASSIVE SOLAR FEATURES/COMPONENTS

The aim was to create a better environment and make energy savings by reducing both thermal transmission losses and ventilation heat losses. This was achieved by converting the central courtyard into an atrium.

Two configurations were examined: a completely transparent roof and transparent atrium walls; and a 55% translucent roof and transparent atrium walls. The latter was cheaper to build and gave a lower atrium temperature, particularly in summer. These benefits far outweighed its reduced fuel savings and it was this configuration which was chosen for the final building.

The atrium roof is made of 10 mm polycarbonate cavity sheet with U-value 3.2 W/m² K, solar transmittance 78% and light transmittance 80%. The transparent atrium walls are single glazed.

In winter, the temperature in the atrium is above that of the external air due to passive solar gains. The air entering the atrium is preheated by heat recovered from the flats' exhaust air by means of heat recovery units on the roofs of each wing of the complex. The efficiency of these units is about 85-90%.

To prevent heat loss from the building structure, a high level of thermal insulation has been incorporated into the outer facades. The U-values are given in Table 2; in brackets are the maximum U-values allowed by the Dutch model building regulations.

external walls	0.33 (0.46)
panels	0.39 (0.46)
floor	0.36 (0.64)
roof	0.26 (0.46)
windows	1.80 (3.30)

Table 2. U-values of materials used in building and (in brackets) maximum values allowed by building regulations (W/m² K)

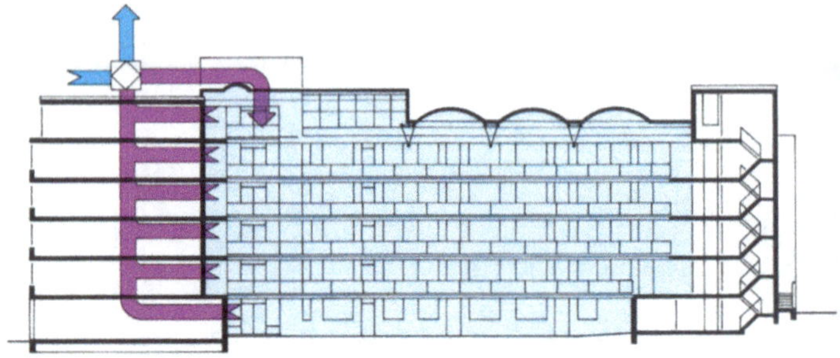

Figure 9. Section through atrium in winter operating mode

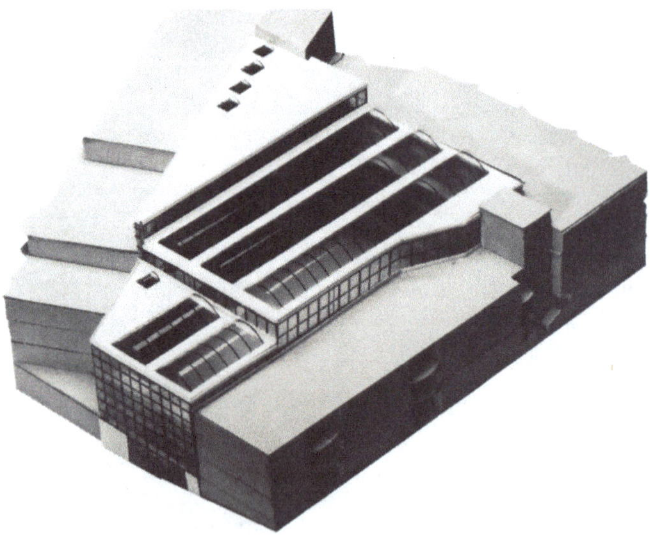

Figure 10. View from the south-west

Operating Mode in Winter

In winter (when the air in the atrium is warmer than that outside), ventilation air for the flats is taken from the atrium through inlets in the atrium wall. All apartment entrances and rooms adjoining the atrium contain these. (During the rest of the year the flats are ventilated with fresh air taken in through vents in the external walls; in winter, these vents are closed.) The flats' exhaust air leaves the building at roof level, where the heat is recovered by exchange units and transferred to the atrium inlet air.

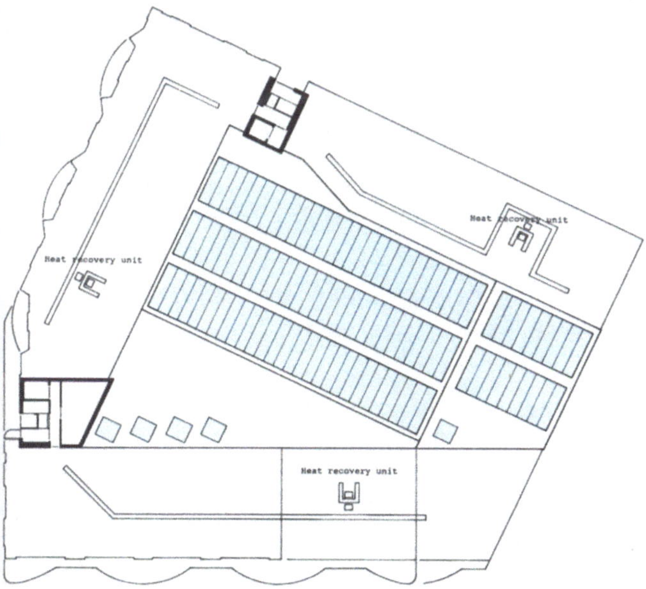

Figure 11. Plan of roof with heat recovery units

Operating Mode in Summer

In summer, the ventilation air for the atrium is fresh air which enters at a rate of 1 air change per hour. The flats' exhaust air is separated from the fresh air by means of a valve and heat recovery from the exhaust air is prevented. In addition, the occupants are asked to close the vents in the entrances and rooms adjoining the atrium so that warm air from the atrium does not enter the building. When the atrium air temperature reaches a certain level, combined ventilation/fire units in the roof open up. There are a number of securable ventilation units and a large door for inlet air in the glazed west facade of the atrium. The door is connected to smoke detectors and opens automatically in case of fire.

Auxiliary Heating

Auxilary space heating is provided by a central heating system with radiators. Hot water is supplied by individual electric water heaters.

ENERGY CALCULATIONS PERFORMED AND DESIGN TOOLS USED

Thermal Performance of Building Without Atrium

The thermal performance of a similar building without an atrium insulated to Dutch model building regulations standards (Case A in Figure 13) was evaluated using a computer program based on the degree day method. The building was divided into three zones. Hour-by-hour simulation runs were carried out for each zone using a weather data file containing mean hourly direct and diffuse solar irradiation and ambient temperature data for each month of the year. The hourly heating load due to transmission and ventilation losses was calculated together with the contribution from solar and internal gains. Assumptions concerning the thermal behaviour of occupants were taken from a study carried out by TPD-Holland. It was found that the occupants' behaviour had a great influence on the ventilation losses and ultimate energy load.

The base heating load due to transmission and ventilation losses was calculated to be over 803,000 kWh/year. The contribution from solar gains was estimated to be 234,000 kWh/year. Thus the net annual load was around 569,000 kWh. This is shown in Case A in Figure 13 and amounted to around 80 kWh/m^2 .

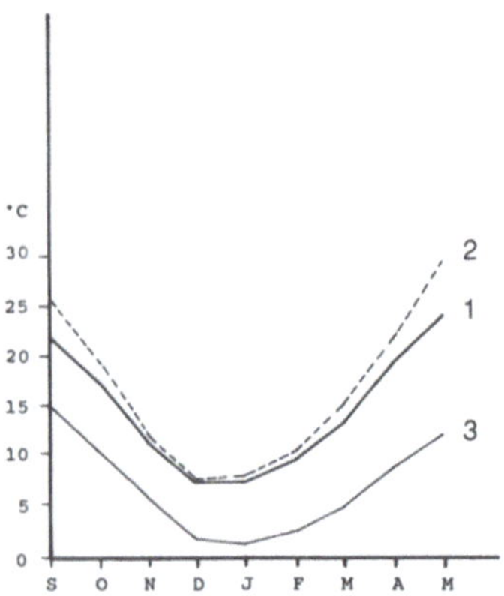

Key: 1 temperature in atrium with 55% translucent roof

2 temperature in atrium with100% transparent roof

3 ambient temperature

Figure 12. Mean monthly temperatures in atria with different roofs

Effect of Energy-Saving Features on Thermal Performance

Atrium Formed by Covering Central Courtyard

The effect on thermal performance in winter-time of forming an atrium by covering the central courtyard with a transparent or 55% translu-cent roof but ventilating the flats and offices with external air (i.e. air not preheated in the atrium) was then assessed (Case B in Figure 13).

It was found that, in winter, the mean temperature of the atrium would be about 5-6° C above the mean ambient temperature (see Figure 12). If the roof is covered with a completely transparent roof, for instance, in December (when the mean ambient temperature is 1.6° C) the mean temperature in the atrium would be about 7.0° C; in January (when the mean external temperature is about 1.5° C) the mean temperature in the atrium would be about 7.3° C. With a 55% translucent roof, the mean atrium temperature would be 6.8° C in December and 7.0° C in January.

The creation of an atrium with a 100% transparent cover would reduce transmission losses by over 101,000 kWh a year, i.e. about 18% of the net energy load for Case A. This is shown in Figure 13.

Replacement of the transparent roof with one which is 55% translucent would have a negligible effect on atrium temperature; the annual reduction in transmission losses would be about 74,000 kWh (i.e. about 13% of the total net energy load for Case A).

Preheating Ventilation Air in Atrium in Winter

The effect on winter thermal performance of taking the flats' ventilation air from the atrium was then assessed (Case C in Figure 13). It was found that this would give a considerable reduction in ventilation losses and a small cut in transmission losses. The annual savings would amount to over 175,000 kWh compared with Case A - a 31% reduction in the total net heating load.

Additional Insulation

It was found that additional savings would result from insulating the building to a much higher level than those given in the Dutch model building regulations. Adding insulation at the level given in Table 2 gave the savings shown in Case D of Figure 13.

Heat Recovery Units

These savings could be improved further by using three heat recovery units (one for each wing of the building) to recover heat from the flats' exhaust air in winter (Case E in Figure 13).

Total Savings

The results of the assessment studies are shown in Figure 13. They show that the total savings from all the above measures - creation of an atrium, preheating the inlet air in the atrium, improved insulation levels and use of heat recovered from exhaust air - would amount to 424,000 kWh a year. This is 74% of the net heating load for the conventional building in Case A - see Figure 14. This would reduce the net energy load from 80 kWh/m^2 to 20 kWh/m^2.

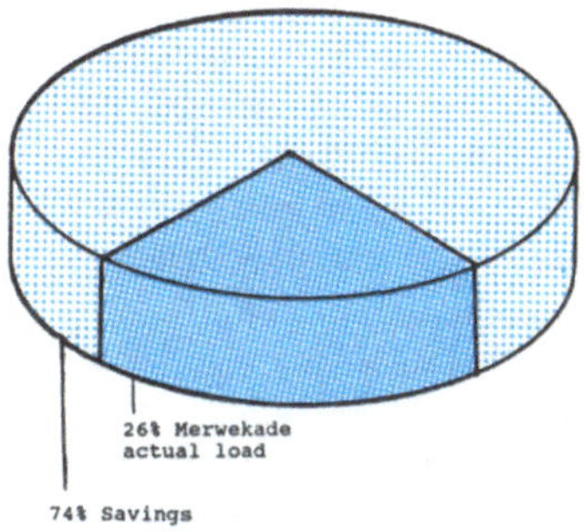

Figure 14. Total energy savings achieved in Case E. (569,000 kWh a year would be required to heat the conventional building without an atrium described in Case A.)

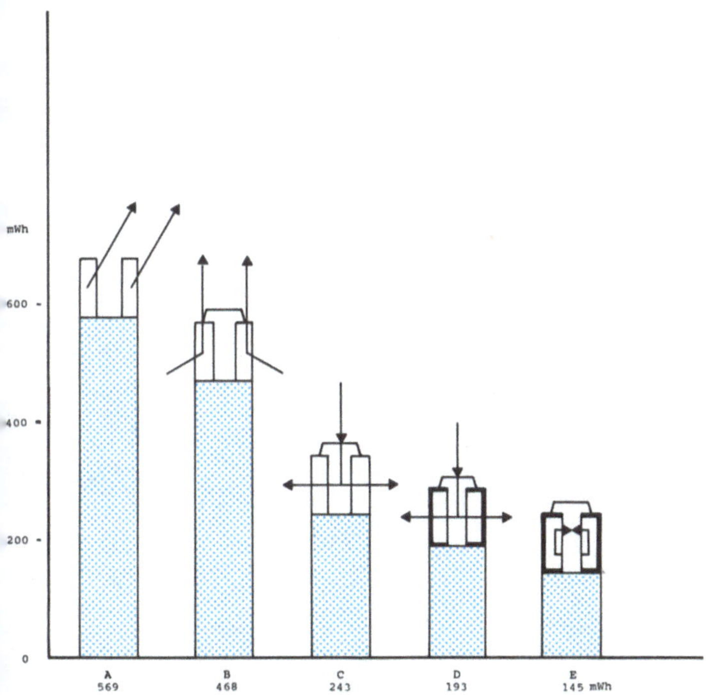

Figure 13. Effect on annual energy load of adding energy-saving features to conventional building. (Cases A, B, C, D and E are described in the main body of text.)

Storage

The effect of short- and long-term storage was also considered.

Short-term storage could have been achieved by withdrawing heat from the top of the atrium and transferring it to (1) a rock bed or (2) a hot water store by means of an air-to-water heat recovery unit. The heat stored in option (1) would not be useful because there is no need for it in summer anyway. Option (2) - which is shown in Figure 15 - could have been interesting from a costs point-of-view if it had been used in conjunction with a communal hot water system. The client did not, however, wish to pursue it.

Long-term storage is theoretically possible because there is a surplus of heat for several months of the year. The small scale of the project, however, made it fruitless to examine the idea in detail.

Prevention of Overheating of Atrium in Summer

Because the atrium is used for traffic, rather than as a sitting area, no air conditioning (heating or cooling) is provided there. In winter, the atrium is perfectly comfortable. In summer, measures have to be taken to prevent overheating. There are three ways of doing this: altering the glazing ratio to reduce the area of transparent surface; use of shading devices; ventilation. Previous experience in The Netherlands had shown that comfortable conditions can be created in an atrium without the use of shading devices if there is sufficient natural ventilation. The VA32 computer program was used to calculate the amount of ventilation needed in this current project. The most important conclusions are shown in Table 3 and Figure 16.

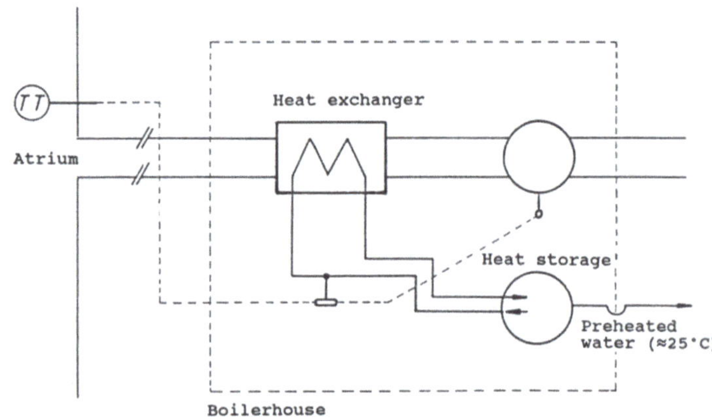

Figure 15. Scheme for short-term storage of surplus atrium heat

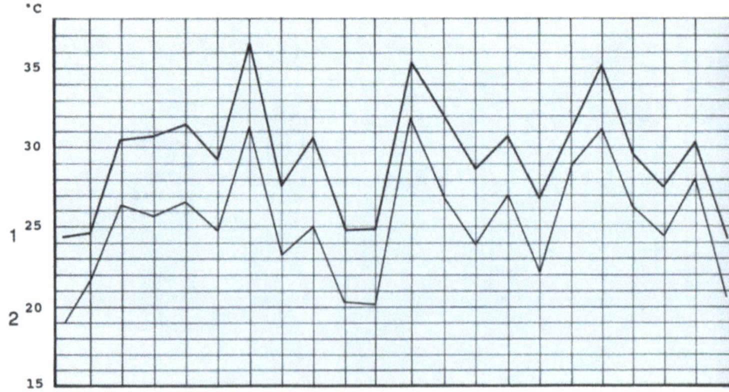

Key: 1 maximum atrium temperature
2 maximum ambient temperature

Figure 16. Maximum weekly temperatures achieved in atrium with glazing ratio of 0.7 and ventilation rate of 6 air changes per hour

With six air changes per hour and a glazing ratio of 0.7, the peak maximum temperature in the atrium is about 6 °C above the peak ambient temperature. In summer, at least four air changes per hour are needed to prevent overheating. This can be achieved by opening 18 m (i.e. 3% of the roof surface) at the top of the atrium. There are 15 m adjustable inlets in the glazed wall on the Hoefijzerstraat. In addition, fresh air can be brought in in summer through the heat exchanger units.

The combined ventilation/smoke openings can be opened or closed manually or automatically. The openings in the atrium walls of the apartments (which provide the latter with preheated air in winter) have to be closed manually by the occupants in summer.

It was left to the client to decide whether or not to introduce shading devices.

air changes per hour	glazing ratio	max. mean	temp. °C peak
3	0.8	34	42
3	0.7	32-33	40-41
6	0.8	31	38
6	0.7	29-30	36-37

Table 3. Effect of ventilation rate and glazing ratio on atrium temperature

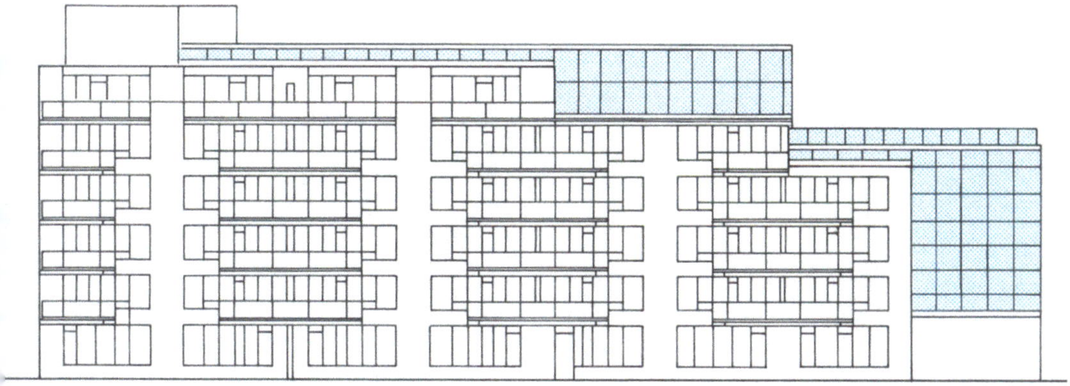

Figure 17. North elevation

GENERAL DESIGN GUIDELINES/POINTS OF INTEREST RESULTING FROM THIS PROJECT

The Architect's Experiences

The idea of using energy saving measures in the Merwekade project was introduced late in the design process. Realization of the idea was therefore more difficult than it might otherwise have been.

In addition, construction of the complex began earlier than expected, due to external factors. Thus a lot of the detailed design of the roof, etc., had to be carried out while building was underway and many decisions had to be made quickly. The consideration of alternatives was out of the question.

Specific problems encountered with the roof included the specification of the steel frame and the junctions with the rest of the building. The roof has three levels and the steel frame had to be connected to three different parts of the building. Care had to be taken, therefore, to adhere strictly to the "fixed points". Relatively large movements occurred in the structure because of the large spans. The architectural details had, therefore, to be designed to accommodate these movements.

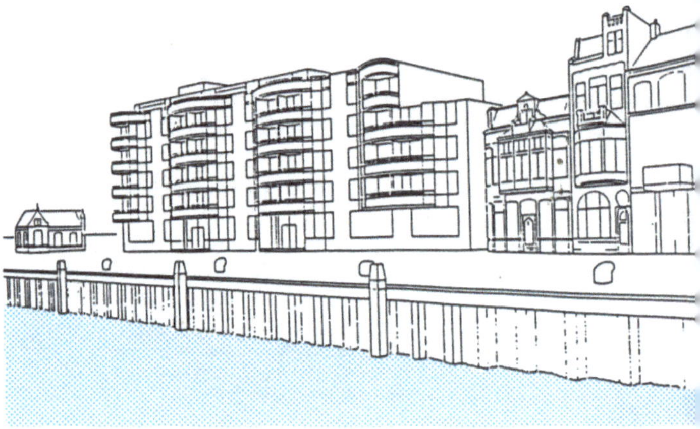

Figure 18. View from the river Merwede

Chronological Survey of Design and Construction of Energy-Saving Measures

March 1987
The client requested examination of the possibility of improving the comfort in the flats, support centre and offices. Proposals submitted consisted of inclusion of an atrium, high levels of thermal insulation and heat recovery.

January 1988
The proposal was made financially possible by a grant from the Finance Department.

June 1988
The municipal authorities granted the building permit on condition that additional work was carried out on the shape of the roof, noise control in the atrium, and fire and smoke control devices.

Concerning the roof, two extra roof levels had to be made in the atrium so that the new building had more visual unity with the adjacent historical buildings.

With regard to noise, the reverberation time in the atrium had to be kept to reasonable levels.

Concerning fire and smoke control, fire and smoke detectors had to be fitted at three levels, instead of one. Also the smoke ventilation openings had to be enlarged.

July 1988
Construction began.

August 1988
To make the building more stable, the steel structure of the roof had to be turned through 90°.

January-March 1989
The finishes had to be simplified to keep the project within budget.

April 1989
The highest point was reached.

May 1989
The steel frame of the roof was erected.

June-October 1989
The final stages of construction were carried out.

October 1989
The building was handed over to the client.

BUILDING 2000 Building 2000 brochures are published by Directorate General XII of the Commission of the European Communities to show how design studies can help architects and other building designers use passive solar principles to the best effect to produce attractive energy-efficient buildings. Each brochure describes studies carried out with the support of the Commission during the design phase of one of thirty-six non-domestic buildings in the EC Member States. The studies were on such topics as daylighting, heating, cooling, ventilation, comfort, control systems and urban design. They were carried out with the help of acknowledged European experts in these fields and drew heavily on lessons learned and techniques developed through the Commision's research and development programme on solar energy applications to buildings.

Commission of the European Communities/Directorate-General for Science, Research and Development

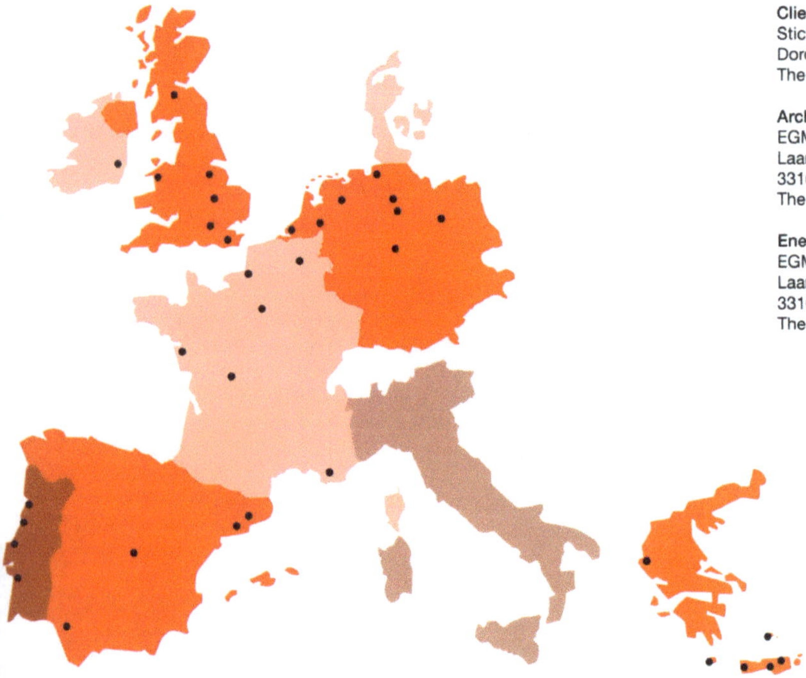

List of Design Team Participants and Advisers

Client
Stichting Wooncentra Ouderen
Dordrecht
The Netherlands

Architect
EGM architecten BV
Laan der Verenigde Naties 9
3316 AK Dordrecht
The Netherlands

Energy Consultant
EGM engineering BV
Laan der Verenigde Naties 9
3316 AK Dordrecht
The Netherlands

This set of **Building 2000** brochures illustrates how architects and other building designers can successfully apply passive solar principles to produce energy-efficient buildings.

BUILDING 2000 Participants

Project Director
Theo C. Steemers

Coordinator
Cees den Ouden

Technical Steering Committee
Dean Hawkes
Nick Baker
Alex Lohr
Jean P. Lepoivre

Regional Liaison Agents
(D) Jörn Behnsen
(E) Vicente Sifre
(F) Michel Raoust
(GB) Alan Hildon
(GR) Matheus Santamouris
(P) Eduardo Maldonado

Further information or copies of the brochures can be obtained from prof. ir. Cees den Ouden, EGM Engineering BV, P.O. Box 1042, 3300 BA Dordrecht, The Netherlands.

BUILDING 2000

Commission of the European Communities

28 units for new businesses plus central reception and secretarial, training and legal support centre

Winter-time heating using direct solar gain and atrium in centre of building

Daylighting of offices through glazing in external facades and atrium walls

Summer comfort using shades, natural ventilation and light-coloured roof surface

CENTRE FOR NEW BUSINESSES

REZE/FRANCE

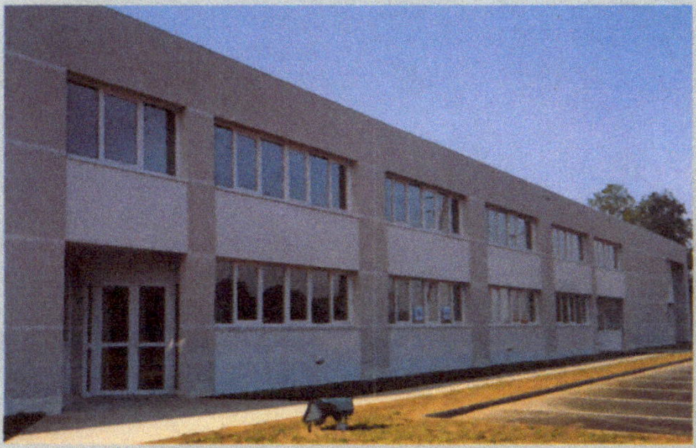

Building 2000 is a series of design studies illustrating passive solar architecture in buildings in the European Community.

ISSUE 04

FEB 1991

PROJECT DESCRIPTION

Building Type

This building has been created by the local authorities to encourage the formation and survival of new businesses in the area. Known as the Technosite Business Development Area, it has been designed with the needs of various types of organization in mind. Examples include those working in computer processing, graphics and metalworking (e.g. silversmiths and goldsmiths). It contains 28 units for individual businesses which have been operating for less than two years plus a central reception/service area where tenants can obtain secretarial, fax, telex, legal, training and other support services.

Twenty schemes of this nature are being developed in France at the moment. The overall objective of the Technosite project was to modify the standard design to produce a cost-effective energy-efficient building which benefits from the specialist knowledge on thermal comfort, daylighting and energy use available through the Building 2000 programme and is capable of being replicated elsewhere.

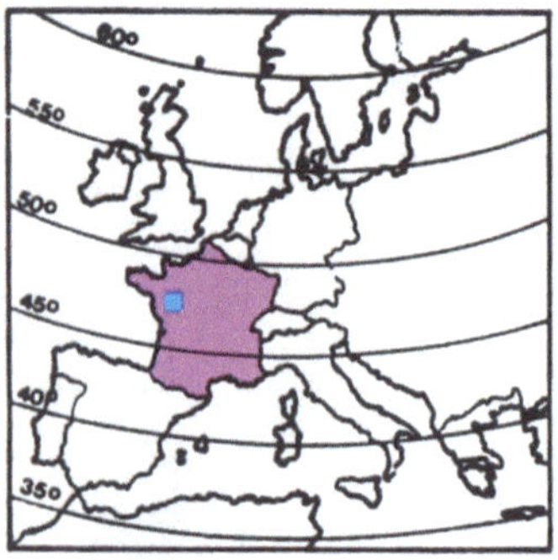

Figure 1. Location

Location

The site is in south Reze, a suburb of Nantes in southern Brittany (Figure 1). It is near the ring road around Nantes and 5 km from the international airport. It is on the edge of a large built-up area and comes within the Praud Industrial Activity Zone. It is 50 km from the Atlantic Ocean and 5 km from the river Loire. The land, which is 10 m above sea level, is flat and is not overshadowed by buildings or other obstacles. The site plan is shown in Figure 3.

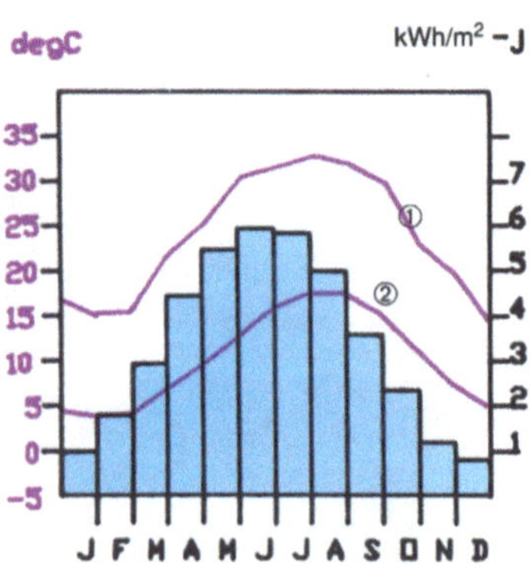

Key: ① maximum external temperature at 2 pm
② average daily external temperature
global irradiation on a horizontal plane

Figure 2. Ambient temperature and solar irradiation data

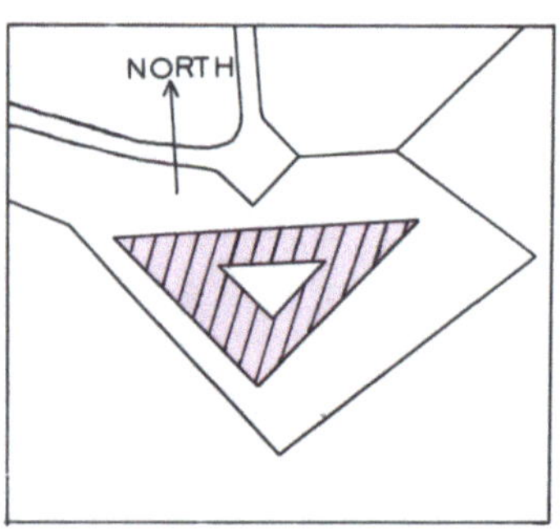

Figure 3. Site plan

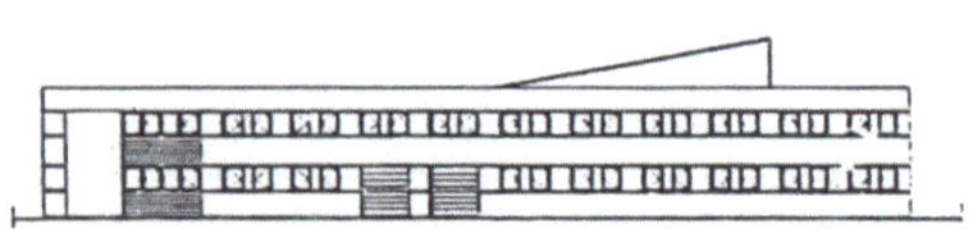

Figure 4. South-west facade

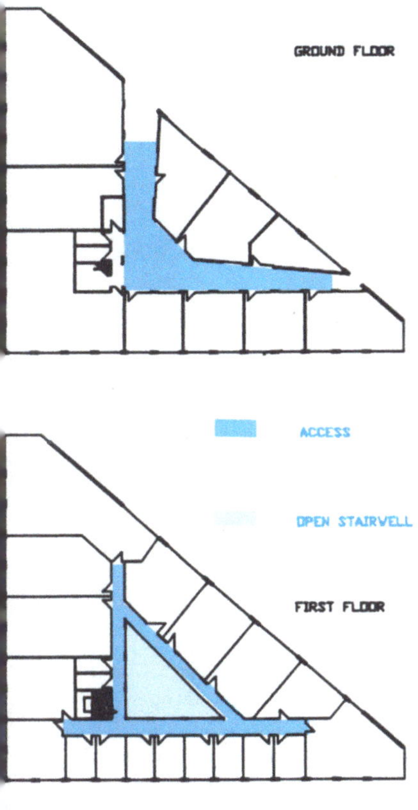

Figure 5. Ground and first floor plans

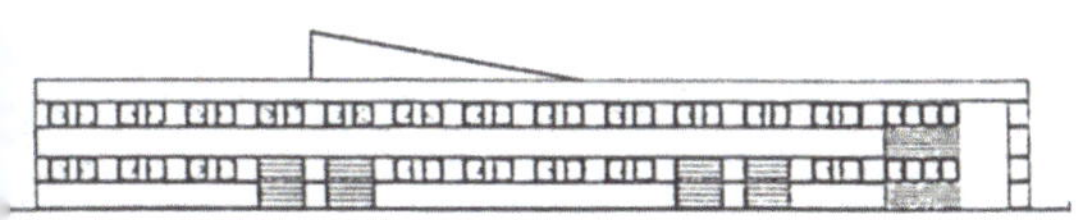

Figure 6. South-east facade

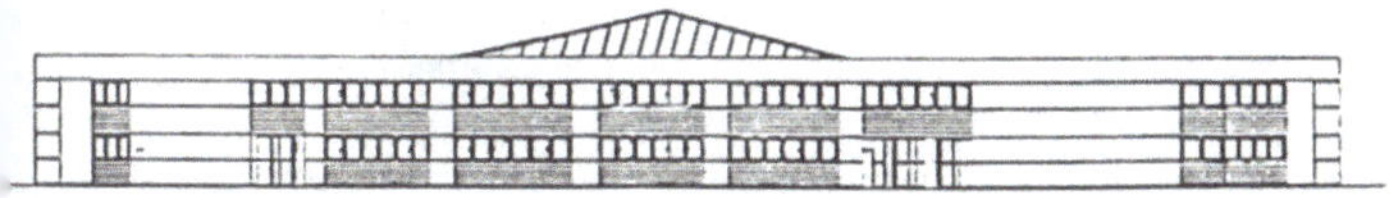

Figure 7. North facade

Site Microclimate

Temperature and solar irradiation data for Nantes are given in Figure 2. The minimum temperature found there is -5° C. In summer, external temperatures at 2 pm can reach 30-32° C. The annual global irradiation on a horizontal plane is 1227 kWh/m² (4417 MJ/m²). The annual global irradiation on a vertical plane is 913 kWh/m² (3287 MJ/m²). There are 2200 sunshine hours a year. The prevailing winds are from the south-west and the north-east. The south-westerly winds are accompanied by heavy rainfall. The north-easterly winds are cold and dry.

Design and Construction Details

The building is on two storeys, the 28 individual office units being arranged around a central atrium. The ground floor and first floor plans are given in Figure 5 and the south-west, south-east and north facades in Figures 4, 6 and 7. The total office space amounts to 2000 m² and the atrium and access areas cover 500 m².

The building is constructed on a platform. The general frame consists of concrete blocks. A 40 mm layer of polystyrene insulates the outer structure. The inside walls have 80 mm polystyrene plus plaster. The U-value is 0.40 W/m² K. The flat roofs (U-value 0.43 W/m² K) are of solid concrete insulated on the outside with 60 mm polyurethane and covered with waterproofing topped with a 40 mm layer of white pebbles. Fittings are made of PVC. There is single glazing on the south-east and south-west facades and double glazing on the north facade. The atrium is made of polycarbonate.
The building cost FF6 M (850,000 ECU) to construct, some 70% of the average cost of a similar conventional building.

DESCRIPTION OF PASSIVE SOLAR FEATURES/COMPONENTS

Priorities in Choice of Passive Solar Systems

In designing the building, the objective has been to provide occupants with summer-time comfort, good daylighting and winter heating in a way which brings savings in capital and operating costs and requires less energy to run than a conventional building of the same type (see Figure 8).

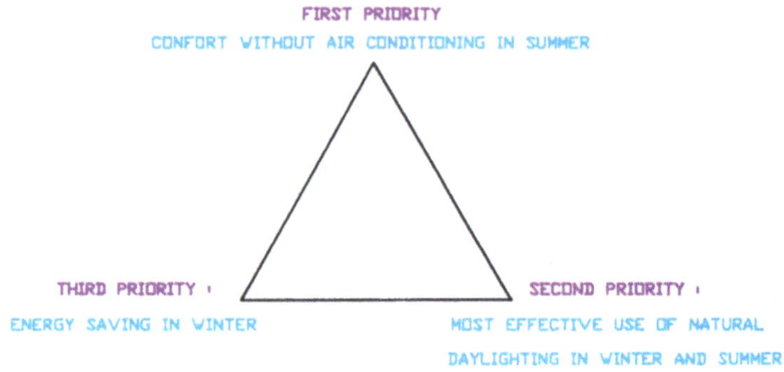

Figure 8. Priorities in choice of passive solar and energy-saving systems

Summer Comfort

In most buildings of this type, summer-time comfort is achieved with individual air conditioning units run on electricity. These are expensive to install (say, FF300,000 (42,000 ECU)) and operate. The objective in the Technosite building was to provide the same level of comfort (i.e. to produce an internal temperture of 28° C when the external temperature is 33° C) without them.

Lighting

Craft and graphics businesses require good quality light. To cut capital costs and save energy, the objective in the Technosite building was to meet as much as possible of the occupants' lighting needs by daylight. Effective daylighting is achieved by good design rather than extra capital cost. The target was to obtain an illuminance of 200 lux in the building when there was 20,000 lux outside.

Winter Heating

In the area of winter-time heating, the objectives were for the building to comply with the French statutory requirements regarding heat loss and to make effective use of simple and inexpensive solar systems to reduce energy consumption. The target was to achieve a 10% contribution to heating requirements from solar energy - it was recognized that a design which gives a higher contribution could lead to overheating in summer.

General Philosophy

The philosophy has been to meet the above objectives by the use of straightforward systems which are cheap to run, use materials and components in common use and do not contain active equipment like air conditioning units. The techniques chosen involve direct solar gain through southerly windows, an atrium, natural lighting of the offices through the atrium walls and glass panelling in the exterior facade, natural ventilation, solar protection using blinds, increase in thermal inertia of the roof, good insulation and judicious use of colour. These features are used in such a way that a correct balance is achieved between summer-time comfort, good daylighting and winter heating. The overall concept, which is described in more detail below, can be adapted to meet the specific requirements of other projects.

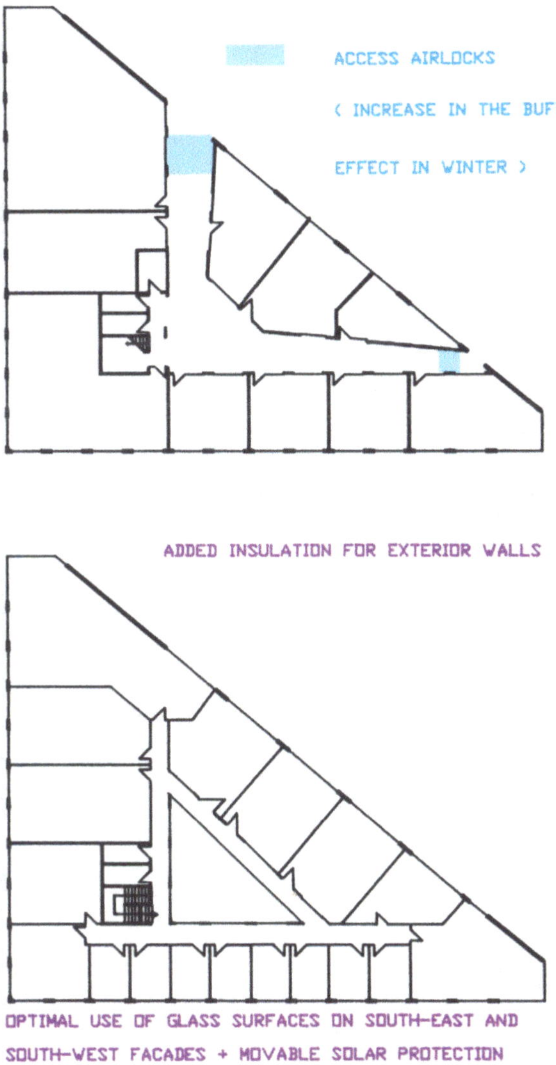

Figure 9. Plan of ground and first floors of building showing construction details aimed at improving thermal performance

Heating and Cooling Strategies

General Principles

The atrium in the centre of the building has a social function - it facilitates business exchange and serves as an access way. The offices are private domains where the core of individual business activity takes place. It is important, therefore, that the offices are maintained at comfortable temperatures. The general principles of achieving this are illustrated in Figure 9.

In winter, there is direct solar gain through the glass windows on the south-east and south-west facades and through the glazing panels in the atrium. Heat loss through the outside walls is prevented by high levels of internal insulation. Heat loss caused by opening and closing exterior doors is minimized by addition of air locks. There is exchange of energy between the offices and the atrium.

In summer, the glazing on the south-east and south-west facades is shaded by external blinds. The atrium is cooled at night by natural ventilation. The good insulation and light surface of the flat roofs prevents solar energy entering the building through them (see Figure 10).

Buffer Effect of Atrium

The thermal resistance of glass walls/doors is ten times lower than that of the solid external walls of the building with their internal insulation. Therefore, the walls between the offices and atrium contain glass partitioning. Because of direct solar gain through the atrium roof and the greenhouse effect, the temperature in the atrium is on average 7-8° C above the external temperature.
Therefore the atrium acts as a buffer and minimizes the amount of heat lost from the offices via the glass partitions. The buffer effect is strengthened by the placing of air locks where there is access from the atrium to the outside of the building and restricting the amount of glass panelling in the atrium to the minimum required for good daylighting.

Auxiliary Heating Systems

The building is open from 8 am to 6 pm, except at weekends and public holidays. It is therefore only occupied 30% of the time. The need for auxiliary heating is therefore intermittent. The auxiliary heating system is managed by a control system which optimizes all available energy sources. The temperature of the offices is reduced to 12° C when they are not occupied.

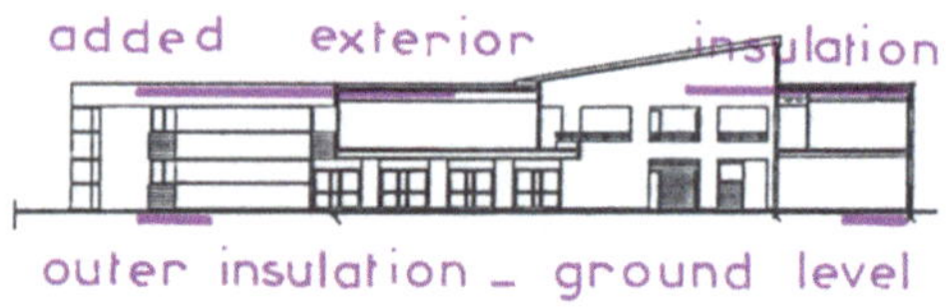

Figure 10. Section of building showing external insulation to roof and concrete base

Ventilation

In a building of this type, all the doors are frequently opened and closed. The opening of the external doors would give a change of air every ten minutes which would make the building uncomfortable and increase energy consumption in winter by 30%. To prevent this, a system of double airlocks has been installed between the atrium and the outside of the building.

Mechanically-operated extract fans remove stale air. During business hours, the air change rate is 1 per hour. Outside working hours, air change rates are reduced to 0.3 per hour.

In summer, the atrium is ventilated at night by means of windows above the airlock access doors and a 6 m^2 opening higher up the exterior wall of the atrium (see Figure 11).

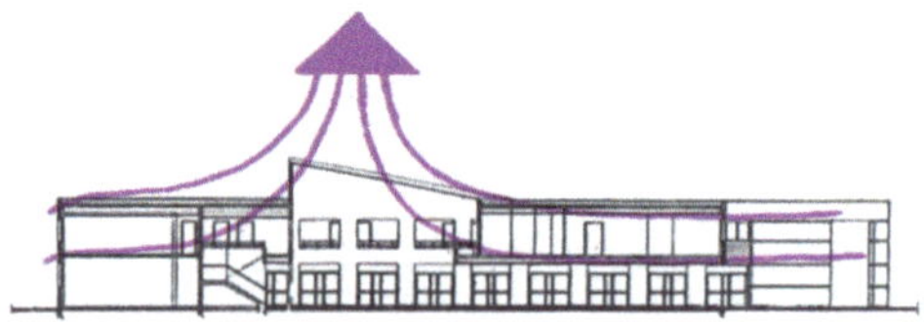

Figure 11. Natural ventilation of the building

Daylighting Strategy

The daylighting strategy is illustrated in Figure 12. Because most of the offices are very deep they have two sources of natural light - light entering directly through the glazing on the exterior facade and light entering indirectly through the atrium.

In winter, direct light through the southerly facades provides most light.

In summer, exterior blinds are used on the southerly facades to prevent overheating and glare from direct solar radiation. Most of the light is provided by the glazing in the atrium walls. Visual comfort can also be improved by the use of curtains on the south wall in the atrium.

The studies carried out through the Building 2000 programme showed that it is the position, rather than the quantity, of the glazing which produces good lighting. It was found that the windows should be unobstructed and positioned high above the ground. The window area was restricted to the minimum required for good daylighting to avoid unnecessary heat loss to the outside.

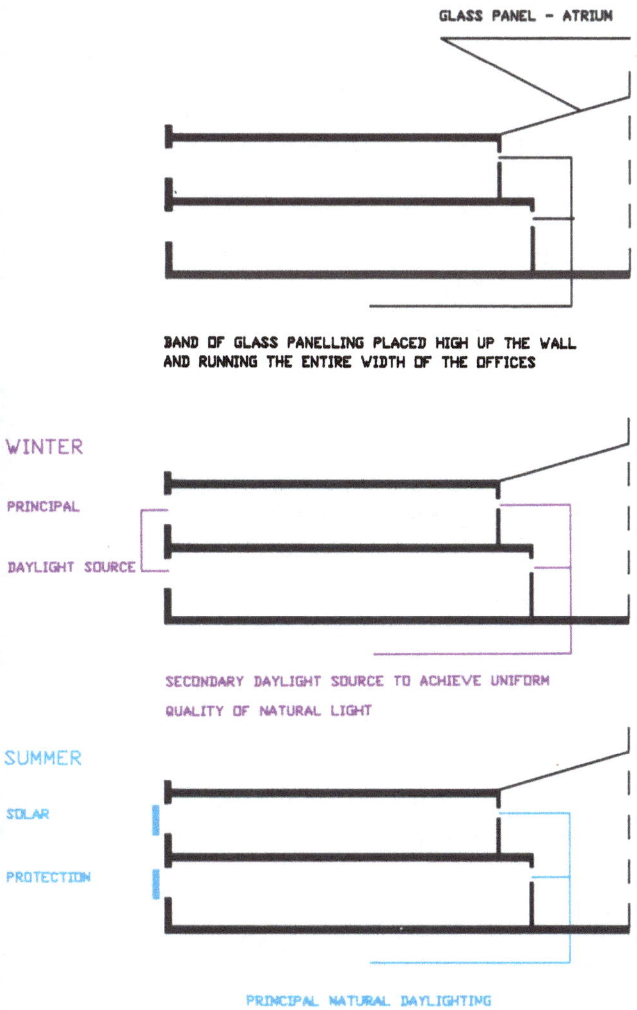

Figure 12. The daylighting systems in winter and summer

ENERGY CALCULATIONS PERFORMED AND DESIGN TOOLS USED

The daylighting, heating and cooling systems were designed with the help of a number of studies carried out with advice from the Building 2000 experts. Some results are given below.

Results of Daylighting Analysis

The daylighting work was carried out using the NATUREL simulation model of the Centre Scientifique et Technique du Bâtiment (CSTB). Daylight factors were calculated for a standard Commission Internationale de l'Eclairage (CIE) overcast sky. (The daylight factor at a point is the ratio of the illuminance at the point due to the light received directly or indirectly from the sky, to the illuminance on a horizontal plane due to an unobstructed hemisphere of the sky. The contribution of direct sunlight to both illuminances is excluded.) The transmittance of the glazing and the reflectance of the interior surfaces were taken into account. It was assumed that the glazing consisted of a single pane (Tdf = 0.85), the ceiling reflectance was >0.6, the wall reflectance was in the range 0.4 to 0.6 and the floor reflectance in the range 0.3 to 0.5.

It was found that all units in the building except the garages had sufficient daylight when there was an external illuminance of 20,000 lux. Some of the rooms are, however, deep - 8.6 m with a ceiling height of 2.6 m. To improve the lighting, therefore, it was decided to add a glazed aperture to the wall in the atrium.

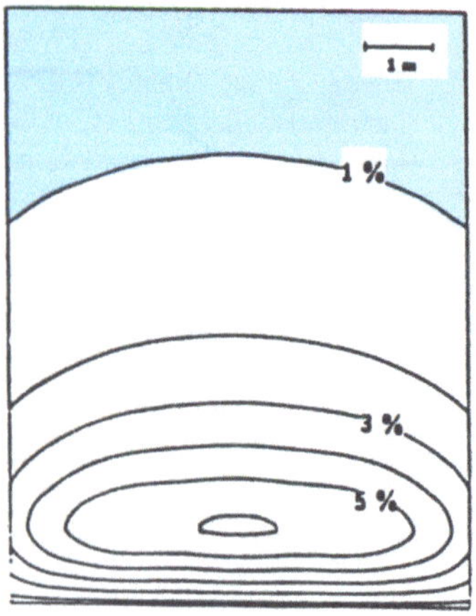

Room area	62 m^2
Glazing area (facade only)	5.5 m^2
Corrected glazing index	7.5%

Figure 13. Daylight factors for first-floor office on north facade without glazing on the atrium

First-Floor Offices

The results of the daylighting analysis of a 62 m^2 first-floor office on the north facade with and without glazing in the atrium wall are given in Figures 13 and 14. In the figures, the glazing index of the room is the ratio of glazing area to room area. The corrected glazing index of the room is the ratio of glazing area to room area multiplied by the transmittance of the glazing.

It can be seen that the addition of glazing to the atrium wall improves the distribution of daylight considerably. When the external illuminance is 20,000 lux, the minimum lux achieved in the room without glazing in the atrium is 150 lux; with glazing in the atrium a minimum of 200 lux is produced in the room. This is in line with the design target.

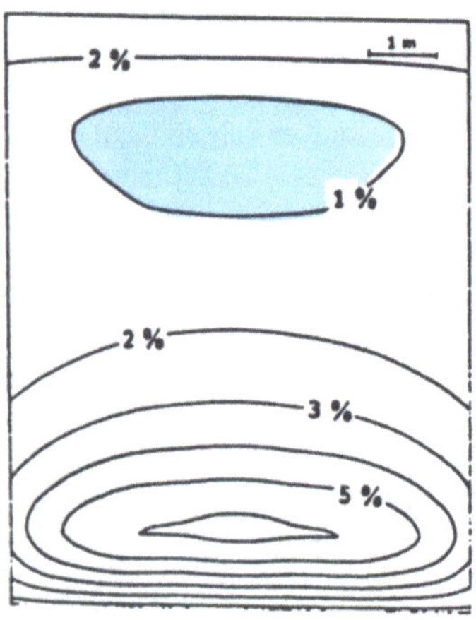

Room area	62 m^2
Glazing area	5.5 m + 3 m^2
Corrected glazing index	11%

Figure 14. Daylight factors for first-floor office on the north facade with glazing on the atrium. (The south aperture of the room is on the atrium.)

Ground-Floor Office

The daylighting analysis showed that, throughout the ground floor, the daylighting of all the offices is good. The results for a 162 m² ground-floor corner office with windows on the south-east and south-west facades are shown in Figure 15. A minimum of 200 lux is achieved with an external illuminance of 20,000 lux.

Conclusions of Daylighting Analysis

The studies showed that placing bands of glazed panelling high up on the facades of each office provided an optimum level of natural lighting. When the windows are placed correctly the total area of glazing can be reduced so that there is a reduction of heat loss in winter and consequent energy saving. There is also a reduction of excess solar gain in summer. Further, there is a saving in construction costs because glass walls are more expensive than solid ones.

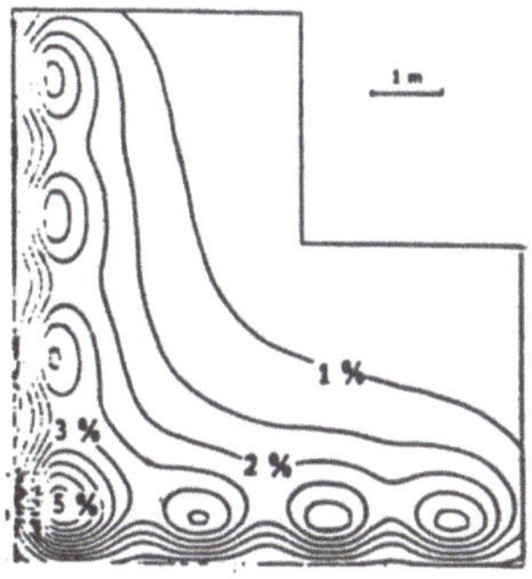

Room area	162 m²
Glazing area	17 m²
Corrected glazing index	9.5%
Reflectances:	
ceiling	0.7
walls	0.5
floor	0.3

Figure 15. Daylight factors for ground-floor corner office on south-west and south-east facades

Results of Evaluation of Summer-Time Comfort Levels

Standard Design

Evaluation of the thermal comfort of the standard building of this type (i.e. one whose performance has not been improved with help from the Building 2000 experts) showed that temperatures reached in the first-floor offices were unsatisfactorily high in summer. They could climb to as much as 35° C when the external temperature was 30-32° C.

Effect of Introducing Special Measures

With the help of the experts, studies were carried out to assess the effect on the standard design of night-time ventilation, shading with external blinds, increasing thermal inertia and insulation of the flat roofs and providing the flat roofs with a light-coloured surface. The results are given in Figures 16-20. The diagrams show the cumulative effect of the improvements so each new Figure includes all the improvements introduced in the previous Figures, plus the new ones.

Figure 19 (where there is an increase in the thermal inertia and insulation of the flat roof) represents a real change in construction technique. The steel panelling and insulation of the standard design are replaced by a concrete slab insulated externally by 60 mm polyurethane.

In Figure 20 the colour of the flat roof is changed from the traditional black to white by the introduction of a 400 mm layer of white pebbles to the top of the roof. This reduces the temperature of the top floor offices by 1° C.

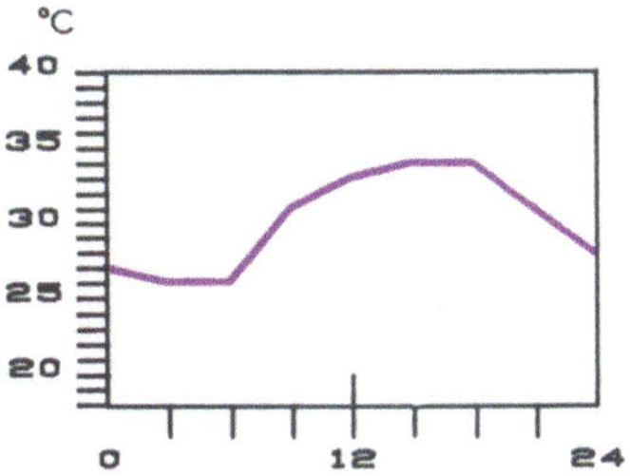

Figure 16. Temperatures reached in first-floor offices in standard building (before introduction of measures to improve thermal comfort)

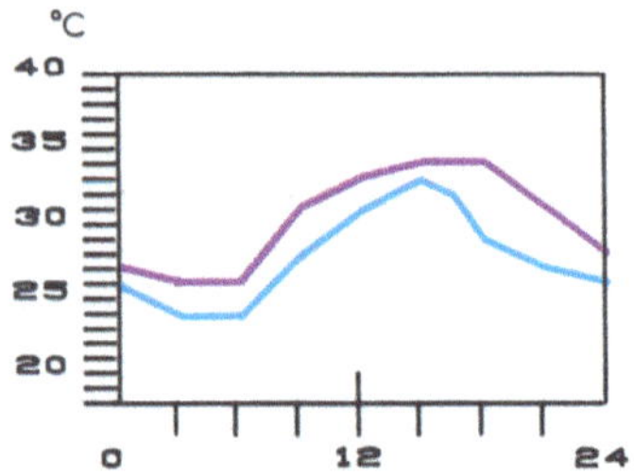

Figure 17. Temperatures reached in first-floor offices with night-time natural ventilation

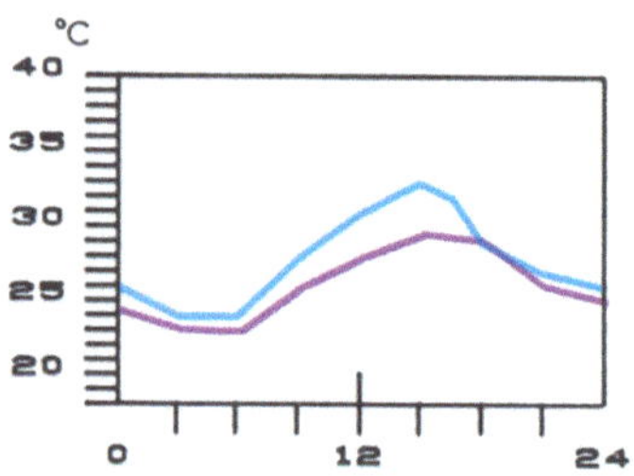

Figure 18. Effect on temperature in first-floor offices of adding exterior blinds

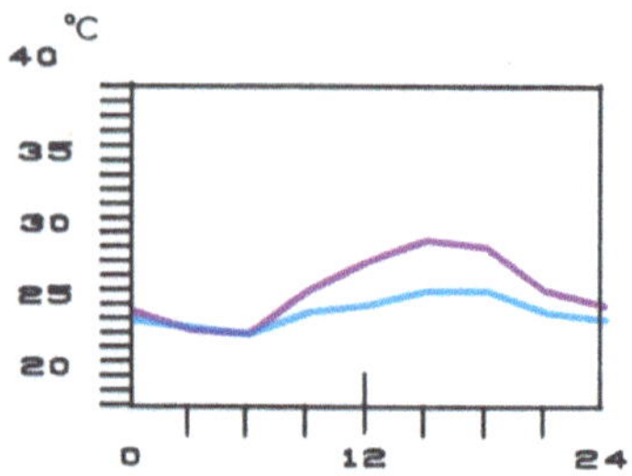

Figure 19. Effect on temperature in first-floor offices of increasing thermal inertia and insulation of flat roof

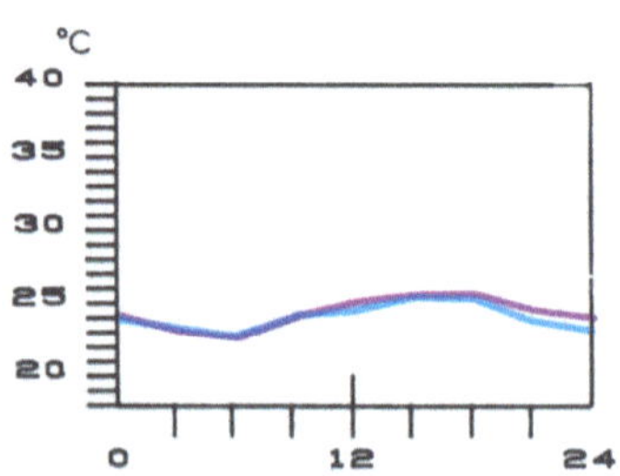

Figure 20. Effect on temperature in first-floor offices of adding light-coloured roof surface

Conclusions of Thermal Comfort Evaluations

The studies showed that, by using the straightforward techniques described, summer-time comfort can be achieved without the use of conventional air conditioning.

The total cost of implementing these recommendations is around FF100,000 (14,300 ECU) without tax. The cost of installing a conventional air conditioning system would be around FF 360,000 (51,000 ECU) exclusive of tax. Thus by using the design developed for the Technosite building it is possible to cut construction costs by FF260,000 (36,700 ECU) exclusive of tax and at the same time avoid paying energy bills for summer-time air conditioning.

Results of Evaluation of Heating in Winter-Time

The standard design required 340,000 kWh per year for heating. The evaluation work showed that the passive solar and other energy-saving techniques incorporated in the Technosite building (i.e. the increased solar gain through the south-east and south- west facades, higher levels of insulation - 60 mm instead of 40 mm polyurethane on the roof and 40 mm polystyrene in the interior walls separating the offices, control of the air change rate, the airlocks and control of the auxiliary heating system) reduced this by 47%. The breakdown is shown in Figure 21.

Overall Results

A breakdown of the annual energy requirements for air conditioning, lighting and heating the standard building (i.e. one whose performance has not been improved with help from the Building 2000 experts) is given in Figure 22. In Table 1 these results are compared with those for the Technosite.

	standard building (kWh)	improved building (kWh)	saving (%)
air conditioning	120,000	0	100
artificial lighting	72,000	43,200	40
heating	340,000	180,200	47

Table 1. Annual energy consumption of standard and improved (Technosite) buildings

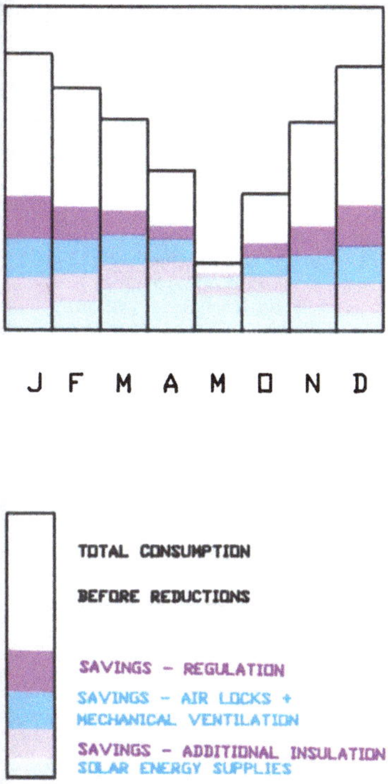

Figure 21. Reduction in annual energy requirements achieved in improved (Technosite) building compared with standard building

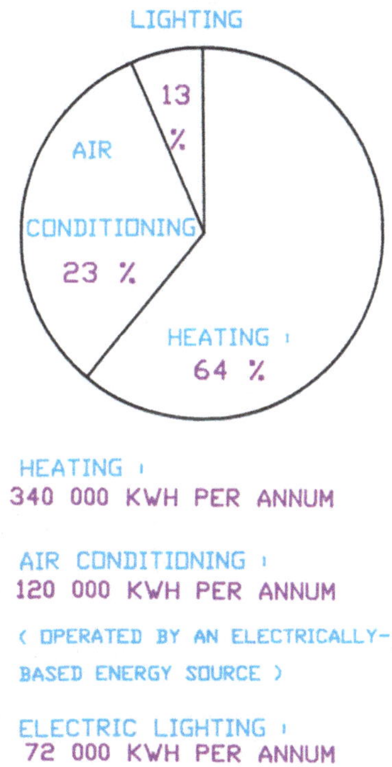

Figure 22. Breakdown of annual energy consumption of standard building

DESIGN GUIDELINES/POINTS OF INTEREST RESULTING FROM THIS PROJECT

The developer's aim has been to produce a building with reduced construction costs, a comfortable environment in summer and low energy requirements. This has been achieved with a simple design and straightforward and well-known energy-saving techniques. Because the techniques needed to produce summer-time comfort, winter-time heating and daylighting are interlinked, it was found necessary to evaluate the interaction of all three systems to develop a properly balanced building.

The use of a conventional air conditioning system (which would have been expensive to buy and maintain) has been avoided.

The type of use to which the building is put made it necessary to provide adequate solar control and to reduce the amount of auxiliary heating in winter.

BUILDING 2000 Building 2000 brochures are published by Directorate General XII of the Commission of the European Communities to show how design studies can help architects and other building designers use passive solar principles to the best effect to produce attractive energy-efficient buildings. Each brochure describes studies carried out with the support of the Commission during the design phase of one of thirty-six non-domestic buildings in the EC Member States. The studies were on such topics as daylighting, heating, cooling, ventilation, comfort, control systems and urban design. They were carried out with the help of acknowledged European experts in these fields and drew heavily on lessons learned and techniques developed through the Commision's research and development programme on solar energ applications to buildings.

Commission of the European Communities/Directorate-General for Science, Research and Developmen

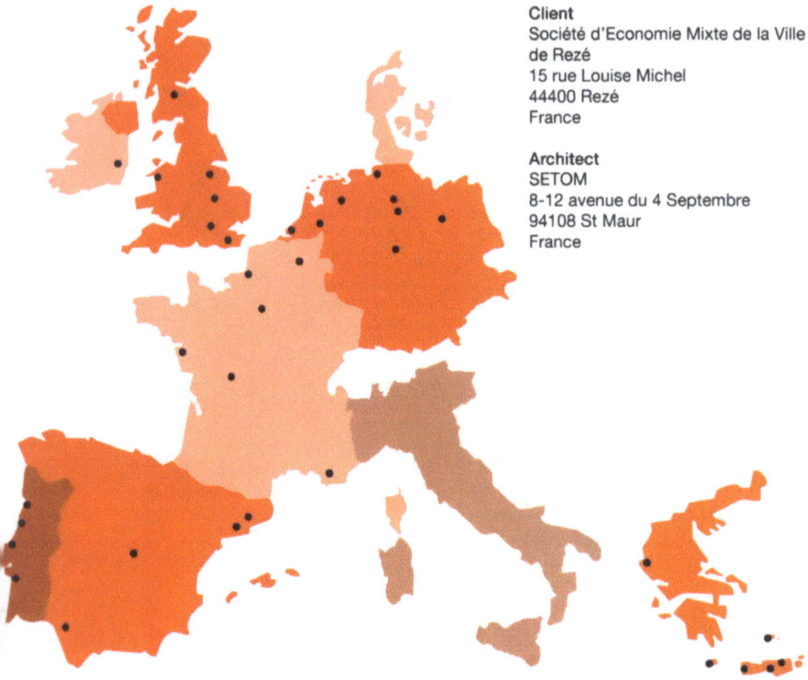

List of Design Team Participants and Advisers

Client
Société d'Economie Mixte de la Ville de Rezé
15 rue Louise Michel
44400 Rezé
France

Architect
SETOM
8-12 avenue du 4 Septembre
94108 St Maur
France
assisted by Euve/Blisson/Saint Gealm
Atelier 3 architectes
13 avenue des 3 peuples
781800 Montigny le Bretonneux
France

Energy Consultant
Gaudin Ingenierie
11 rue de la Butte de Praud
44400 Reze
France

Daylighting Consultant
Mme Chauvel
Centre Scientifique et Technique du Bâtiment (CSTB)
11 rue Picherit
44000 Nantes
France

Summer-Time Comfort Consultant
M Grelat
Centre Experimentale de Recherches d'Etudes du Bâtiment et Travaux Publ
- Fédération Nationale du Bâtiment (CEREBTP-FNB)
Domaine de St Paul
78470 St Remy les Chevreuses
France

Builder
Vezin (Groupe Quillery)
Le Croisy
BP 4
44708 Orvault
France

This set of **Building 2000** brochures illustrates how architects and other building designers can successfully apply passive solar principles to produce energy-efficient buildings.

BUILDING 2000
Participants

Project Director
Theo C. Steemers

Coordinator
Cees den Ouden

Technical Steering Committee
Dean Hawkes
Nick Baker
Alex Lohr
Jean P. Lepoivre

Regional Liaison Agents
(D) Jörn Behnsen
(E) Vicente Sifre
(F) Michel Raoust
(GB) Alan Hildon
(GR) Matheus Santamouris
(P) Eduardo Maldonado

Further information or copies of the brochures can be obtained from prof. ir. Cees den Ouden, EGM Engineering BV, P.O. Box 1042, 3300 BA Dordrecht, The Netherlands.

BUILDING 2000

ommission of the European Communities

Renovation of two nineteenth century buildings for use as artists studios, offices and dwellings.

Addition of transparent nsulation system to outer wall o reduce heat loss and enable solar gains to contribute to space heating.

Optimizing building use by mprovement of daylighting.

INNERCITY STUDIOS AND OFFICES

WEST BERLIN/ FEDERAL REPUBLIC OF GERMANY

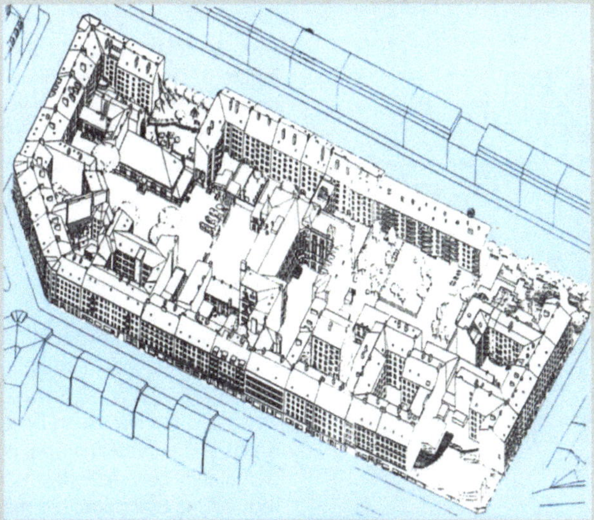

Building 2000 is a series of design studies illustrating passive solar architecture in buildings in the European Community.

ISSUE 05

FEB 1991

PROJECT DESCRIPTION

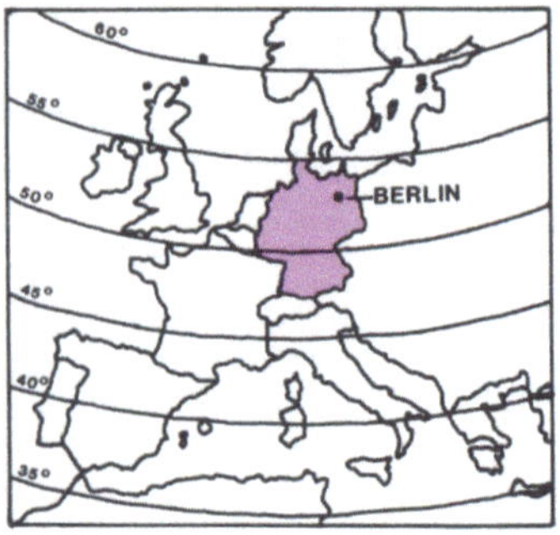

Figure 1. Location of Berlin

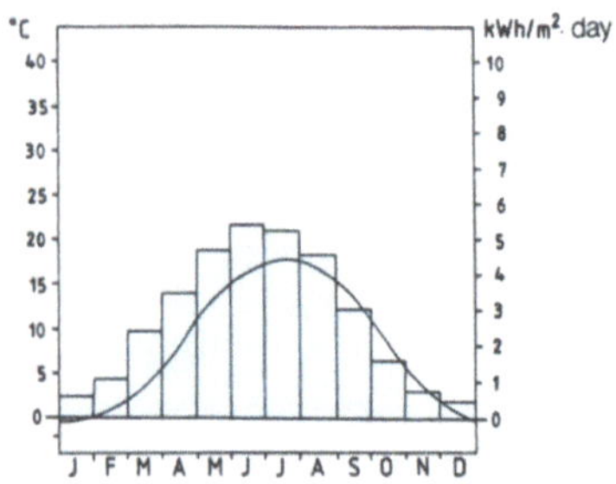

Figure 2. Mean solar irradiation and temperature data

Building Type

This project concerns the renovation of two five-storey inner city buildings which were originally constructed in the middle of the nineteenth century. The completed buildings will contain a mix of residential accommodation, artists' studios and offices for small businesses, trade organizations and social institutions.

Location

The buildings are at 80 Naunynstrasse and 9 Oranienstrasse in Block 103 in Kreuzberg in West Berlin (latitude 52.4 ° N). Kreuzberg, part of central Berlin, contains a high density of residential, industrial and commercial buildings dating from the nineteenth century. Many of the buildings are in urgent need of repair and Block 103 is part of a large inner city renewal area.

Site Microclimate

Average solar irradiation and temperature data for Berlin are given in Figure 2. The average annual temperature is 8.5 ° C. In January the daily average temperature is -0.6 ° C; in July it is 18 ° C. The average annual global irradiation is 1024.3 kWh/m^2 (3647 MJ/m^2). At individual sites, however, climate data can differ very considerably from these mean values. In the city centre, for instance, the building density and road surfaces causes thermal warming. In addition, the compact design of the buildings does not allow for much air change. Local temperatures can, therefore, be well above the Berlin average.

Air pollution is a problem throughout the city. Pollutants comes in from other places. In addition, many of the older buildings in the city are heated with fossil fuels so that smog occurs regularly in winter and it is often necessary for temporary limits to be put on motor traffic and industrial production. The situation in Kreuzberg is particularly bad; pollution reaches the city's peak levels and is regularly above permitted maxima. Only if these maxima are exceeded in several districts of the city at the same time are restrictions put on industrial production and traffic.

Design and Construction Details

The Original Buildings

Both buildings are located in Block 103 (see Figure 3) and are typical of central Berlin's five-storey older buildings. Each consists of a front section facing the street and a side wing running into the inside of the block (see Figures 4 and 5). Separate staircases give access to the front section and side wing upper floors.

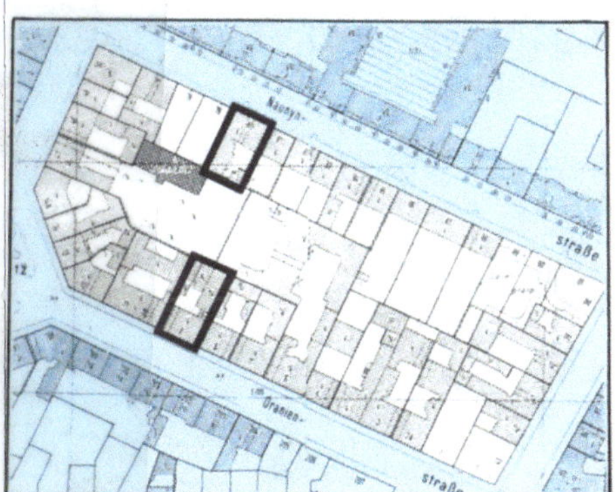

Figure 3. Location of 80 Naunynstrasse and 9 Oranienstrasse in Block 103

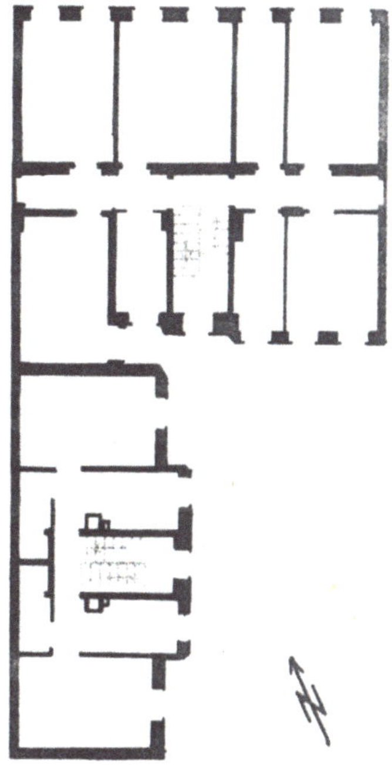

Figure 4. Ground floor plan of 80 Naunynstrasse

80 Naunynstrasse was built in 1874 and is on the north-eastern side of Block 103. The whole building has a volume of 3315 m^3 , of which 2153 m^3 are in the front section and 1162 m^3 in the side wing. The front section was originally part of a terrace. It is now joined to the building to its south-east. The building to the north-west of the front section, however, was destroyed during the war and there is now a space in that position. The buildings originally joined to the side wing are no longer there either.

Before renovation, 80 Naunynstrasse was still heated by individual coal- fired stoves. Because of its age and poor maintenance it was not energy efficient and was in urgent need of improvement. The outer walls (640 mm thick at ground floor level reducing to 250 mm on the top floor) had poor U-values (1.0-1.9 W/m^2 K). The windows were in bad condition and the side wing end walls were unplastered in places and penetrated by damp.

9 Oranienstrasse was originally built in 1862 and is on the south-western side of Block 103. The front section forms part of a terrace, all of which is intact. The building to the north-west is five storeys high so the main courtyard of 9 Oranienstrasse is badly lit. The inner yard is only open to the north-west. There, it borders the inner area of Block 103 where there is some greenery. Most of the side wing of 9 Oranienstrasse is free of adjacent buildings. There is, however, a single-storey building on the south-eastern end wall. The rooms in the side wing are particularly badly lit because their windows face north-west.

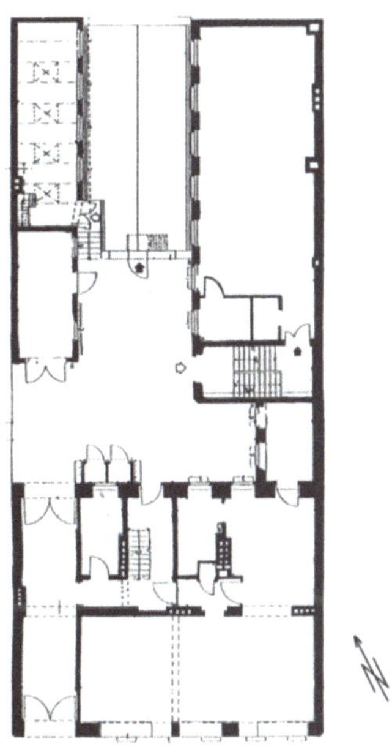

Figure 5. Ground floor plan of 9 Oranienstrasse

Use of Renovated Buildings

80 Naunynstrasse is to be used as offices for trade organizations and social institutions.

The front section of 9 Oranienstrasse currently contains residential accommodation and small businesses; the cellar on the yard side and a former coach house are artists' studios. These uses will continue once renovation is complete.

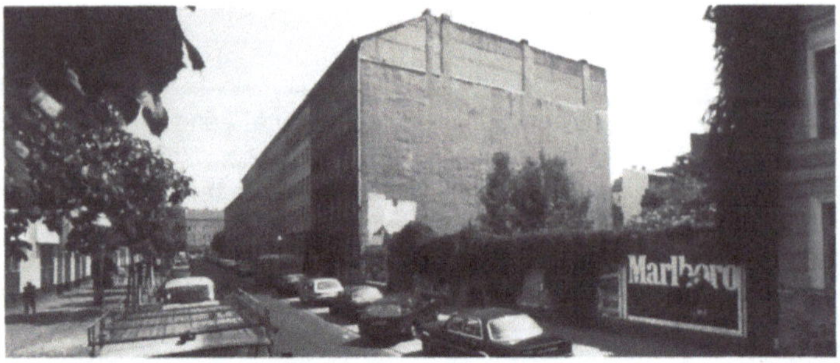

Figure 6. View of 80 Naunynstrasse from the south-west

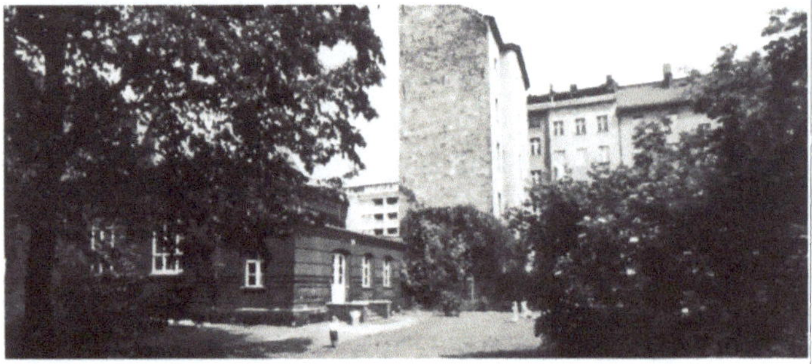

Figure 7. View inside Block 103 and side wing of 80 Naunynstrasse from the south-east

The Renovation Programme

The renovation of Block 103 is being carried out as part of a publiclyfinanced urban renewal programme where prototype ecological features are incorporated in the buildings and building users are heavily involved in the planning and implementation process. In Block 103 it has been decided that a rational, environmentally-friendly energy technology should be used. The basic heating system for the Block has already been installed. The general level of insulation has been improved. In addition, 80 Naunynstrasse and 9 Oranianstrasse make use of some passive solar features.

In 80 Naunynstrasse, a transparent insulation system (TIS) has been installed on the side wing end wall and will be used in conjunction with a gas-fired central heating system to heat the building. This is the first installation of its kind in Berlin. It is hoped that, if the initial tests are successful, it will be used in a number of buildins and make a useful contribution to energy savings and environmental improvement.

At 9 Oranianstrasse, a daylighting scheme is being tried out. This aims to save energy by reducing the need for artificial light and improve living and working conditions. In the studios, in particular, there is a need for even light without harsh shadows.

DESCRIPTION OF PASSIVE SOLAR FEATURES/COMPONENTS

Transparent Insulation System (TIS) at 80 Naunynstrasse

Principles of the TIS

The idea of the TIS was suggested by B.u.L. Energieplan and developed further by the Weidlich engineering company. It consists of a transparent layer of insulation which is fitted to a southerly-facing end wall of a building. There is a gap between the insulating material and the wall which is sealed off from the outer air. The insulation slows down heat loss from the adjacent rooms. In addition, the layer of air between the insulation and the wall of the building warms up when the sun shines and can be used for heating the building.

The system has three variations. In the simplest (the solar wall without convection - see Figure 8), the wall of the building serves as the medium for transporting the radiant heat from the sun. The heat of the warmed air in the gap is absorbed by the wall which acts as a thermal store, heating up the rooms adjacent to it. To prevent overheating, the insulated wall must be shaded once the heat requirements of the building have been met. This variation of the TIS is not as effective in using the available solar radiation as the other two versions described below.

In the solar wall with convection (Figure 9) direct use is made of the air warmed by the sun. Vents in the building wall permit warm air to be conveyed into the building with the assistance of fans so that it can heat the rooms. Cooler air is directed back into the air gap behind the insulating facade by means of door cracks and return vents in the end wall. It is then warmed up again. Like the solar wall without convection, this version of the TIS requires a shading facility as a protection against overheating. The advantage of the open loop system is that, with suitable piping, more rooms can be heated than just those adjacent to the insulated facade. However, a means of cleaning the inside of the insulation must be provided in the open version because impurities can get into the gap between the wall and the insulating facade.

In the third version of the TIS (the flat collector system with transparent insulation - see Figure 10), water is the heat transfer medium. This is warmed in flat collectors located between the wall of the building and the insulating layer and collected in a storage tank which, by means of heat exchangers, preheats the water for the conventional domestic hot water and space heating systems. The presence of the storage tank smooths out the minute-by-minute changes to the solar radiation incident on the building facade. Solar energy is used more effectively in this version of the TIS than in the others. On the other hand, the system is more complex, has higher capital costs and requires supplementary energy for circulation of the water.

Choice of TIS for 80 Naunynstrasse

It was decided to install the simplest version of the TIS (the solar wall without convection) at 80 Naunynstrasse because of its low capital costs. It was felt that if it could be showed that transparent insulation was competitive with conventional opaque insulation, future use of this system in other old buildings would be encouraged.

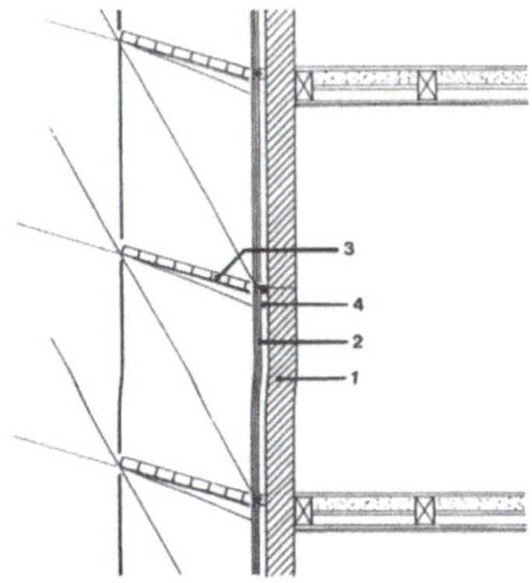

Figure 8. TIS without convection

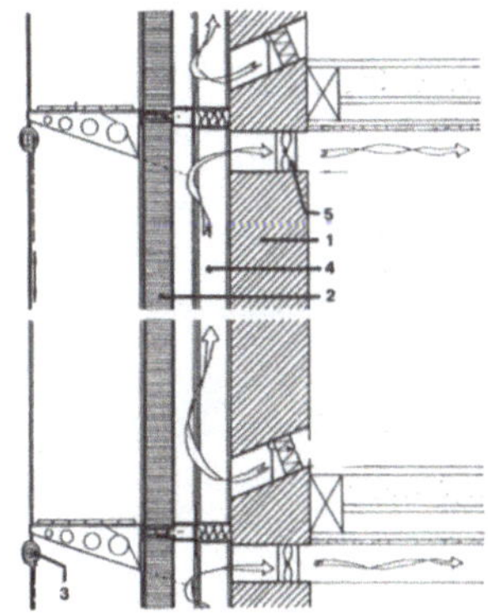

Figure 9. TIS with convection

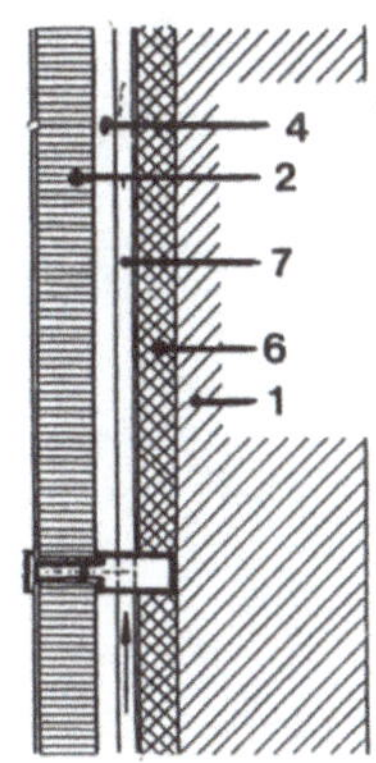

Figure 10. TIS with flat collector system

Key to Figures 8-10:

1. End wall
2. Transparent insulating layer
3. Shading element
4. Air gap
5. Air pipe with fan
6. Opaque insulating layer
7. Flat collector

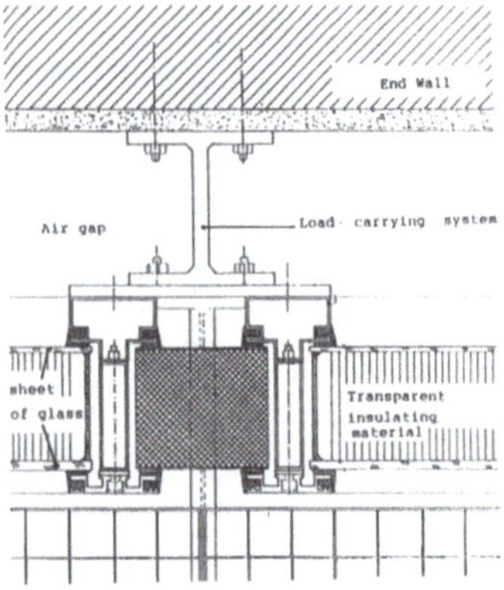

Figure 11. Horizontal section of the TIS facade

Installation of TIS at 80 Naunynstrasse

The insulating facade has been installed between the first and top storeys on the south-west end wall of the side wing - see Figure 12. The insulating elements consist of light-permeable homeycomb or capillary 80 mm polycarbonate sheets protected from contamination on both sides by sheets of glass. They are mounted on a load-bearing system of thermally-separated sections at a distance of 150-200 mm from the outer wall of the building - see Figure 11.

To maximize the energy savings achieved by the TIS, the rest of the building is well insulated. The other exterior walls of the side wing are insulated with conventional materials. This also prevents tensions being set up and the building damaged from the temperature changes in the end wall. All windows have heat-protective glazing and the ceilings of the upper storey and cellar are also insulated to reduce heat loss.

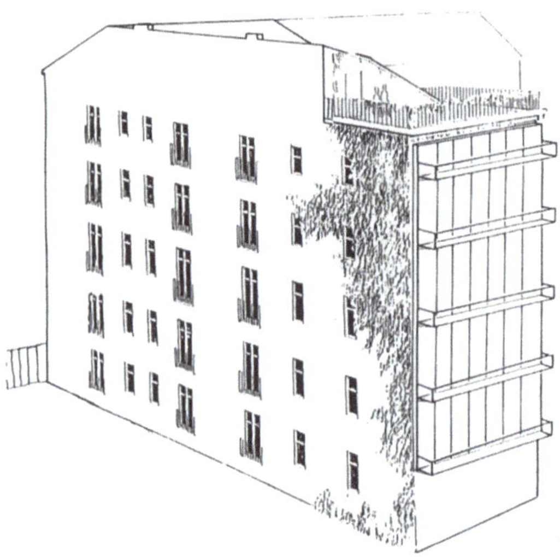

Figure 12. The TIS facade on side wing of 80 Naunynstrasse

Use of the TIS at 80 Naunynstrasse

The main use of the TIS is to insulate the end wall to reduce winter-time heat losses. In addition, if sunshine conditions are right, heat produced by the sun is stored in the wall of the building and provides basic heating for adjacent rooms in the side wing. The local climate conditions are such, however, that it is not possible to heat the rooms by the TIS alone - the solar system has to be augmented by a gas-fired low temperature central heating system.

In summer, no heating is generally required in the building. Therefore the solar wall will as a rule be shaded. At the time of writing this brochure the possibility of using solar collectors for this purpose was being considered.

Daylighting System at 9 Oranienstrasse

Objective of Daylighting System

The objective is to improve the daylighting of 9 Oranienstrasse so that optimum working conditions are provided for the artists using these rooms. A high level of illumination is needed with uniform light distribution. Direct sun penetration which could cause harsh shadows is to be avoided.

Original Daylighting at 9 Oranienstrasse

As indicated earlier, the daylighting in the original building was very poor because of the density of the surrounding buildings and the orientation and position of the windows, which produced sidelighting - see Figures 13-16. Lighting levels were low, especially on the lower floors. The distribution was typical of side-lit spaces - relatively high illumination levels near the windows and very little light at the rear of the rooms. There were glare problems due to the contrast between the bright sky visible from the workplace and the dark surroundings.

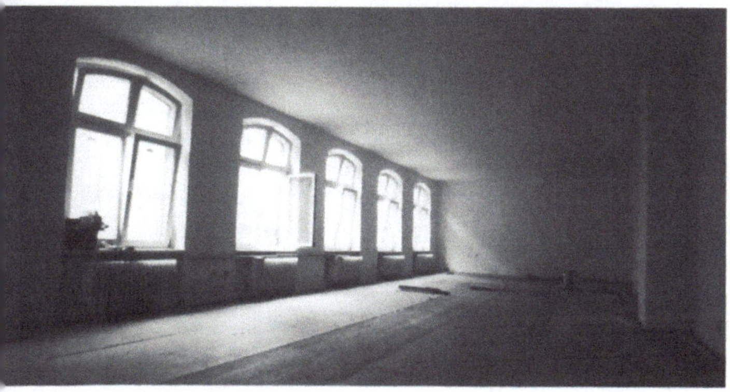

Figure 13. Side wing of 9 Oranienstrasse showing artist's studio with windows facing inner yard

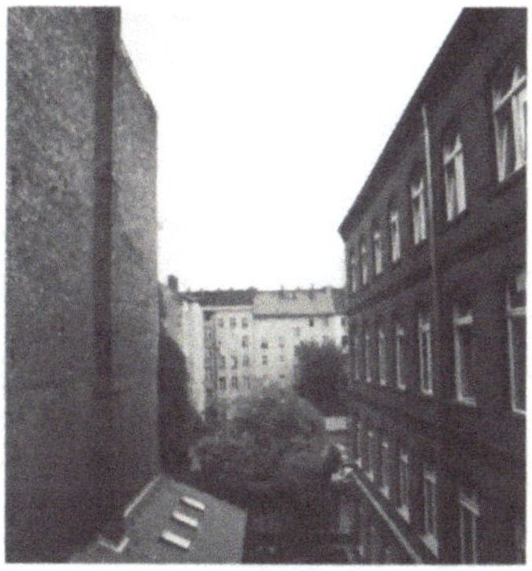

Figure 14. Inner yard of 9 Oranienstrasse

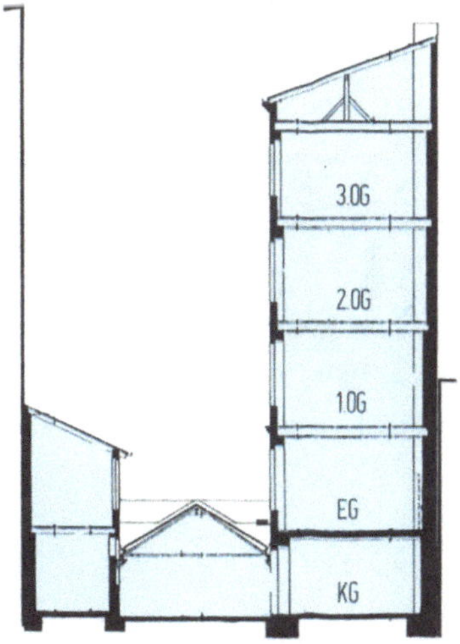

Figure 15. Vertical section of side wing of 9 Oranienstrasse

Options for Improving the Daylighting

Several options were available for improving the daylighting both quantitatively and qualitatively:

- the illumination conditions in the courtyard could be improved by painting the courtyard walls white;
- sidelighting from two sides could be achieved by adding window openings to the rear (fire) wall of the rooms;
- light-directing elements could be used to increase the lighting levels in the dark zones of the rooms.

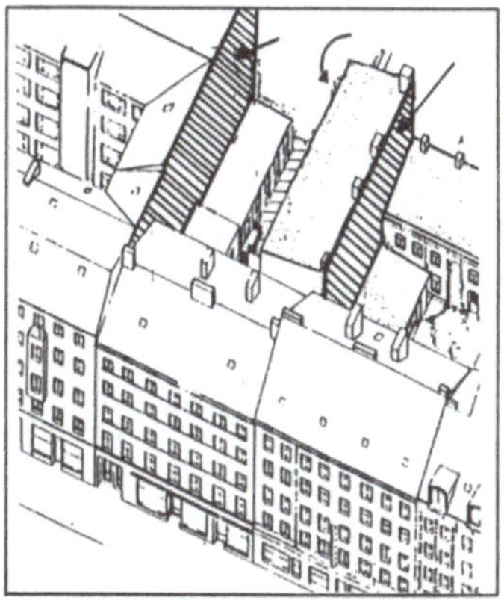

Figure 16. Areas affected

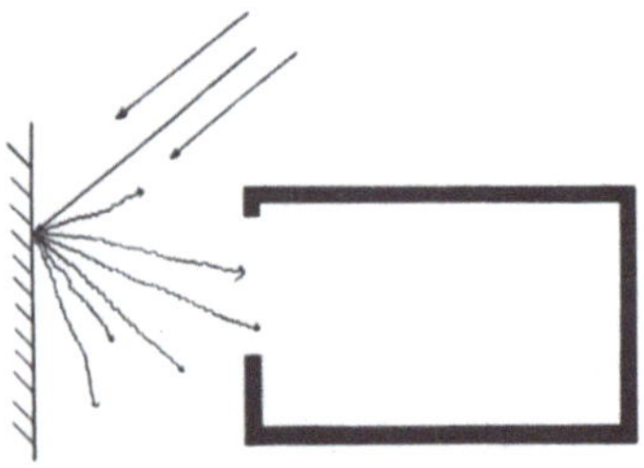

Figure 17. Effect of painting courtyard walls white

The use of light-coloured paint on the courtyard walls would affect the inter-reflection of daylight in the courtyard - see Figure 17. Daylight striking the walls would be reflected more strongly because of the higher reflectance of the walls. Therefore, more diffuse daylight would penetrate the rooms, thus enhancing working conditions from the artist.

In general, double sidelighting provides a relatively high illumination level while reducing the dark zones in a room - see Figure 18. Deepening a room under these circumstances can lead to an increase in the useful area. In the case of 9 Oranientstrasse, introduction of a second window opening could increase the illumination level for a given room depth and at the same time reduce the generation of shadows, thus providing more suitable conditions for an artist. The position, size and design of the opening in the south-east wall would have to be considered very carefully, however, if entry of direct sunlight was to be avoided throughout the day. An additional restriction is that the south-east wall is a fire wall, where normally no openings are allowed.

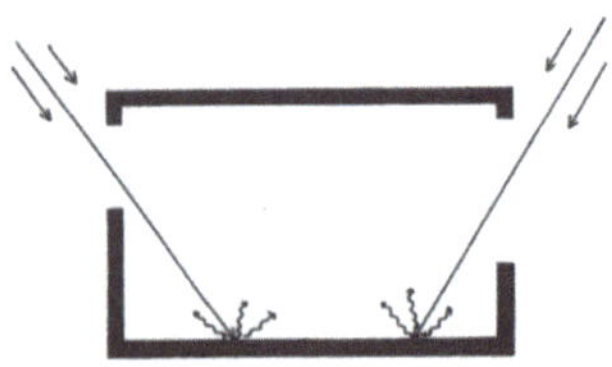

Figure 18. Effect of adding a second window

Location of control elements (such as a mirrored plate inclined at a specific angle in front of the window) at the level of the window transom (i.e. above eye height) would affect daylight distribution in the room - see Figure 19. The amount of light near the window would be reduced and the amount in the rear of the room increased because of reflection of light onto the ceiling. These elements would also act as shading devices and prevent entry of direct sunlight into the room.

Choice of Daylighting System for 9 Oranienstrasse

As described in the energy calculations/design tools section below, the above options were tested using scale models. The tests showed that the following actions would be technically appropriate for 9 Oranienstrasse:

(1) paint the dark brick walls white in order to change their reflectivity;

(2) add windows to the side wing end wall if permission can be obtained from the building regulations authority on the grounds that the new windows are required to improve the lighting level in the rooms;

(3) install reflective light-directing elements in front of the windows.

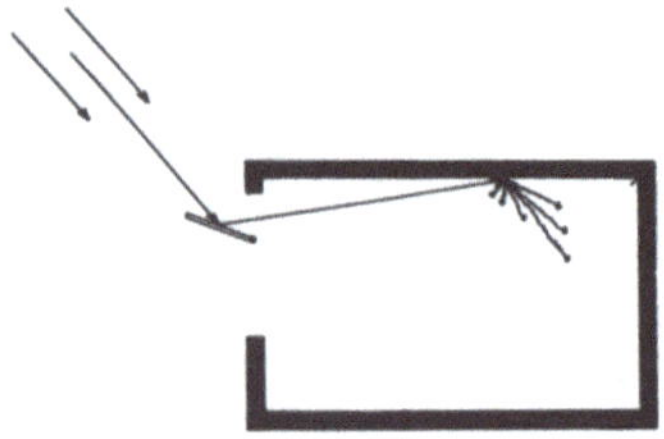

Figure 19. Effect of placing a mirrored plate in front of window

It was decided to go ahead with recommendation (1) immediately. Implementation of action (2) was dropped for the time being on the grounds that it would be time-consuming (because of the need to enter into discussions with the building regulations authority) and bring the costs of the renovation above the budget limit. Installation of the reflector elements can be effected at any time and it was decided to postpone implementation of recommendation (3) until the artist confirmed that there was a real need for improvement of the quality of light in the studio and the money had been made available.

ENERGY CALCULATIONS PERFORMED AND DESIGN TOOLS USED

Introduction

The thermal performance characteristics of 80 Naunynstrasse were established using various computer modelling and energy calculation procedures. The various daylighting options for 9 Oranienstrasse were evaluated using scale models. These are described below.

Thermal Performance Evaluation of 80 Naunynstrasse

Results

The calculations showed that the annual effective energy consumption of the side wing with a conventional heating system and thorough insulation would amount to 33,800 kWh a year. Assuming that the efficiency of the heating system was 0.85, the final energy consumption would be 39,800 kWh/year.

With the TIS as described, the effective energy consumption would be 27,100 kWh/year and the final energy consumption 31,900 kWh/year. Thus the TIS would save 7,900 kWh final energy a year - some 20% of the energy required for a conventionally heated and insulated side wing.

Because only a little experience has been gained to date in the field of transparent insulationm the cost effectiveness of the TIS compared with a conventional system can only be estimated with difficulty. At an energy price of DM 0.10 DM/kWh, the annual savings on heating amount to DM 800. The additional cost of installing the TIS compared with traditional opaque insulation is estimated to be approximately DM 50,000. The Land of Berlin is paying a subsidy for the 80 Naunynstrasse installation which reduces the chargeable investment costs to DM 18,000. This will enable the system to pay for itself in about 22 years.

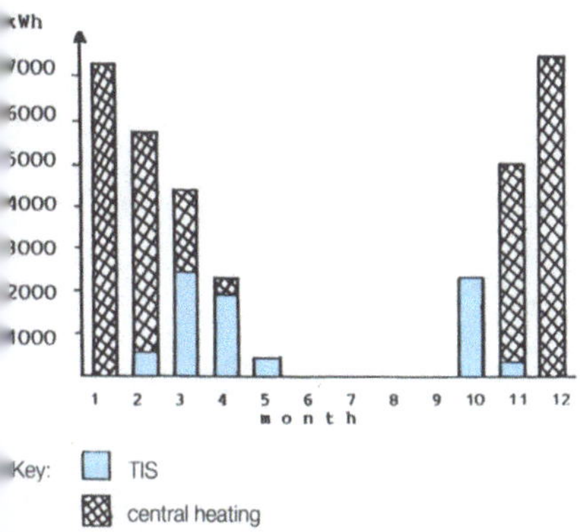

Figure 20. Monthly contribution of TIS and gas-fired central heating to heating side wing

Method of Evaluation

To establish the thermal performance characteristics of 80 Naunynstrasse, the ISFH dimensioning program for architects and the table calculation procedure of the Büro für energiegerechtes Bauen (Cologne) were used. The energy balance for the side wing was calculated with (a) the TIS as described and (b) a conventional variant consisting of a conventional heating system plus opaque insulation of the outer walls, insulation of the ceilings of the upper storey and cellar and heat-protective glazing on the windows.

The evaluations were carried out assuming that room temperatures were set in the range 18-20 C, internal heat gains were 40 kWh/day and the average air change rate was 0.8 volumes/hour.

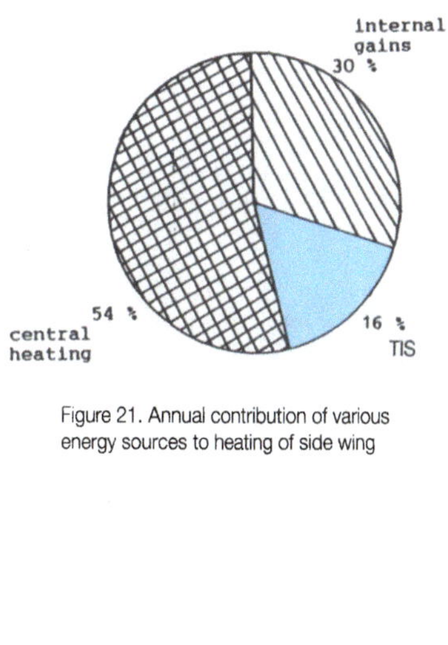

Figure 21. Annual contribution of various energy sources to heating of side wing

Method of Evaluation

Because of the complexity of the light path from the light source (i.e. the sky) to the artists' worktops, simple design tools could not be used to predict the daylight distribution on the different floors of 9 Oranienstrasse. Only scale model studies would provide the needed quantitative and qualitative information on the proposed daylighting options. Therefore a 1:50 model was built (see Figure 22), fitted with a number of photometric sensors and tested under overcast sky conditions in the artificial sky facility at RWTH Aachen. (Testing under clear sky conditions was not felt to be necessary because the density of the surrounding buildings at 9 Oranienstrasse prohibits much of the relatively small amount of sunshine experienced in Berlin entering the rooms.) The options were tested separately and in various combinations and compared with the results for the original building when the courtyard walls had low reflectivity - see Figure 23.

Performance of Daylighting Options for 9 Oranienstrasse

Results

The tests showed that the greatest quantitative improvement is achieved by changing the reflectance of the courtyard walls from 0.15 (as it would be with dark brick walls) to 0.70 (white paint). When this is done, 86% more light on average enters the rooms on all floors of the building and the light distribution becomes more uniform (and therefore more suited to an artist) because a greater amount of diffuse light reaches the rear of the rooms.

If in addition new windows are added to the end wall of the side wing, there is a 119% increase in the light entering the rooms compared with the original situation. The dimension and location of these windows is different on each storey. On the upper floors, horizontal openings of small height give the best result. On the second storey, narrow vertical windows perform best. (The single-storey building attached to the rear wall of the side wing prohibits additional windows being added to the ground floor.)

The greatest improvement from the qualitative point of view can be achieved by using light-directing elements on both the window openings of the rooms. When this is done, the daylight factor at the rear of the fourth storey room is increased from 1.2% for the original situation (i.e. dark brick courtyard walls and one side window) to 5.1% and the light level is almost uniformly throughout the room. There is also the possibility of redirecting the reflected light at the ceiling so that it falls on the artist's workplace.

It was found that if the reflectivity of the courtyard walls is low, light-directing elements can increase the amount of light entering the space by 10-30%. If, however, these elements are used in combination with white courtyard walls then the amount of light entering the rooms is reduced but the uniformity of the illumination is increased to a high degree.

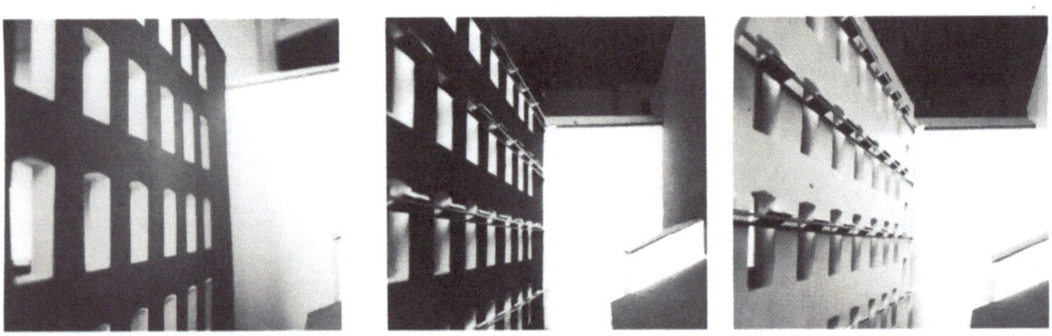

Figure 22. Scale model of 9 Oranienstrasse used for testing different daylighting options

GENERAL DESIGN GUIDELINES/POINTS OF INTERST RESULTING FROM THIS PROJECT

Transparent Insulation Systems (TIS)

So far, there has been little experience of the use of TIS in inner city areas under climate conditions similar to those found in Berlin. The studies described in this brochure have shown that:

- the installation of a TIS in older inner-city buildings is possible from the town planning, architectural and technical points of view;

- the fossil fuel savings achieved by use of a TIS can make a real contribution to the local environment;

- at present, the TIS is only economically acceptable if public sector financing can be obtained for system development and installation. For the system to become more widely marketable, simple inexpensive versions would have to become available.

Further research on the installation of TIS in older inner city buildings is needed. Monitoring is necessary to optimize the performance of the system in use.

Daylighting

The studies carried out for improving the daylighting at 9 Oranienstrasse have shown that:

- incorporation of various daylighting features in a building in a densely built up area can lead to such an improvement in lighting levels that the building can be put to a greater use;

- considerable improvements in daylighting can be achieved by relatively inexpensive measures such as use of light-coloured paint on end walls and facades;

- lighting requirements depend on the type of use to which the building is being put. Therefore daylighting measures must bear in mind specific user needs if they are to produce genuine improvements.

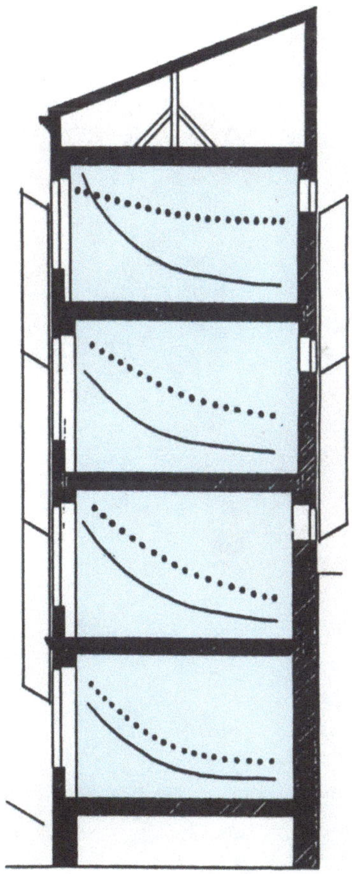

Figure 23. Daylight distribution in each storey

BUILDING 2000 Building 2000 brochures are published by Directorate General XII of the Commission of the European Communities to show how design studies can help architects and other building designers use passive solar principles to the best effect to produce attractive energy efficient buildings. Each brochure describes studies carried out with the support of the Commission during the design phase of one of thirty-six non-domestic buildings in the EC Member States. The studies were on such topics as daylighting, heating, cooling, ventilation, comfort, control systems and urban design. They were carried out with the help of acknowledged European experts in these fields and drew heavily on lessons learned and techniques developed through the Commission's research and development programme on solar energy applications to buildings.

Commission of the European Communities/Directorate-General for Science, Research and Development

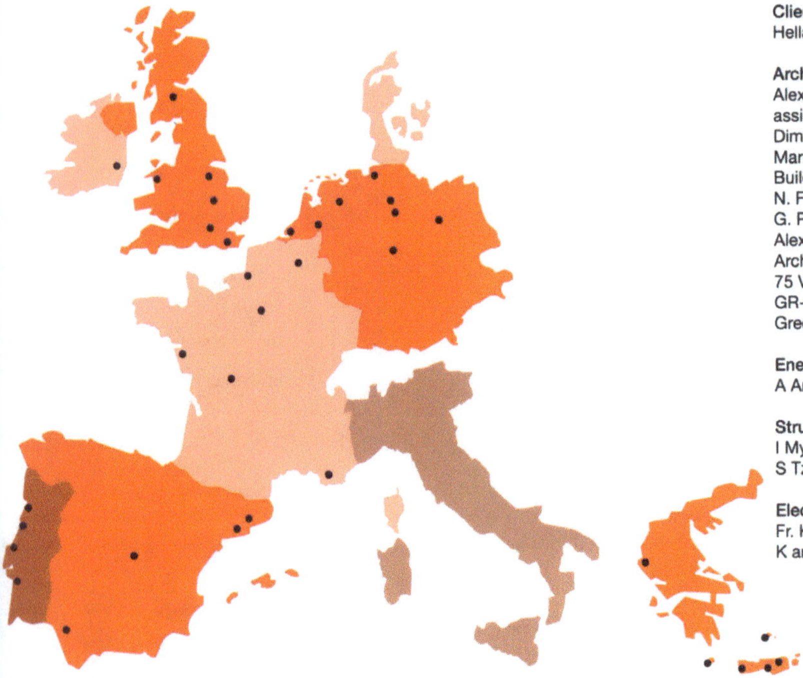

List of Design Team Participants and Advisers

Client
Hellas Holiday Hotels

Architect
Alexandros N Tombazis
assisted by
Dimitris Potiropoulos, architect
Maria Spyridaki, architect for the Building 2000 aspects of the project
N. Fletorides, L. Nella-Potiropoulou and G. Panetsos, design team architects
Alexandros N Tombazis and Associates, Architects
75 Vasilisis Sofia Avenue
GR-11521 Athens
Greece

Energy Consultant
A Argiriou

Structural Engineer
I Mylonas
S Tzivanakis

Electrical and Mechanical Engineer
Fr. Krispis
K and A Alexopoulos

This set of **Building 2000** brochures illustrates how architects and other building designers can successfully apply passive solar principles to produce energy-efficient buildings.

BUILDING 2000 Participants

Project Director
Theo C. Steemers

Coordinator
Cees den Ouden

Technical Steering Committee
Dean Hawkes
Nick Baker
Alex Lohr
Jean P. Lepoivre

Regional Liaison Agents
(D) Jörn Behnsen
(E) Vicente Sifre
(F) Michel Raoust
(GB) Alan Hildon
(GR) Matheus Santamouris
(P) Eduardo Maldonado

Further information or copies of the brochures can be obtained from prof. ir. Cees den Ouden, EGM Engineering BV, P.O. Box 1042, 3300 BA Dordrecht, The Netherlands.

Commission of the European Communities

BUILDING 2000

PUBLIC BUILDINGS

P1
Magistrates Courthouse, Rotherham, UK
P2
Community Centre for Cultural Activities, Le Vigen, France
P3
Public Library, Newbury, UK
P4
Community and Educational Centre, Steyerberg, Germany
P5
Town Hall, Ilhavo, Portugal

Building 2000 is a series of design studies illustrating passive solar architecture in buildings in the European Community.

VOLUME 2 / **KLUWER ACADEMIC PUBLISHERS**

BUILDING 2000

Commission of the European Communities

- Magistrates' courthouse containing 10 courthouses, public waiting areas and other facilities
- Orientation allows high proportion of south-facing glazing
- Spread-out building form incorporating central courtyard allows much of the building to receive daylight and natural ventilation
- Provision of direct gain spaces two storeys high
- Preheating of ventilation air for courtroom by heat extracted from exhaust from direct gain spaces
- All courtrooms receive direct and/or borrowed daylight

MAGISTRATES' COURTHOUSE

ROTHERHAM/UNITED KINGDOM

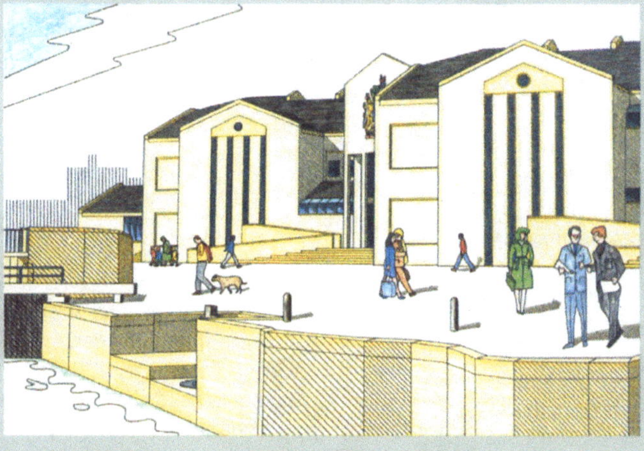

Building 2000 is a series of design studies illustrating passive solar architecture in buildings in the European Community.

ISSUE

P 1

FEB 1991

PROJECT DESCRIPTION

Building Type

This project concerns a new magistrates' courthouse for the Rotherham District in the County of South Yorkshire, UK. The building contains 10 courthouses plus public waiting areas and other public facilities, magistrates' facilities including retiring rooms, solicitors', police and press facilities, public counters for payment of fines and administration offices.

Location

The building is being constructed on the edge of Rotherham town centre (latitude 53 ° 26 ' N) in the industrial valley of the River Don, much of which is now being redeveloped for commercial use (see Figure 1). The site is 25 m above sea level and is roughly triangular in shape, bounded by a canal, a railway line and the local police headquarters (see Figure 2). In winter, the latter causes overshadowing between 10.00 am and 1.00 pm.

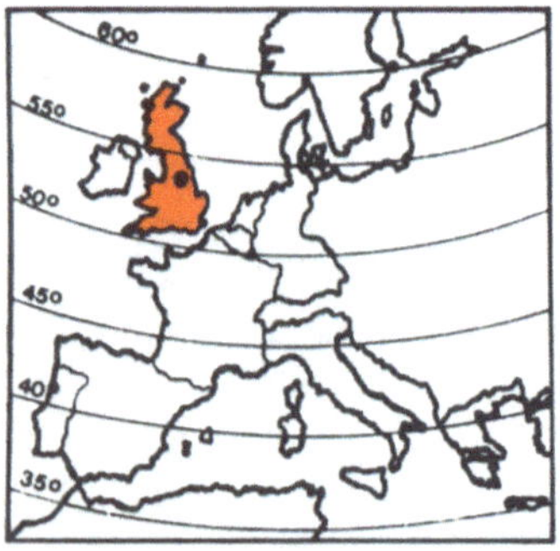

Figure 1. Location of Rotherham

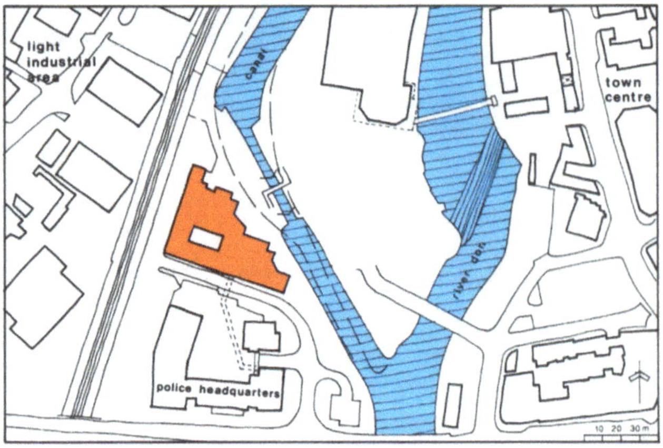

Figure 2. Site layout

Site Microclimate

The climate is maritime in character with relatively mild winters and fairly cool summers. Passing frontal systems travelling east from the North Atlantic make the weather changeable and bring cloud and rain. Their effect is moderated by the presence of the Pennine hills to the west of Rotherham. The total annual rainfall is 808 mm. Winds are predominantly from the south-west but in winter there is the occasional severe east wind. Monthly mean solar irradiation, degree day, sunshine hour and temperature data are given in Table 1.

Month	Solar irradiation on horizontal plane global	diffuse	Degree days base 18 ° C	Sunshine hours	Air temperature
	MJ/m^2	MJ/m^2	° C days	hours	° C
January	144	70	447	38	3.4
February	248	99	411	54	3.5
March	483	166	390	92	5.5
April	637	220	307	129	8.3
May	816	237	215	162	11.3
June	851	207	125	183	14.5
July	829	246	88	160	16.0
August	712	233	89	142	15.7
September	510	187	137	111	13.9
October	347	128	237	84	10.7
November	181	73	347	43	6.6
December	104	46	421	35	4.6
Total	5863	1912	3214	1233	

Table 1. Solar irradiation, degree day, sunshine hour and temperature data

Design and Construction Details

Design Objectives

The design objectives were as follows:

- to reflect the authority of the courts;
- to reflect the open and public nature of the administration of justice;
- to segregate the circulation routes of the different categories of people using the building (magistrates, defendants, adults, juveniles);
- to provide a humane and comfortable internal environment for all buildings, relying as much as possible on natural ventilation and daylighting and exploiting the potential for passive solar heating;
- to make a positive contribution to the town centre, not only by creating an important public and civic building but also by providing well-designed outdoor spaces around and within the complex which relate to the canal and pedestrian approaches;
- to achieve a building with low running costs;
- to provide security and deter vandalism.

Design Details

Plans, sections and elevations, etc., are shown in Figures 3-11 and design data are given in Table 2. The courtrooms are arranged in two staggered two-storey blocks separated by the main entrance. Together with the L- shaped administration block they enclose an open landscaped courtyard in the middle of the complex. One block contains juvenile courtrooms; the other contains formal and informal adult courtrooms. The large, high- security courtroom is on the lower ground floor next to a secure holding area for defendants in custody. Both courtroom blocks have a public waiting/circulation space, two storeys high, along the southern edge. In both blocks, vertical circulation for magistrates and defendants in custody are kept separate, the horizontal connections being made on the lower ground floor. A sidewalk between the building and the canal provides a pedestrian through route across the site and connects with a bridge across the canal which provides the major approach to the site from the town centre. At intervals the canal walk widens into larger public spaces. The three public entrances to the building open off these.

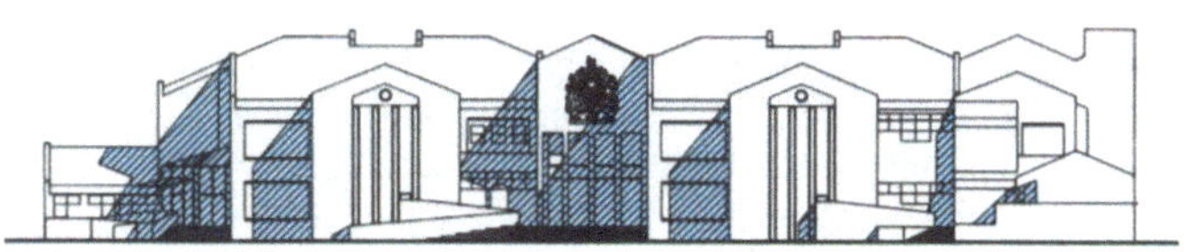

Figure 3. North elevation

Construction Details

The building has a heavy structure for security reasons and to provide sound insulation. This takes the form of reinforced concrete framework and floor slabs with brick cavity-walling and dense concrete block internal partitions. The external walls have 100 mm brick outer leaf, a cavity filled with mineral fibre insulation and 140 mm dense concrete block inner leaf. The partly-buried external walls of the lower ground floor and all concrete floor slabs in contact with the ground or exposed to open air are also insulated with mineral fibre. The roof is a timber and steel structure insulated with mineral fibre quilt and clad with slate. There is double glazing throughout. In highly glazed walls the inner pane is of toughened glass - as is the outer pane in all glazing on the bottom floor. All materials and constructional detailing are robust to resist damage from vandals. U-values are given in Table 3.

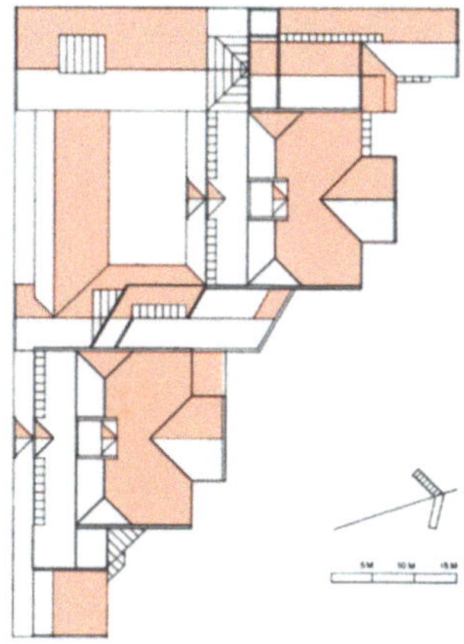

Figure 4. Roof plan

Floor area	4800 m^2
Roof area	2300 m^2
External wall area	
(including all glazing)	3650 m^2
Glazing areas	
total	750 m^2
south facing	400 m^2
Budget cost (1989) £6,750,000	

Table 2. Design data

Roof	0.23
Glazed roof	3.20
Floor	0.33
External walls	0.36
Double-glazed curtain walling	3.20
Windows	3.20

Table 3. U-values in $W/m^2 K$

DESCRIPTION OF PASSIVE SOLAR FEATURES/COMPONENTS

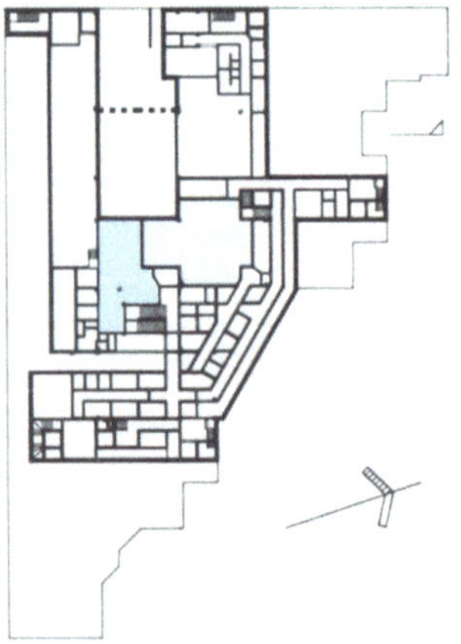

Figure 5. Lower ground floor plan

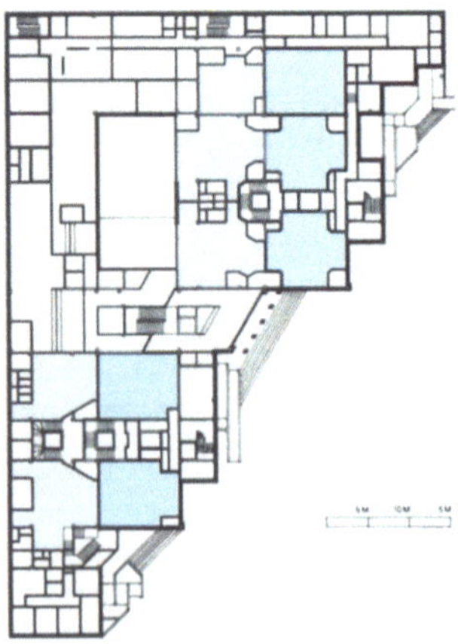

Figure 6. Ground floor plan

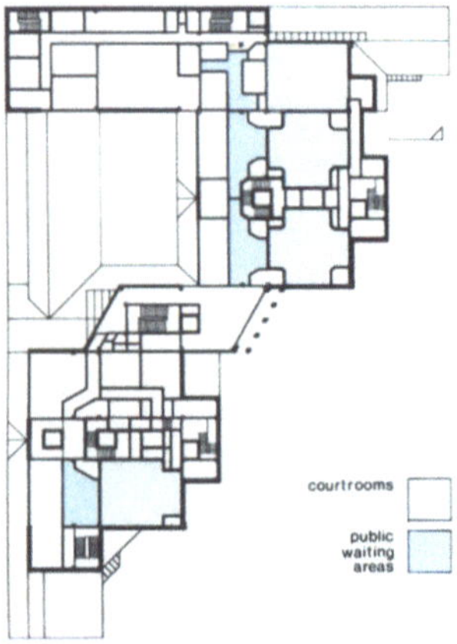

Figure 7. First floor plan

Introduction

When the brief was formulated, the magistrates expressed a strong desire for daylighting and natural ventilation (as opposed to artificial lighting and air conditioning) to be used as much as possible in the new building. This is in line with current UK trends away from deep-plan air conditioned buildings. Since running costs are important in this project, it was decided to follow up the magistrates' preferences and explore the use of design features which allow natural light and air to penetrate the building and permit passive solar energy to contribute to its heating requirements.

Passive Solar Features

Orientation

The building has been oriented north-south to enable a high proportion of glazing to be located on south-facing walls. The north, east and west glazing has been minimized.

Building Form

The building has been spread out over the whole site and arranged around an open courtyard, rather than made compact and monolithic in form. This allows more rooms to be located on external walls and, therefore, to be daylit and naturally ventilated. It also increases the proportion of south- facing glazing. The part of the building to the south of the courtyard has been kept low to minimize overshadowing of the courtyard.

Direct Solar Gain

Early in the design development it was recognized that the public waiting/circulation areas, especially, had the potential to use direct solar gains to offset heating costs. Accordingly, it was decided to locate these spaces on the south side of the courtrooms and give them large amounts of glazing. Subsequent decisions resulted in these spaces becoming two storeys high. Solar heated air will be extracted at high level from these spaces and redistributed or used to pre-heat ventilation air for the courtrooms. Locating the waiting areas on the south side has the added advantage of making them bright and sunny spaces - which could help to relieve stress and tension in those waiting for court hearings. The large amount of glazing will also make it easier to supervise these areas.

Insulation

Although the spread-out design increases the potential for daylighting, natural ventilation and solar gain, it also increases the possibility of heat loss. This is offset by incorporating more insulation into the building envelope than that required by the new UK Building Regulations.

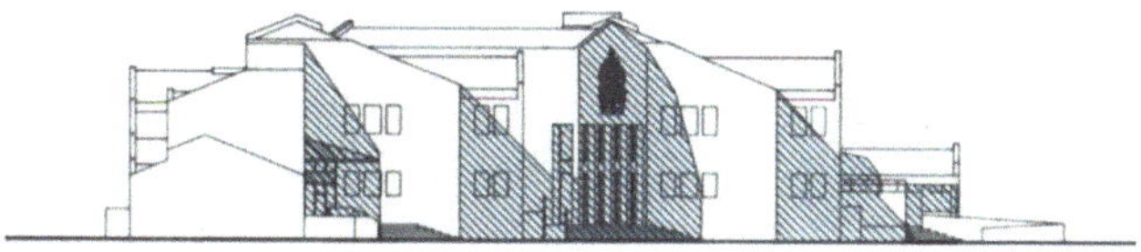

Figure 8. East elevation

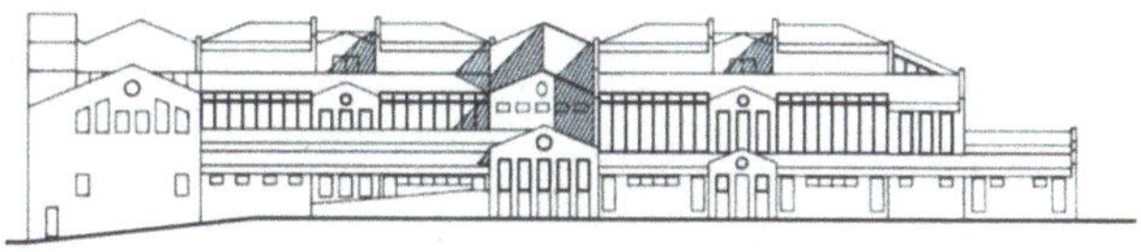

Figure 9. South elevation

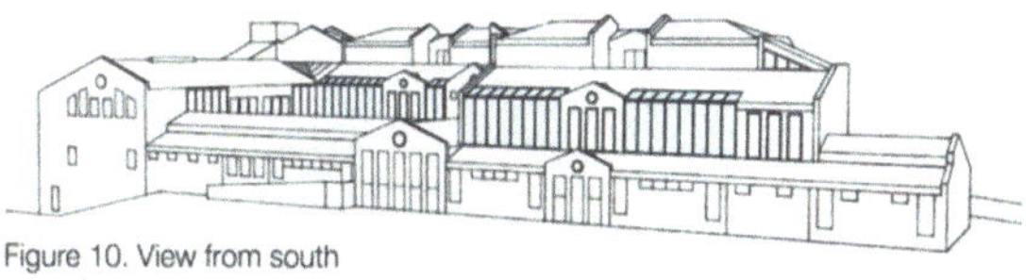

Figure 10. View from south

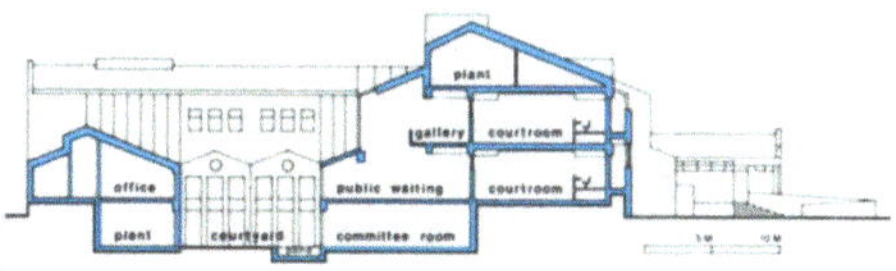

Figure 11. Section

Thermal Mass

The heavy structure required for security and sound insulation reasons will also provide sufficient thermal mass for heat storage and enough thermal inertia to help prevent summer overheating.

Ventilation/Cooling

Wherever possible, ventilation is provided by natural means (i.e. opening windows) and controlled by building occupants. Where greater environmental control is required, such as in the courtrooms, a system is installed to supply fresh air mechanically. The input air will be pre-heated, if necessary, in the direct gain spaces. In the six courtrooms with direct daylight the magistrates will be able to switch off the mechanical air system and open the windows if they wish.

In summer, the air flows generated by temperature stratification and natural ventilation will keep the public waiting areas cool and ameliorate the effect of overheating in these highly-glazed spaces.

Shading

Automatically-controlled retractable blinds will be installed on all south- facing roof glazing and some vertical glazing.

Control

Although occupants will be able to control their own environment in much of the building, a building management system will be installed to control and record all aspects of the building's energy consumption and use.

Daylighting Features

The spread-out design permits daylight penetration into a greater proportion of the internal spaces.

The central courtyard allows much of the lower ground floor to be daylit.

Usually, courtrooms are located in the interior of buildings. In this project, six out of the ten courtrooms have been placed on an external wall and thus receive direct natural light. In addition, borrowed daylight has been introduced into nine out of the ten courtrooms via the highly-glazed two-storey public waiting/public circulation spaces.

Stairwells allow daylight to penetrate the lower reaches of the building.

A pond in the open courtyard enlivens the daylighting in the Magistrates' Committee Room on the lower ground floor.

Operation of Passive Solar Systems

Basic Principles

The public waiting areas are used as direct gain sunspaces. As these areas are occupied by members of the public wearing outdoor clothes, the temperature in these spaces is allowed to float. It is felt that a minimum temperature of 16 ° C is acceptable in winter and a maximum of 26 ° C in summer. The predicted temperatures reached in these areas in the different seasons are given in Figures 12-14.

Conditions in the courtrooms are more critical and will be controlled when required to 19-21 ° C. When the temperature in the waiting areas is above the minimum set point in winter, warm air is extracted at a high level and passed through a heat exchanger to preheat the fresh air input to the courtrooms. The preheating takes place in recuperators built within the main air handling plant.

Blinds are fitted to all glazing in the public waiting area to limit solar gains in summer. To reduce any overheating which does occur, low and high vents are opened to allow outdoor air to pass through these spaces by the stack effect.

The courtrooms are also ventilated using a displacement air system. Fresh air is introduced at a low level and at a temperature slightly lower than the temperature required in the space. The system provides pleasant and healthy air because it avoids the mixing of stale and fresh air and the recirculation of used air. However, because 100% fresh air is used, heat recovery has to be incorporated into the plant.

Auxiliary Heating

Auxiliary heating in the courtrooms and public waiting areas is by low-pressure hot water radiators or perimeter heating convectors.

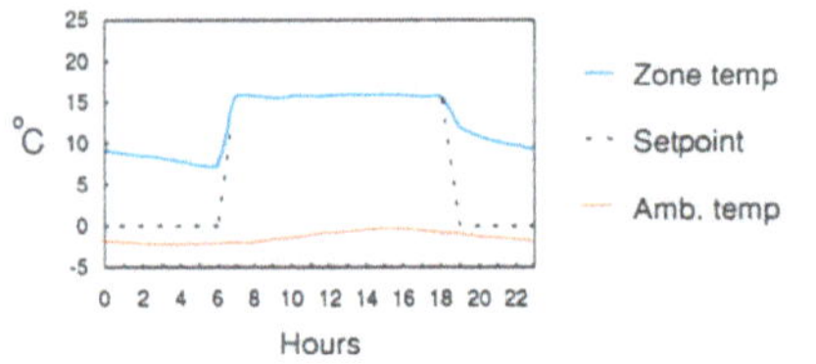

Figure 12. Temperatures in public waiting area on a winter day

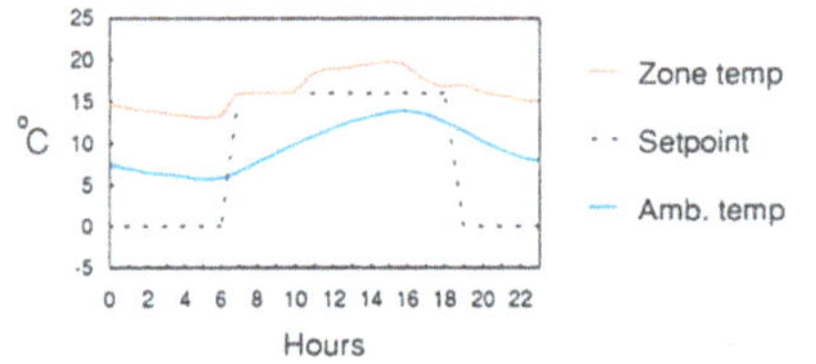

Figure 13. Temperatures in public waiting area on a spring day

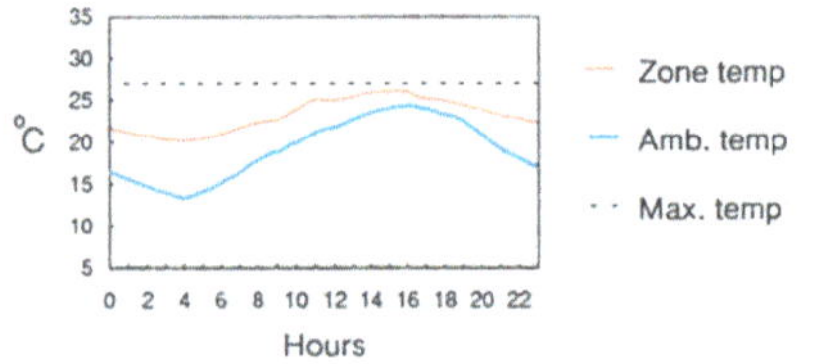

Figure 14. Temperatures in public waiting area on a summer day

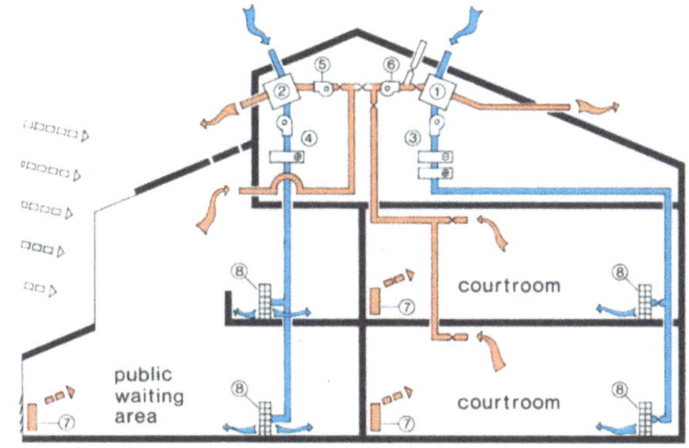

Figure 15. Mode of operation in winter

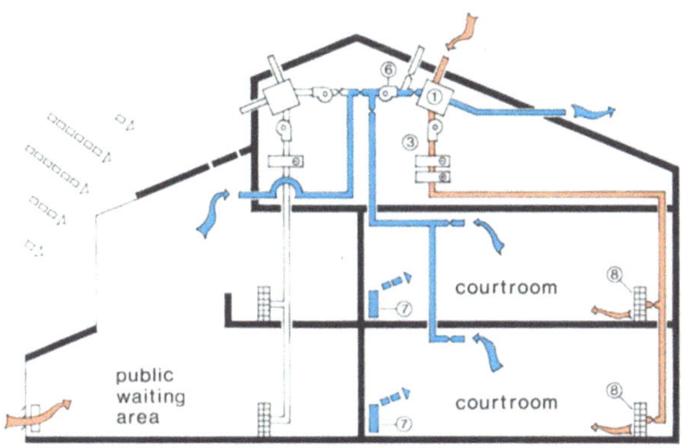

Figure 16. Mode of operation in spring/autumn

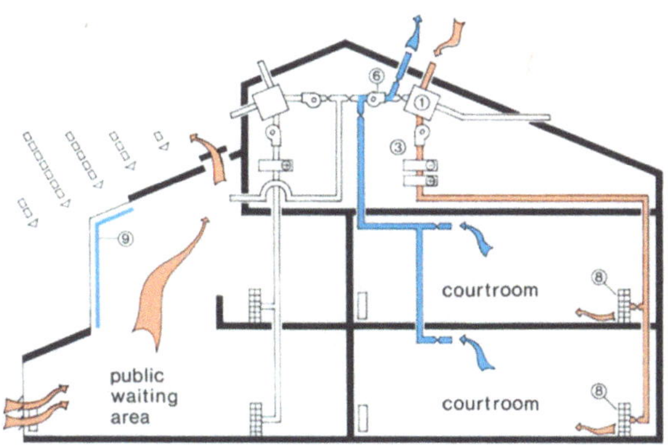

Figure 17. Mode of operation in summer

Key to Figures 15-17:

1. courtroom supply recuperator
2. public waiting area supply recuperator
3. courtroomplant
4. public waiting area plant
5. public waiting area extract
6. courtroom extract
7. heating system
8. displacement ventilation input units
9. blinds

Modes of Operation

Winter

In the public waiting areas, the auxiliary heating is on when required to maintain a minimum temperature of 16 ° C. The ventilation system gives a minimum air change rate. The vents are closed. Used air passes to the public waiting area recuperator.

In the courtrooms, the auxiliary heating is on, maintaining a temperature of 19 ° C. The ventilation system is on - or windows are opened if desired. Used air passes to the courtroom recuperator.

Spring/Autumn (Bright, Cold Days)

In the public waiting areas, there are solar gains and the heating is on when required to maintain a temperature of 16-25 ° C. The ventilation system is off. The low level vents are regulated to provide fresh air input while maintaining the minimum temperature. Used air passes to the courtroom recuperator to preheat fresh air.

In the courtrooms, the heating is on if required. The ventilation system is on or the windows opened. The fresh air input is preheated by extract from the public waiting areas and courtrooms.

Summer (Hot Days)

In the public waiting areas, the heating and mechanical ventilation systems are off. The low level and high level vents are open and the blinds closed.

In the courtrooms, the heating is off. The mechanical ventilation system is on - or windows opened. Used air from the courtrooms goes direct to waste, not via the recuperator. In really hot weather the cooling plant is run to reduce the fresh air input temperature. If the extract air from the courtrooms is at a lower temperature than the fresh air then the recuperator is brought into use to restore cool conditions.

ENERGY CALCULATIONS PERFORMED AND DESIGN TOOLS USED

Introduction

The passive solar and daylighting systems were designed using various design tools. The SERI-RES computer model was used to determine the thermal performance in the public waiting areas and courtrooms. GNOME, a program developed in the School of Architecture of Humberside College of Higher Education, UK, was used to assess whether glare would be a problem in the courtrooms.

Results of Thermal Performance Studies

Design of Glazing in Public Waiting Rooms

The preliminary design for the public waiting areas incorporated fully-glazed roof areas in an attempt to maximize both daylight and solar gain. The thermal performance evaluations, however, showed that, if this was done, temperatures in the public waiting areas could rise to 28 ° C. To limit the excess heat gains to an amount which could be used to preheat the input air for the courtrooms and to reduce overheating in the public waiting areas, the roof glazing was reduced to the minimum which would allow daylight to reach the courtroom via the window in the partition wall between the waiting area and courtroom. By doing this, the maximum temperatures predicted were reduced to 25-26 ° C (see Figure 18).

Savings Made by Using Solar Gains to Preheat Ventilation Air

Using direct gains in the public waiting areas to preheat ventilation air for the courtrooms typically saved 90 kWh/day for a spring/autumn day and 60 kWh/day for a winter day. This gave annual savings of 23,000 kWh for the whole building. Very little additional plant is required for this recovery technique. The recuperators need to be slightly larger to enable them to cope with additional air volume from the waiting areas and more ductwork and dampers have to be installed to complete the air circuit.

The above savings represent the difference between using spare heat from the waiting areas and a similar system operating on full fresh air, controlled using the existing energy management facility. Savings in capital plant and running costs due to incorporation of heavy mass and insulation into the building and glazing optimization are additional to this. Compared to a conventional air-conditioned building with air conditioning, the overall capital and operational costs of this building are low.

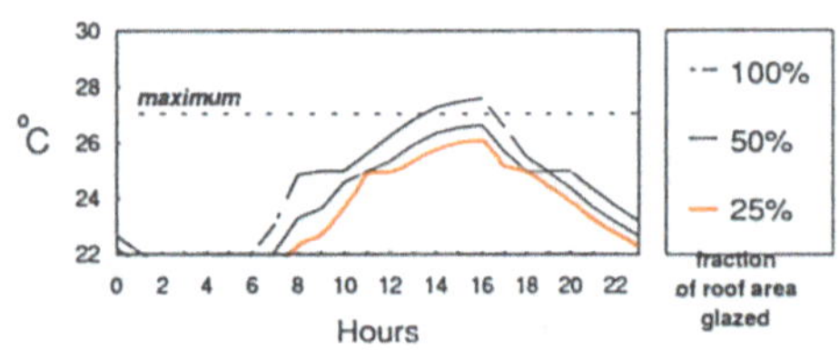

Figure 18. Temperatures reached in a public waiting area on a summer day for different amounts of roof glazing

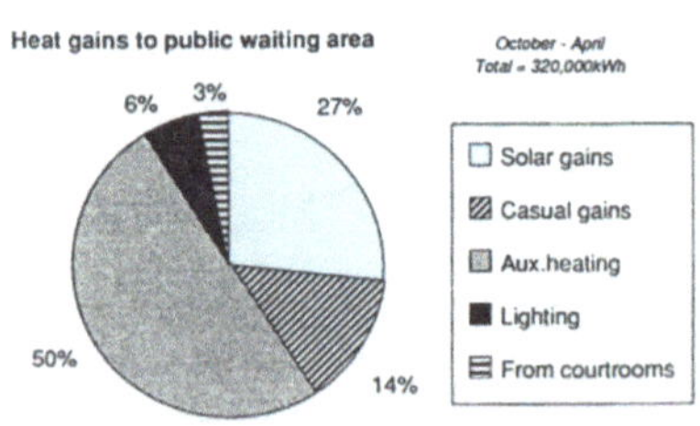

Figure 19. Breakdown of heat supply to public waiting area from October to April. (Total usage, October to April, is 320,000 kWh.)

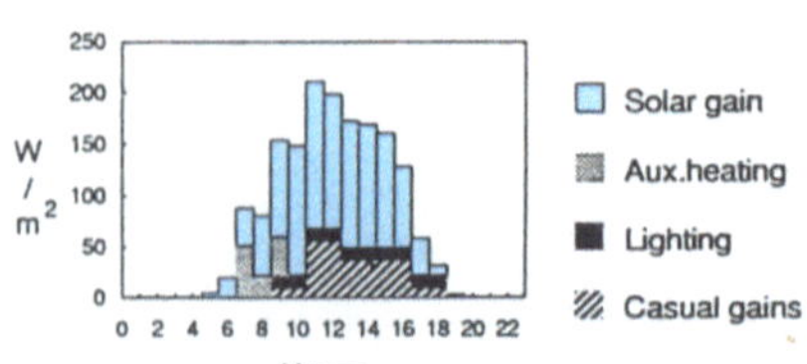

Figure 20. Supply of heat in public waiting area over a spring day

Results of Courtroom Daylighting and Solar Glare Studies

Introduction

The building has been planned so that every courtroom receives daylight, direct and/or borrowed. The borrowed daylight is introduced to nine of the ten courtrooms from their highly-glazed public waiting areas. The partition wall between the courtroom and the waiting area has a long high-level window fitted with acoustic double glazing. This provides a noticeable level of borrowed daylight in the courtroom and some contact with the outside world - especially important in those courtrooms which have no direct daylight.

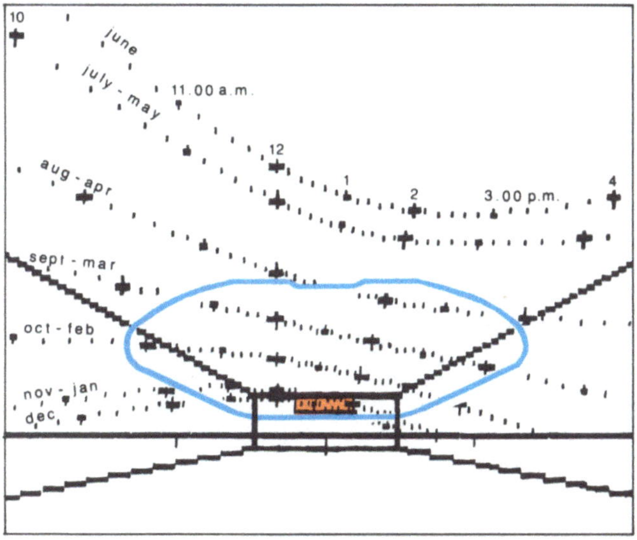

Figure 21. Plot of sunpath perspective as seen from magistrates' bench produced by GNOME computer program

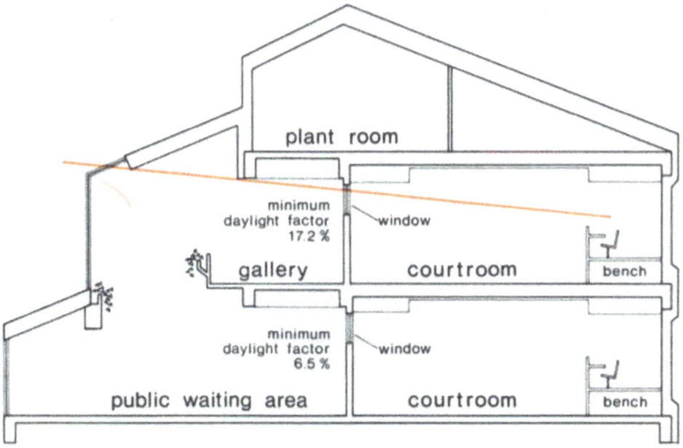

Figure 22. Section showing point where solid roof in waiting area should end if daylighting in courtroom is not to be impaired

Glare Studies

The windows in the partition wall are oriented 18 ° west of south and are directly opposite the magistrates' bench. To assess whether there could be a problem of glare for the magistrates, the GNOME computer program was used to produce a wide-angle interior perspective showing the window and a plot of the monthly sunpath perspectives at 10 minute intervals (see Figure 21). The sun will be visible from the chosen viewpoint when it is within the window outline. The direction within which complaints of solar glare can be expected is also outlined and so the season and duration of complaints can be estimated.

The study showed that there would be a problem of glare at the magistrates' bench in the early afternoon in winter. It was decided to install movable blinds or curtains for control by the courtroom ushers.

Design of Roof Glazing in Waiting Areas

As indicated above, to optimize the thermal performance of the public waiting areas, it was decided to reduce the roof glazing in the latter as much as possible without affecting the borrowed daylight in the courtrooms. A line was therefore drawn on section from eye level at the magistrates' bench on the first floor through the head of the internal window (see Figure 22). The point at which this line intersects the waiting area roof is the point below which solid roofing would restrict the view of the sky from the magistrates'. bench, thus indicating the optimum proportion of solid to glazed roofing.

Design of the Public Waiting Areas

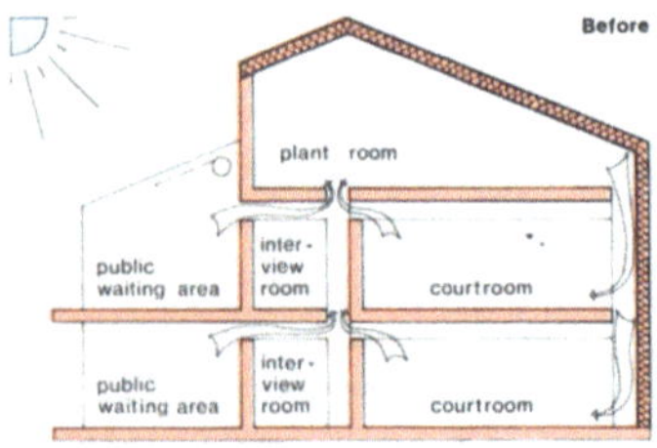

Figure 23. Preliminary design of public waiting area

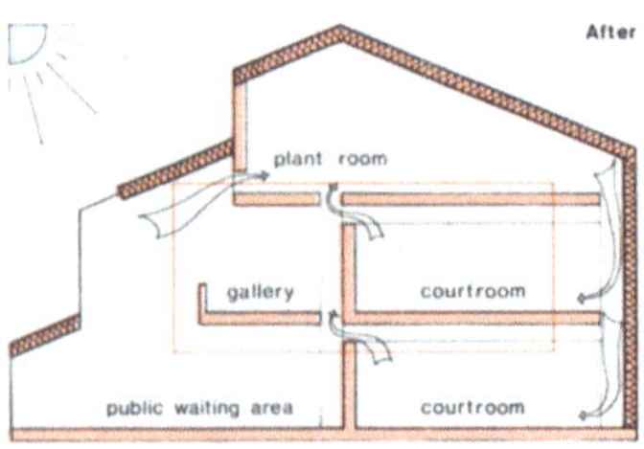

Figure 24. Final design of public waiting area

In the preliminary designs (see Figure 23), the public waiting areas were simple single-storey conservatories/sunspaces separated from the courtrooms by a row of ancillary rooms. The idea was to take air warmed by solar gains from them to the plant room where it could be used to preheat the ventilation air for the courtrooms. However, after discussing these proposals with one of the Building 2000 consultants, it was realized that, if the public waiting areas were designed in this way, they would be too hot in summer and too cold in winter. It was further recognized that rigid environmental control was not required in these spaces because they are used by people wearing outdoor clothing. They will, however, be occupied all the year round and by people who might be there for considerable lengths of time. They were, therefore, redesigned as well-insulated direct gain spaces which provide a greater degree of protection from outside conditions than a sunspace but allow a greater degree of temperature fluctuation than is normal for a completely internal environment. The new design is shown in Figure 24.

The major changes are as follows:

- increase in the height of the spaces to two storeys. This increases the potential for natural ventilation due to the stack effect (particularly useful in summer to prevent overheating) and provides more spacious and attractive spaces.

- provision of a gallery waiting area at first floor level. This provides a shaded area where people may escape from direct sunlight if they wish.

- reduction in the area of roof glazing. The roof glazing does not contribute much to the direct gains. In the winter, with the sun low in the sky, the vertical glazing is far more important; the roof glazing is mainly a source of heat loss. In the summer, with the sun high in the sky, the roof glazing allows solar gain when it is least required. It is an important source of daylight and can be a key design feature. The roof glazing was, therefore, reduced to the minimum compatible with achievement of good daylighting.

- relocation of some of the ancillary accommodation away from the courtrooms to the outer edge of the ground floor waiting area. This allows the direct gain spaces (the public waiting areas) to abut the courtrooms, enabling daylight to penetrate to the courtrooms themselves. In addition, the courtrooms benefit thermally from being adjacent to the direct gain spaces by the storage and passage of heat through the partition wall.

- allowing one of the waiting areas to open onto the external courtyard, providing a continuation of internal into external space, especially in summer. This has a psychological advantage in that the internal space becomes semi-external and people will tolerate greater extremes of temperature, as they do outside.

- provision of a variety of space types in the waiting areas , as indicated above. This gives people the choice of moving to the types of space they find most comfortable.

GENERAL DESIGN GUIDELINES AND POINTS OF INTEREST RESULTING FROM THIS PROJECT

General Principles of Passive Solar Heating in a Courtroom

The study described in this brochure showed that:
- Only spaces in which temperatures can be allowed to float between much wider limits than is normally acceptable should be designed as sunspaces. If the sunspace is allowed to reach higher temperatures in spring/autumn, more useful heat can be recovered. Acceptance of lower temperatures in winter reduces the required heating load.

- Roof glazing tends to cause overheating in the spaces below because it allows the rays of the summer sun to enter the building. Vertical glazing is more useful because it admits the low winter sun but does not allow entry of the overhead summer sun.

- Use of a displacement ventilation system means that higher supply temperatures can be used even when there is a net cooling load, thus making maximum use of fresh air preheated by the extract air from the sunspaces.

Daylight in Courtrooms

Although, in order to achieve a more comfortable and pleasant internal environment, some daylight should be introduced into courtrooms, it is not practical to try to achieve workable levels of light from natural lighting alone. The courtrooms are the key spaces in a courthouse and it is essential that they function satisfactorily in every respect. The courtroom's environment must, therefore, be strictly controlled and certain standards, including lighting standards, must be attained. Ideally, courtroom lighting should be unobstrusive, providing a subtle hierarchy of illuminances to signify the importance of different parts of the courtroom. This can be achieved most consistently by artificial lighting but a measure of daylighting should be introduced to provide a very necessary and psychologically important contact with the outside.

Introducing daylight into courtrooms can be difficult because they tend to be surrounded on at least three sides by circulation areas for the public, the magistrates and defendants in police custody. This leaves only one side free which could be on an external wall and have windows. The building must, therefore, be planned from the outset to allow daylight into the courtrooms, if this is required.

In this project, the plan has been arranged and the circulation organized so that six of the ten courtrooms have an external wall with windows. These walls have been oriented so that dazzle and glare from direct sunlight will not occur during court hours.

Top lighting through roof glazing over the courtrooms can be an option if the building is planned to allow it. However, daylighting of this sort is not entirely suitable for courtrooms, especially over the magistrates' bench, and the problem of down draughts has to be dealt with.

Care must be taken over placement of windows in a courtroom. It is especially important that there are no windows in the wall behind the magistrates' bench. If there were, the magistrates would be seen in silhouette by the courtroom occupants. Ideally, the wall behind the magistrates' bench should be of intermediate brightness and colour to provide a visually comfortable background to the magistrates.

A Note on Reflecting Ponds

Water reflects only 4% of the light shining on it and so a reflecting pond cannot really be regarded as a daylighting feature, although the percentage reflection can be increased if light coloured surfaces are used to line the pond. Perhaps the main daylighting effect of a reflecting pond is to affect the nature of the daylight in adjacent spaces by producing a pleasant ripple effect on the ceiling. However, the problem of glare from upward reflected light must be considered.

BUILDING 2000 Building 2000 brochures are published by Directorate General XII of the Commission of the European Communities to show how design studies can help architects and other building designers use passive solar principles to the best effect to produce attractive energy-efficient buildings. Each brochure describes studies carried out with the support of the Commission during the design phase of one of thirty-six non-domestic buildings in the EC Member States. The studies were on such topics as daylighting, heating, cooling, ventilation, comfort, control systems and urban design. They were carried out with the help of acknowledged European experts in these fields and drew heavily on lessons learned and techniques developed through the Commision's research and development programme on solar energy applications to buildings.

Commission of the European Communities/Directorate-General for Science, Research and Development

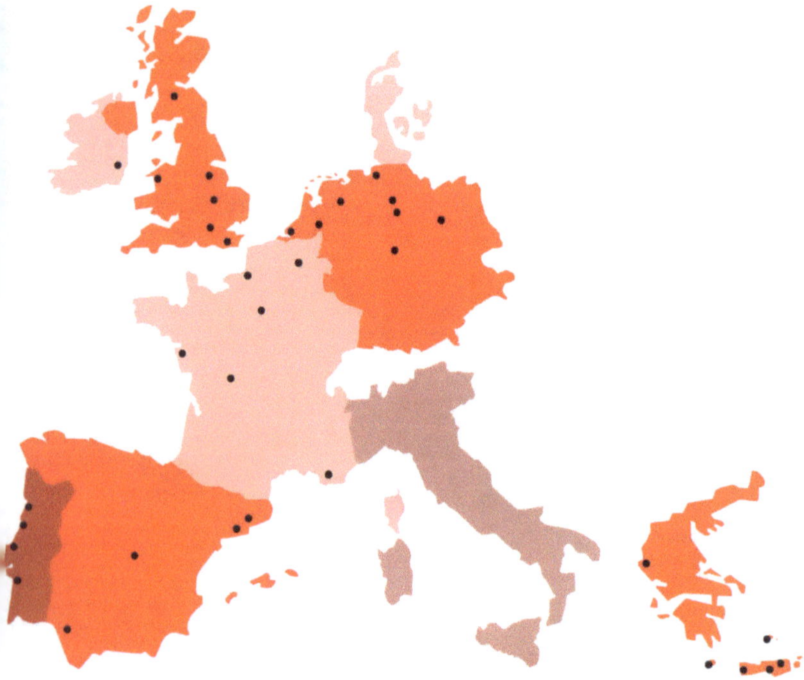

List of Design Team Participants and Advisers

Client
Rotherham Borough Council
Rotherham Magistrates' Courts Committee

Architect and Energy Consultant
Director of Architecture
Rotherham Borough Council
Norfolk House
Walker House
Walker Place
Rotherham S65 1AN
UK

Building 2000 Consultants
Cambridge Architectural Research Limited
Cambridge CB2 2EB
UK

YARD
233 High Holborn
London
UK

J Lynes
Humberside College of Higher Education
School of Architecture
Hull HU2 9BT
UK

This set of **Building 2000** brochures illustrates how architects and other building designers can successfully apply passive solar principles to produce energy-efficient buildings.

BUILDING 2000 Participants

Project Director
Theo C. Steemers

Coordinator
Cees den Ouden

Technical Steering Committee
Dean Hawkes
Nick Baker
Alex Lohr
Jean P. Lepoivre

Regional Liaison Agents
(D) Jörn Behnsen
(E) Vicente Sifre
(F) Michel Raoust
(GB) Alan Hildon
(GR) Matheus Santamouris
(P) Eduardo Maldonado

Further information or copies of the brochures can be obtained from prof. ir. Cees den Ouden, EGM Engineering BV, P.O. Box 1042, 3300 BA Dordrecht, The Netherlands.

BUILDING 2000

Commission of the European Communities

- Local community centre for exhibitions, theatrical performances, slide shows and private functions
- Winter-time heating using direct solar gain through bay windows, heat recovered from exhaust air and radiant heating panels
- Computer control of heating and ventilation systems and window shutters
- Good daylighting using carefully placed apertures and wellchosen surface materials

COMMUNITY CENTRE FOR CULTURAL ACTIVITIES

LE VIGEN/HAUTE/FRANCE

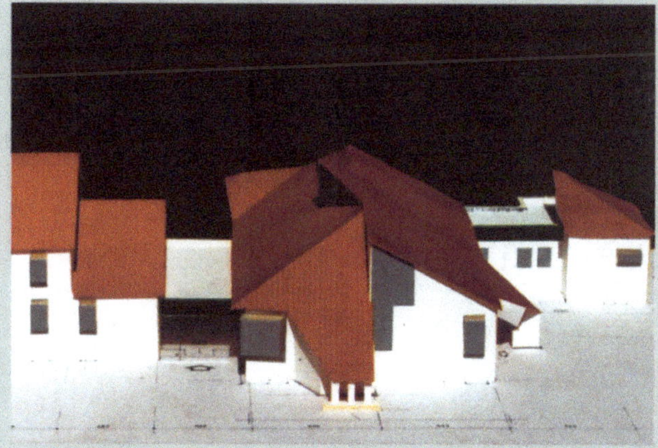

Building 2000 is a series of design studies illustrating passive solar architecture in buildings in the European Community.

ISSUE

P 2

FEB 1991

PROJECT DESCRIPTION

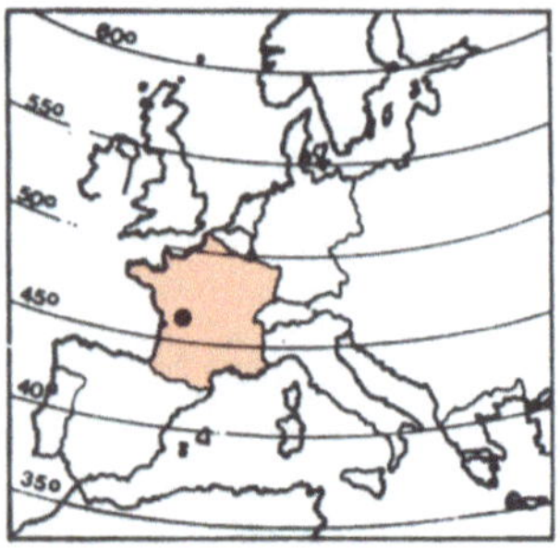

Figure 1. Location of Haute-Vienne department

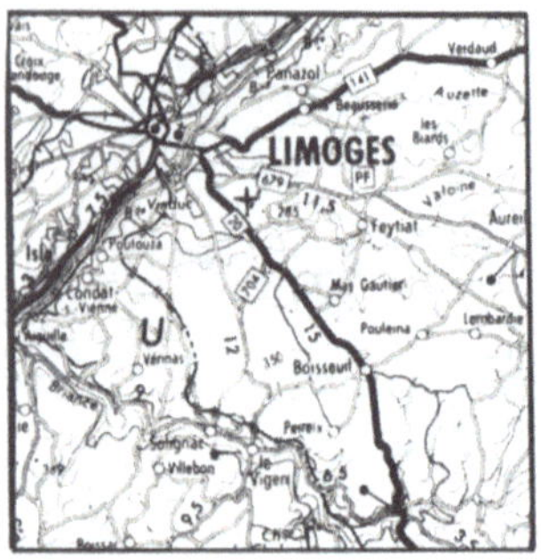

Figure 2. Location of Le Vigen

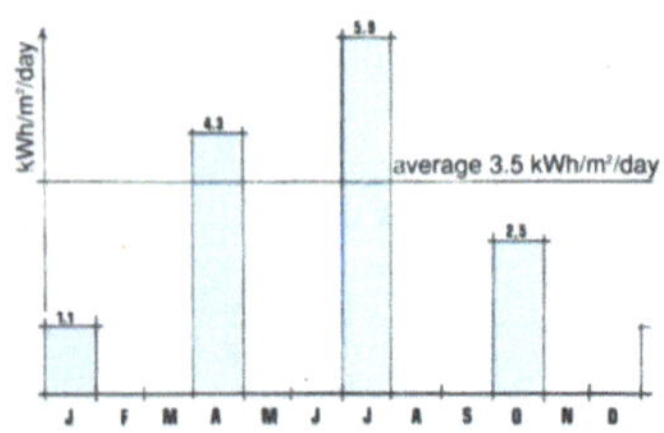

Figure 3. Mean solar irradiation data

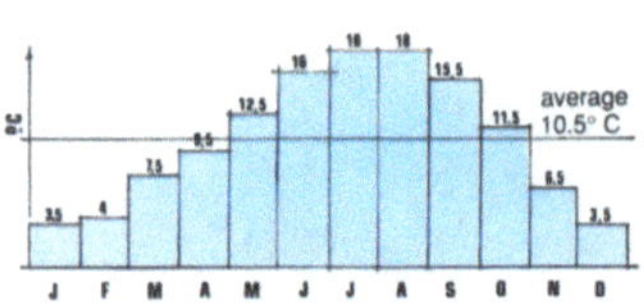

Figure 4. Mean temperature data

Building Type

This building has been developed by Le Vigen town council to provide the community with a centre for cultural activities (exhibitions, slide shows, theatrical performances, etc.), private functions (banquets, wedding receptions, parties, etc.) and so on. The centre aims to be financially self-sufficient and is therefore open to as wide a range of associations and private individuals as possible both within and outside the immediate Le Vigen area.

Location

Le Vigen is a small town of about 1500 inhabitants near Limoges in the department of Haute-Vienne in the Limousin region of south-west France (see Figures 1 and 2). The site (latitude 45° 55'N, longitude 1° 11'E) is in the town centre, near the town hall and the church. It is 283 m above sea level and is completely open. The site plan is shown in Figure 5.

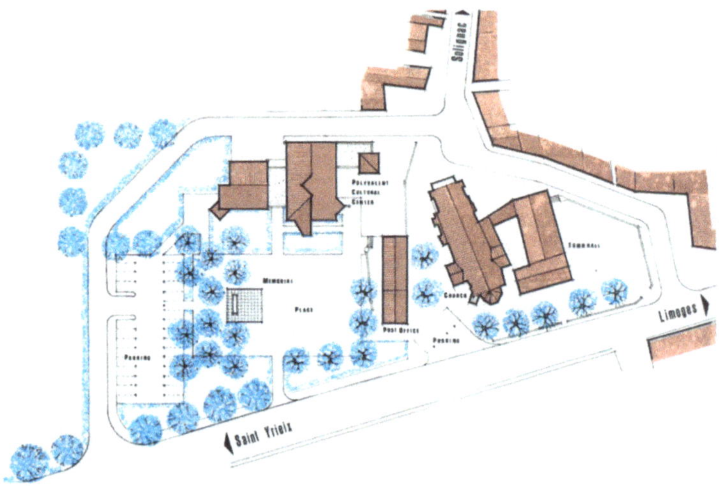

Figure 5. Site plan

Site Microclimate

The climate in this part of France is influenced by the sea: the prevailing winds are from the west, bringing mild rainy weather. On average, there are 1853 sunshine hours a year. Solar irradiation and temperature data are given in Figures 3 and 4.

The winters, however, can be severe. Limousin is on the edge of the Massif Central; this has a hard climate and in winter large temperature swings can be observed within a few hours. Haute-Vienne is classified into climate zone category A where - 10° C is the reference temperature for sizing heating systems. As can be seen from Figure 6, temperatures in the region - 10° C to - 6° C are reached 6 days a year on average.

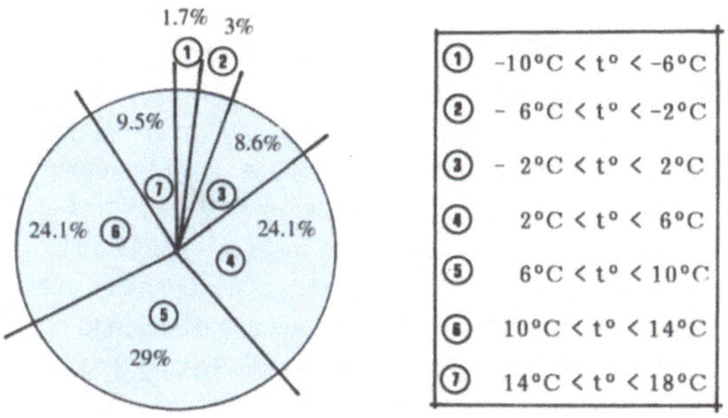

Figure 6. Frequency of days when temperatures reach various levels

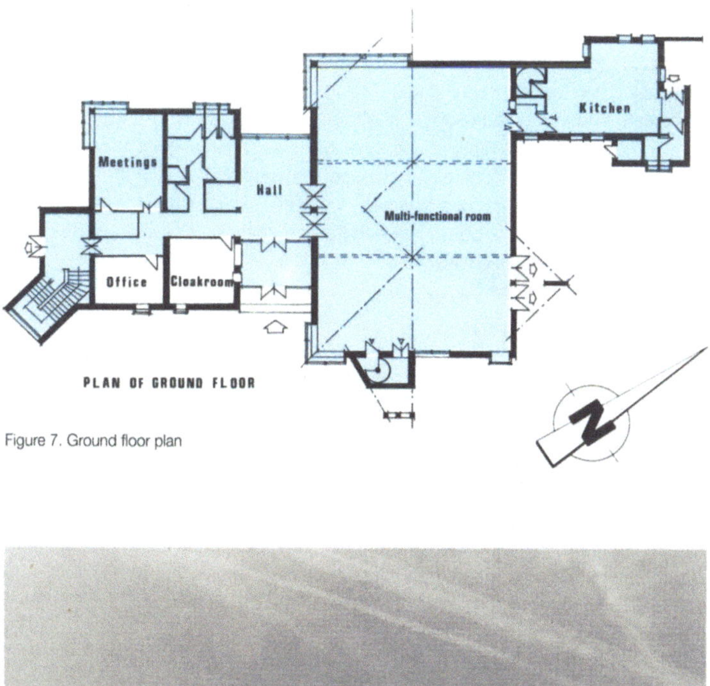

Figure 7. Ground floor plan

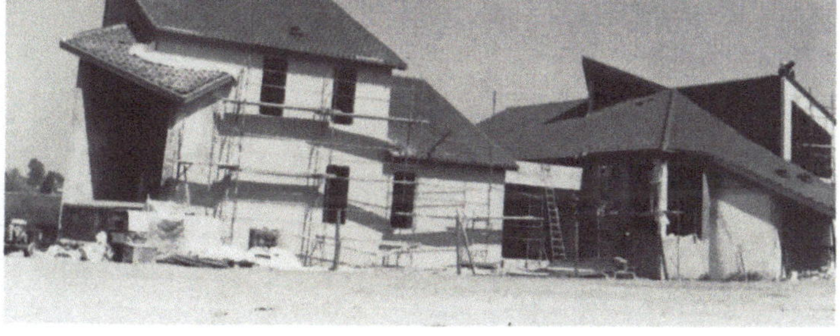

Figure 8. View of the south-east facade (under construction)

Design and Construction Details

The building is on three floors (each about 45 m by 20 m) and contains spaces of different heights and floor areas. The client wanted the design to take advantage of the view of the valley to the north-west. The building has therefore been designed along a SW-NE axis with the meeting rooms facing the north-west. To minimize the effect this orientation had on thermal performance a number of passive solar and other energy-saving features were introduced which are described below. The ground floor plan is given in Figure 7.

The roofs are sloped and tiled, in response to a planning requirement that they should be in keeping with the 13th century Romanesque church nearby.

To keep costs low, construction materials are standard/traditional. Floors and walls are of concrete. There is wood carpentry, aluminium window frames and glass-wool insulation. Double glazing was chosen in preference to polycarbonate for the windows. U-values of some of the building elements are given in Table 1.

For security reasons, the outside of the building is lit at night by means of electric light emitted from the rooflight in the multi-functional room.

roofs	
main tiled	0.24
first floor	0.18
kitchen	0.24
other (terraces)	0.20
floor	0.36
crawl space (kitchen)	0.32
walls (average)	0.28
windows	3.70

Table 1. U-values of some building elements ($W/m^2\,K$)

DESCRIPTION OF PASSIVE SOLAR FEATURES/COMPONENTS

Introduction

The orientation of the main rooms towards the view to the north- west meant that, in the original design, direct solar gains could not contribute much to the energy requirements of the building. Studies were therefore carried out with the help of the Building 2000 experts to establish how various passive solar and energy-saving features could be introduced to minimize this effect and otherwise improve the thermal performance. In the final design, a greater area of window faces in a southerly direction to increase the amount of direct solar gain and improve daylighting. The building spaces have been rearranged and the insulation of the massive parts of the structure increased to improve the collection and storage of solar gains and reduce heat loss. Light-coloured surfaces help daylighting. Incoming fresh air is preheated with heat recovered from the exhaust air. Radiant ceiling panels complete the heating. Window shutters prevent overheating or unwanted heat loss, particularly when rooms are not in use. The building is divided into zones according to occupancy and room type. Ventilation rates, the opening and closing of shutters and operation of the heating systems in each zone is controlled automatically according to external temperature, type of use, solar gains and internal gains from occupants, artificial lighting and other equipment. As a result of these passive solar and other energy-saving measures (which are described in more detail below) the power installed in each room for auxiliary heating is much lower than was envisaged originally.

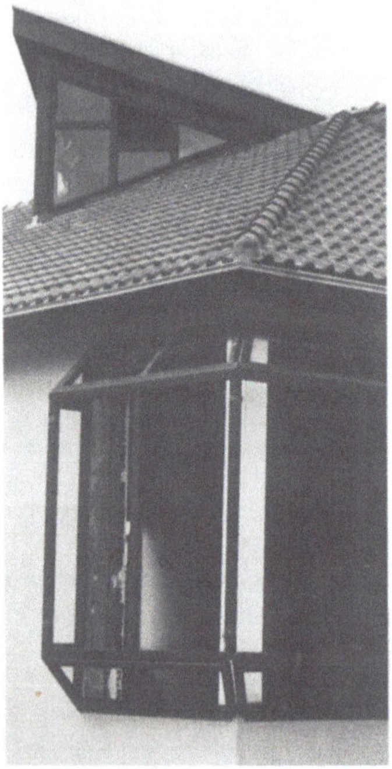

Figure 9. Bay window detail

Direct Solar Gains

Bay windows have been incorporated into a number of the angles of the building. These project out from the facades (see Figures 9 and 10) and make possible the collection of direct solar gains from both the south-west and south-east. Thus those meeting rooms which look out onto the north-western view are able to have the benefit of some direct solar gains. In the large multi-use room the south-west window apertures are larger than required for daylighting so that solar gains can heat up the major part of the room, which is mainly occupied in the late afternoon/evening (see Figure 11). There are no windows on the north-east facades (Figure 12).

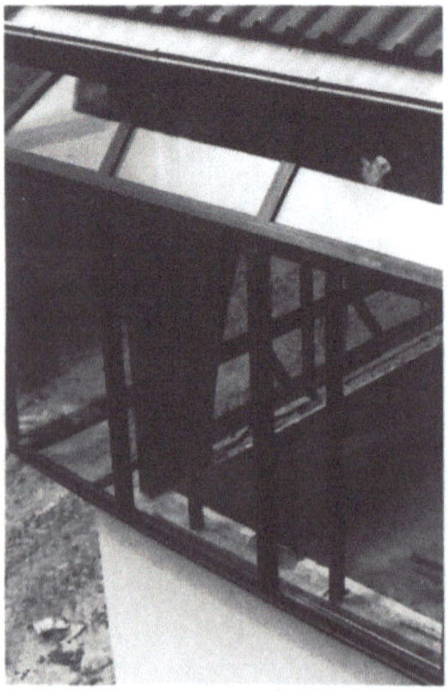

Figure 10. Bay window detail

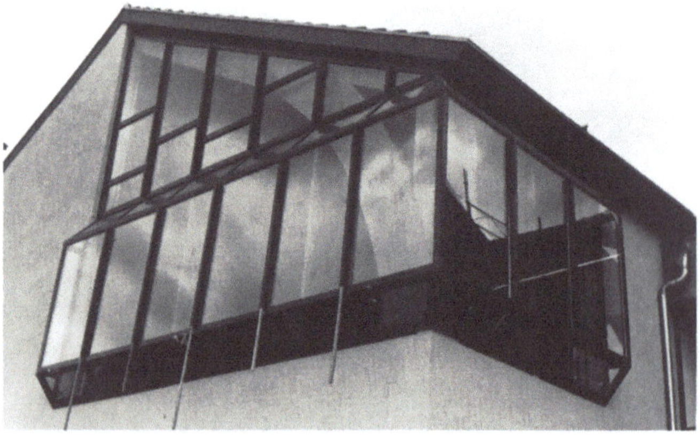

Figure 11. South-west bay window

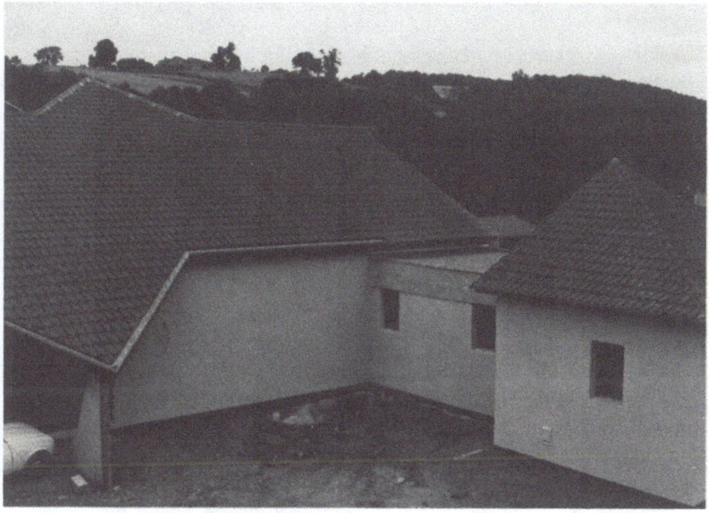

Figure 12. View of north-east wall with no windows

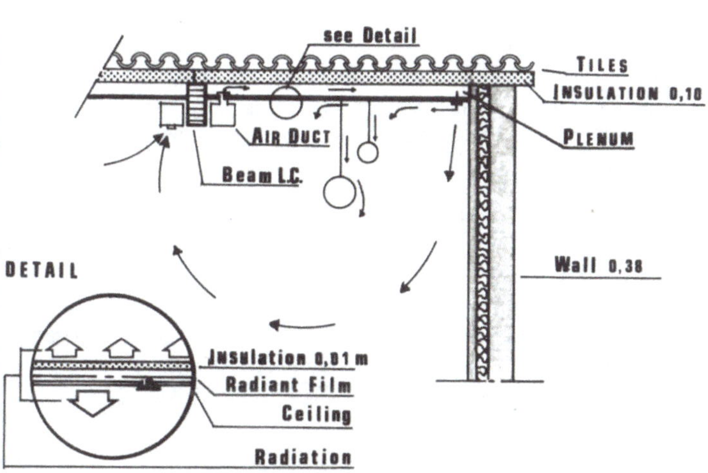

Figure 13. Heat recovery and radiant heating system

Thermal Storage

The concrete floor and wall masses are carefully insulated with FIBRALITH FY 200. Each massive part of the structure receives direct solar radiation from the bay windows. (The idea of using a pebble-or water-filled energy storage system was rejected on the grounds that it would not be cost-effective because of local climate conditions and the building size and type.)

Heat Recovery and Radiant Heating System

The heat recovery and radiant heating system is illustrated in Figure 13. Fresh air and exhaust air are transported at air speeds of less than 4 m/s through insulated pipes to the heat recovery units. These are double flow plate units each having a power of 2 x 0.37 kW. Their heat recovery efficiency is 50%. In spring and autumn any useful excess heat is recovered from the exhaust air and used to preheat the incoming fresh air before it enters the room. The treated air flow is 800 m3 /h in the office and meeting rooms and 2000 m3 /h in the large multi-use room. When the ambient temperature is - 10° C, then the temperature of the fresh air entering the rooms through this system is 5° C. The energy losses from the radiant panels completes the heating of the air. Air introduced at 8° C, say, will slowly move into the main part of the rooms at 16° C using the Coanda effect. Installed electrical power is used to heat the radiant panels when necessary. The installed power is 20 kW in the office and small meeting rooms and 39 kW in the large multi-use room. Fire safety is incorporated into the heating system by means of auto-fuse heating films on the ceiling.

Control System

Automatic control of ventilation rates, radiant heating system and opening and closing of window shutters is achieved by means of a computer control system. The building is divided into three zones: the large multi-use room; the offices and meeting rooms; and an additional zone, which is used in conjunction with either or both of the other zones, which contains the access hallway and cloakrooms.

Each zone is controlled in two ways, by a central system using the French VISIDEL 2000 computer and by local sensors with reduced thermostat settings and non-freezing positions.

In the central control system, data on the external temperature, room temperatures (measured by sensors in the different zones), thermostat settings, the heating system (on or off), solar radiation (measured by sensors) and use of electric power for lighting/heating are collected, displayed on Minitel screens and/or computer visual display units and printed out. The computer makes a real-time analysis of the data and controls the heating systems, ventilation rates and shutters. Control can be made at any time at frequencies of between 1 minute and 67 hours. The system can be managed according to user/activity type and level of occupation. The control system can also trigger off alarms if mechanical or electrical faults occur in the system or if window panes are broken.

The local control systems are for times when the rooms are not in use.

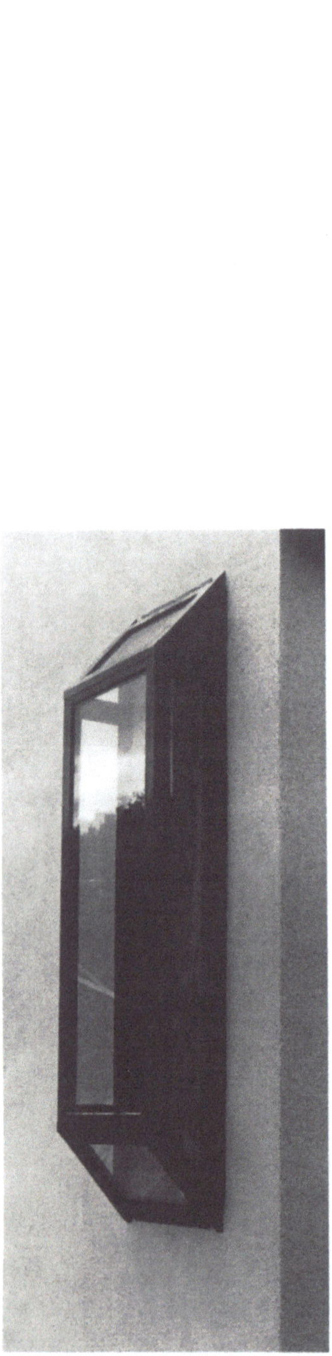

Figure 14. Detail of a window aperture

Daylighting

Natural lighting is available in most rooms in the building (see Figure 14). To improve the daylighting, surfaces are light-coloured, e.g. floor tiles are pale grey.
Because of its size and varied use, particular attention was paid to designing the daylighting of the multi-functional room. This is to be used for exhibitions, banquets and various kinds of shows - but not for sports. For most purposes, therefore, the lighting needs to be lively and pleasant and control of glare is not of paramount importance. To accommodate these requirements, natural light enters the room from several directions through apertures of various shapes and sizes (see Figures 15 and 17).

For some activities, however, such as slide shows and theatrical performances, the room needs to be blacked-out. At such times an average illuminance of up to 30 lux is acceptable and this is provided by a small rooflight (equivalent to 1.4% of the floor area) and the light leaking from behind the curtains. (All the curtains are fully opaque.)

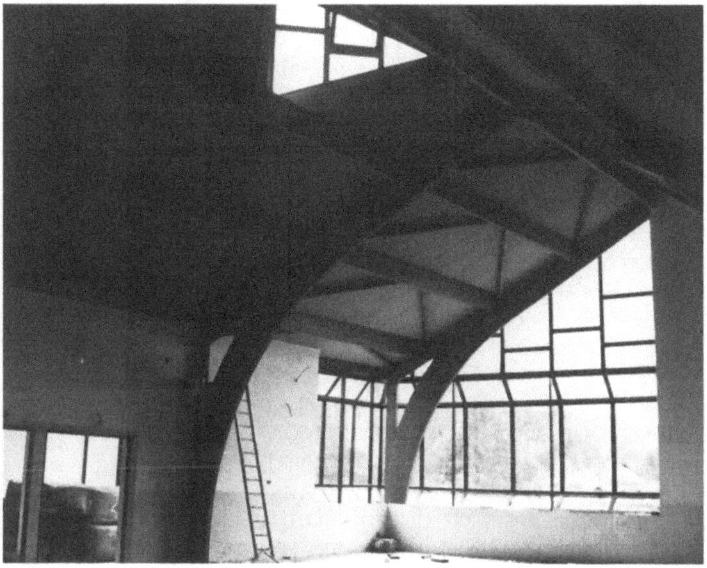

Figure 15. Interior of large multi-use room

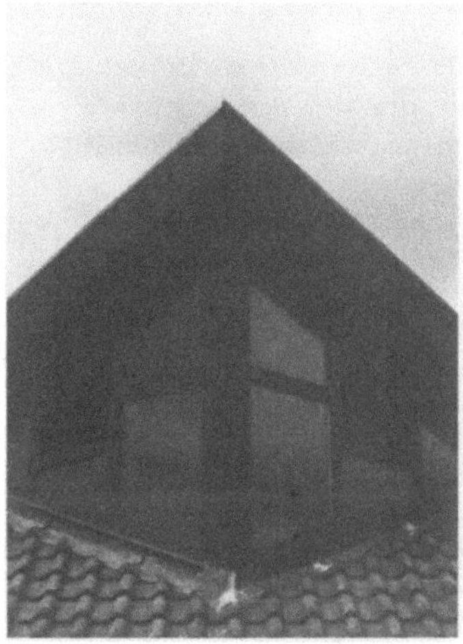

Figure 17. Roof aperture

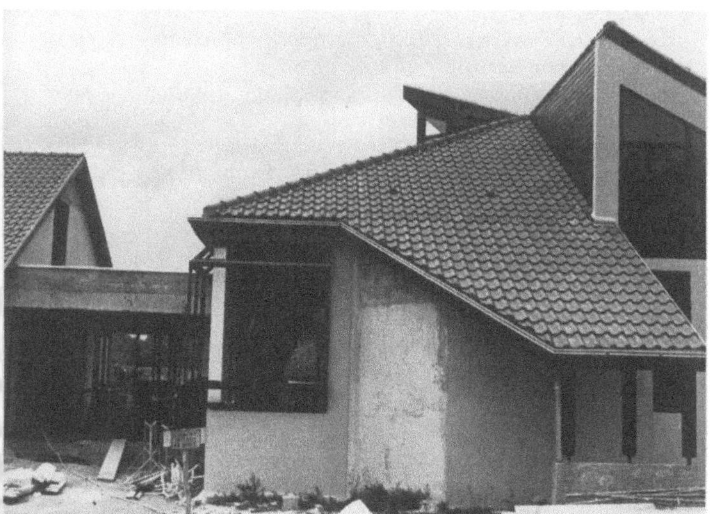

Figure 16. Detail of south-east facade (under construction)

ENERGY CALCULATIONS PERFORMED AND DESIGN TOOLS USED

Introduction

The passive solar and other energy-saving systems were designed with the help of a number of thermal performance and daylighting evaluation studies. Some results of this work are given here.

	offices & meeting rooms	multi-use room
comfort thermostat set point	18° C	18° C
reduced set point	12° C	9° C
actual heating hours (Hh)	1650h	693 h
total heating season (SHh)	5568h	5568 h
coefficient (Hh/SHh)	0.30	0.12

Table 2. Basic data on use and operation of building

Results of Thermal Performance Evaluations

Some basic data on the use and operation of the two main zones of the building - the offices and meeting rooms and the large multi-use room - are given in Table 2. The comfort set point is the set point temperature when the room is in use. The reduced set point is for periods when the room is not in use.

Tables 3 and 4 compare the thermal performances of the initial design of the building (before any modification had taken place to improve its energy performance) and the final design. They show that introduction of the passive solar and other energy-saving features has brought about an annual saving in energy consumption of 12,800 kWh (46,000 MJ) in the offices and meeting rooms and 21,800 kWh (78,500 MJ) in the large multi-use room. Total savings, therefore, amount to 34,600 kWh (124,500 MJ) a year - 62% of the energy requirements of the original design. This includes an 18,500 kWh (66,600 MJ) contribution from direct solar gains but does not take into account any internal gains from occupants, electric lighting and other equipment.

	offices and meeting rooms		large multi-use room	
	initial design	improved design	initial design	improved design
installed power (kW)	28	20	50	39
energy consumption during comfort hours (kWh/year)	10,000	3,5000	12,100	4,200
energy consumption during set back period (kWh/year)	7,500	1,800	14,200	2,500
energy consumption during temperature regain period (kWh/year)	2,300	1,700	10,000	7,800
total consumption (kWh/year)	19,800	7,000	36,300	14,500

Table 3. Comparison of thermal performance of initial and final building designs

	inital design	improved design
total solar gains (kWh/year)	8,600	18,500
total energy consumption (kWh/year)	56,100	21,500
savings achieved by improved design		
(%)		62
(kWh/year)		34,600
(electrical tonnes oil equivalent)		7.70

Table 4. Overall thermal performance of initial and final designs

Evaluation of Daylighting in Multi-Use Room

Introduction

The daylight evaluation studies were carried out on the large multi-use room. Light penetration is directly related to the areas of the aperture and the space to be lit. The ratio of the aperture area on each of the facades to the total floor area of the multi-use room is given in Table 6.

Three different tests were conducted. First, photometric testing of samples of the floor, walls and ceiling coatings was carried out to determine reflection coefficients (R) and the dispersion of light during the reflection process. The Building 2000 experts then carried out computer simulations to identify the daylight factor distribution using the initial selection of surface coatings and various other options. Second, results of the analyses were discussed at a meeting of the architects, engineers and client. Control of glare and sunlight penetration were at the centre of the discussions. The talks led to definition of a set of specifications. Third, on-site measurements were conducted to test the daylight penetration into the multi-use room during successive stages of construction. Measurements were carried out: before windows had been put in and walls and floor consisted of raw concrete; with windows, but walls and floor remained raw concrete; with windows, with plaster on the walls; with windows and the final coatings.

location of apertures	no.	total aperture area (m^2)	ratio of aperture area to floor area(%)
NW facade	0	0	0
NE facade	2	27.3	11.3
SW facade	2	1.8	4.9
SE facade	5	17.0	7.0
roof	2	3.4	1.4

Table 6. Ratio of aperture area to total floor area in multi-use room. (Total floor area is 241.3 m^2.)

Results of Photometric Testing of Surface Materials

The reflectance coefficients (R) and dispersion of light during the refection processes as determined by the photometric testing are shown in Table 5.

Floor - light grey/ochre
R = 36%
Dispersion (when new) = medium to high
Dispersion (after a year's use) = medium

Walls (small dimension) - dark red carpet
R = 6%
Dispersion = high

Walls (large dimension) - beige carpet
R = 50%
Dispersion = high

Ceiling
R in cavity is taken to be equivalent of 40% because of beams and structure

Table 5. Results of photometric testing of floor, wall and ceiling surface materials to determine reflection coefficients (R) and light dispersion

Never less it can be said that more than 80% of the floor area is properly lit, i.e. has a daylight factor of more than 4%. The presence of corner apertures improves the balance of the lighting in the room.

DIAGRAM 1

18 January 1989
Sunny day
No windows installed
No coatings on concrete
HZDIF = 9000 Lx

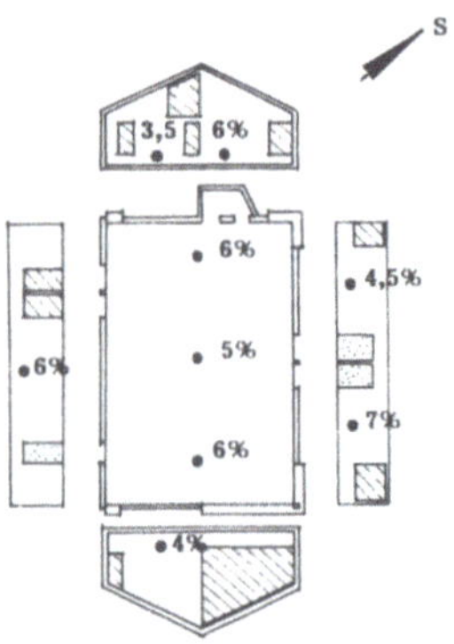

Figure 18. Daylight factors in multi-use room determined by on-site measurements on 18 January 1989

DIAGRAM 2

7 July 1989
Cloudy day
Windows installed
Plaster on walls
HZDIF = 11800 Lx

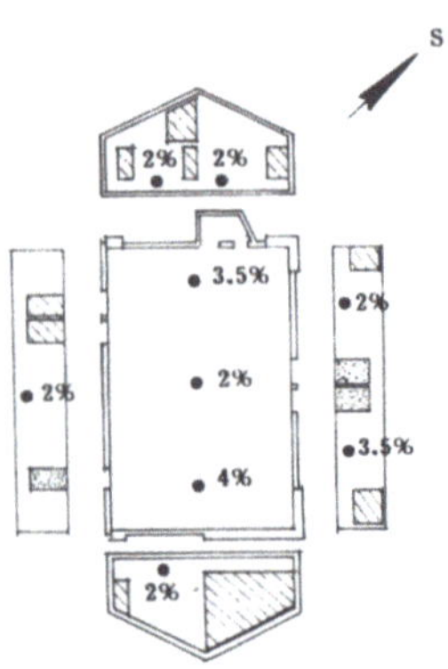

Figure 19. Daylight factors in multi-use room determined by on-site measurements on 7 July 1989

Results of Simulation Studies

The three-dimensional simulation of light penetration into the room showed the position of the lighter and darker zones on the indoor surfaces. The results of the simulation under overcast conditions using the initial selection of surface coatings are given in Figure 20. Correction factors for window frames or dirt deposits on the glazing have not been taken into account. Never less it can be said that more than 80% of the floor area is properly lit, i.e. has a daylight factor of more than 4%. The presence of corner apertures improves the balance of the lighting in the room.

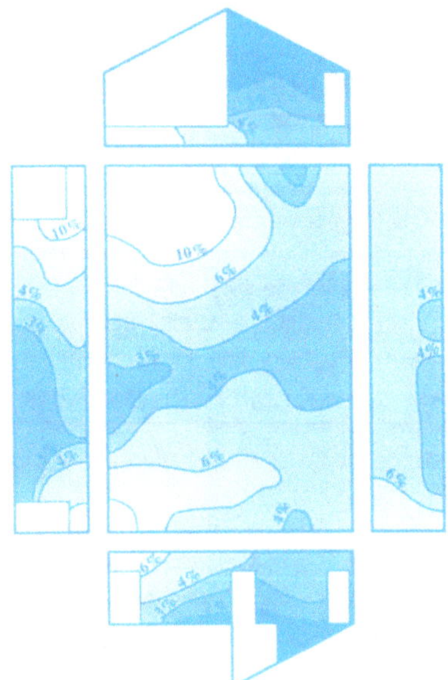

Figure 20. Isolux diagram of multi-use room resulting from simulation studies. This shows indoor illuminances as percent of outdoor illuminance under overcast conditions

Results of On-Site Measurements

Figures 18 and 19 show the results of on-site illuminance measurements on the walls and on a horizontal plane a meter above floor level. Figure 18 presents the distribution of daylight factors (i.e. the ratio of the indoor measured illuminances to the simultaneous exterior illuminances on the horizontal plane) in the room on 18 January 1989 (a sunny day) before any windows had been installed and when there were no coatings on the concrete. Figure 19 is for 7 July 1989 (a cloudy day); here, windows had been installed and the walls plastered. The diagrams show the evolution of daylighting performance as construction proceeds and finishes and glazing are added.

GENERAL DESIGN GUIDELINES/POINTS OF INTEREST RESULTING FROM THIS PROJECT

Some key modifications which were introduced into the design as a result of the Building 2000 studies include: increasing the area of glazing facing south-east and south-west to increase the direct solar gains; control of ventilation rates according to building occupancy; use of heat recovered from exhaust air to preheat the incoming fresh air. Innovative aspects of the design include a heating system which uses radiant panels in the ceiling; incorporation of fire-safety into the heating system by use of auto-fuse heating films on the ceiling; computer control of ventilation and heating systems and window shutters.

In the improved design useful solar gains amount to 18,500 kWh (66,600 MJ) a year. Total savings brought about by all the improvements amount to 34,600 kWh (124,500 MJ) a year.

Other advantages include a quick response to the need to increase temperature at the beginning of an occupancy period; noise-free operation, long-range control and an efficient anti-burglar system. Problems associated with air mixing (such as condensation) are avoided. The air in the rooms does not stratify into different temperature zones. In addition, the heating syste
any pollution problems in the neighbourhood.

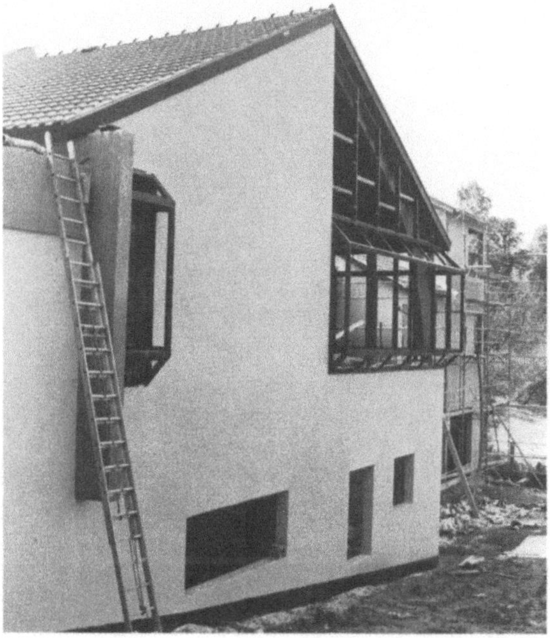

Figure 21. Detail of north-west facade (under construction)

BUILDING 2000 Building 2000 brochures are published by Directorate General XII of the Commission of the European Communities to show how design studies can help architects and other building designers use passive solar principles to the best effect to produce attractive energy-efficient buildings. Each brochure describes studies carried out with the support of the Commission during the design phase of one of thirty-six non-domestic buildings in the EC Member States. The studies were on such topics as daylighting, heating, cooling, ventilation, comfort, control systems and urban design. They were carried out with the help of acknowledged European experts in these fields and drew heavily on lessons learned and techniques developed through the Commision's research and development programme on solar energy applications to buildings.

Commission of the European Communities/Directorate-General for Science, Research and Development

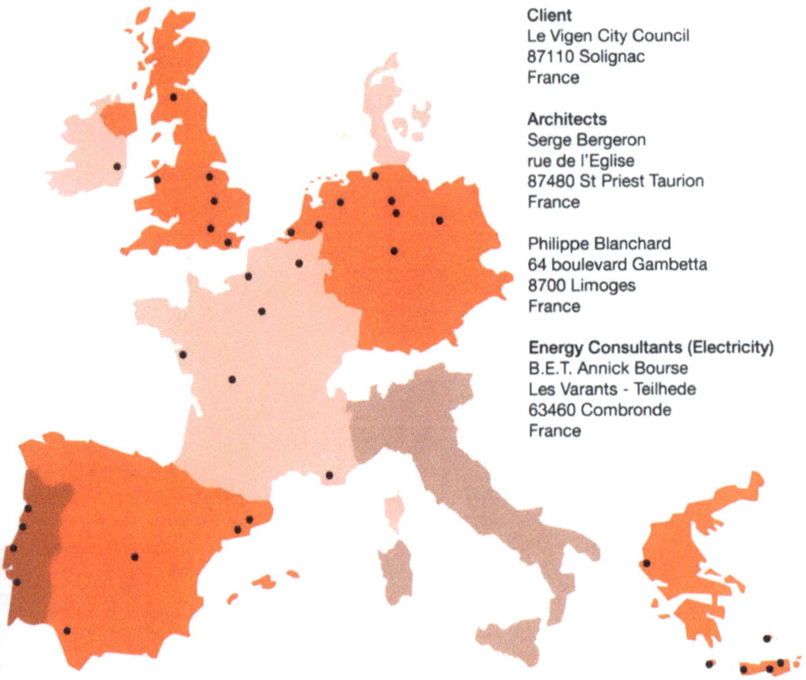

List of Design Team Participants and Advisers

Client
Le Vigen City Council
87110 Solignac
France

Architects
Serge Bergeron
rue de l'Eglise
87480 St Priest Taurion
France

Philippe Blanchard
64 boulevard Gambetta
8700 Limoges
France

Energy Consultants (Electricity)
B.E.T. Annick Bourse
Les Varants - Teilhede
63460 Combronde
France

Energy Consultants - Solar Technolo
B.E.T. Guy Soulier
63870 Orcines
France

Daylighting Consultant
Marc Fontoynont
Laboratoire Sciences Habitat (LASH)
Ecole Nationale Travaux Publiques
69120 Vaux en Velin
France

Consultant
De Mercey
43 rue de la Victoire
75009 Paris
France

Additional Support
Computer programs - Visidel France
Electrical systems - ESWA France an
EdF Limoges

This set of Building 2000 brochures illustrates how architects and other building designers can successfully apply passive solar principles to produce energy-efficient buildings.

BUILDING 2000
Participants

Project Director
Theo C. Steemers

Coordinator
Cees den Ouden

Technical Steering Committee
Dean Hawkes
Nick Baker
Alex Lohr
Jean P. Lepoivre

Regional Liaison Agents
(D) Jörn Behnsen
(E) Vicente Sifre
(F) Michel Raoust
(GB) Alan Hildon
(GR) Matheus Santamouris
(P) Eduardo Maldonado

Further information or copies of the brochures can be obtained from prof. ir. Cees den Ouden, EGM Engineering BV, P.O. Box 1042, 3300 BA Dordrecht, The Netherlands.

BUILDING 2000

Commission of the European Communities

- Two-storey public library building containing shelves for 100,000 books, exhibition space, meeting rooms, staff rooms and other ancillary facilities
- Optimized daylighting systems with automated, daylight-responsive switching of natural lighting
- Ventilation rates controlled to minimize heat loss in winter
- Direct solar gains and solar preheating of ventilation air contribute to winter-time heating
- Movable shades prevent glare and overheating in summer but maximize entry of diffuse light when required

PUBLIC LIBRARY

NEWBURY/UNITED KINGDOM

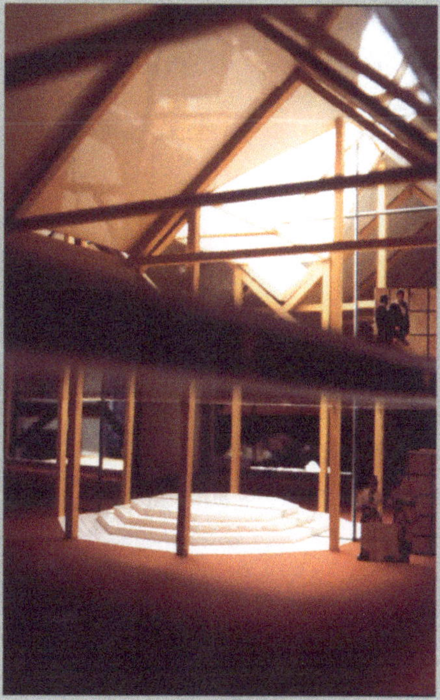

Building 2000 is a series of design studies illustrating passive solar architecture in buildings in the European Community.

ISSUE **P 3**

FEB 1991

PROJECT DESCRIPTION

Building Type

This project concerns a building to be constructed for housing public library facilities consisting of shelves for 100,000 books, a foyer, exhibition space, meeting rooms, staff workrooms, two flights of stairs and a lift.

Location

The site is in the centre of Newbury (latitude 51 ° 24 ' N), a provincial town in Berkshire, UK - see Figure 1. The site is flat, fairly sheltered and 70 m above sea level. On the northern boundary it is obstructed by a two- storey building with a tall, steeply-pitched roof. The rest of the site is not overshadowed. As well as the library building, the site will also accommodate car parking, an area for servicing a mobile library and a public footpath on the northern boundary. The site plan is shown in Figure 3.

Site Microclimate

The climate is typical of the temperate, low-land type. Details are given in Figure 2 and Table 1.

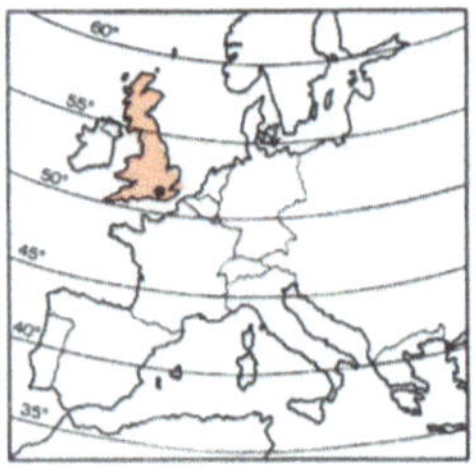

Figure 1. Location of Newbury

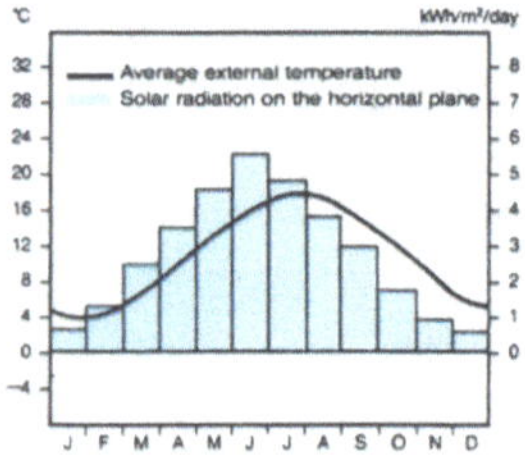

Figure 2. Mean solar irradiation and temperature data

Oct-Apr:	
Degree day (base 15.5 ° C)	2104
Sunshine hours	
total	681
mean daily in hours	3.2
mean daily as % of daylight hours	32%
Global irradiation on horizontal	
total in kWh/m^2	315
(total in MJ/m^2	1134)
mean daily in kWh/m^2	1.5
(mean daily in MJ/m^2	5.4)
Minimum average Jan day temperature	1.5 ° C
Mean hourly illuminance (diffuse + clear sky) from 9.00 to 17.00	
min.	0.4 klux
max.	47.5 klux
May-Sep:	
Sunshine hours	
total	949
mean daily in hours	6.2
mean daily as % of daylight hours	43%
Maximum average Jul day temperature 21.5 ° C	
Mean hourly illuminance (diffuse + clear sky) from 9.00 to 17.00	
min.	4.8 klux
max.	62.2 klux

Table 1. Some climate data

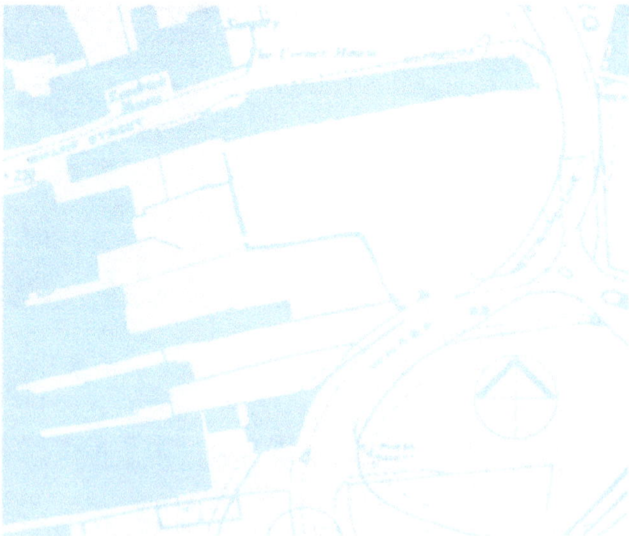

Figure 3. Site plan

Design and Construction Details

Design Details

The building is on two floors with a total floor area of 1150 m^2. The volume is 3500 m^3. The ground and first floors are shown in Figures 4 and 5. Because the height of the building has been restricted by local planners, the upper floor has been left open to the roof to maximize the internal volume. The building entrance consists of a two-storey atrium.

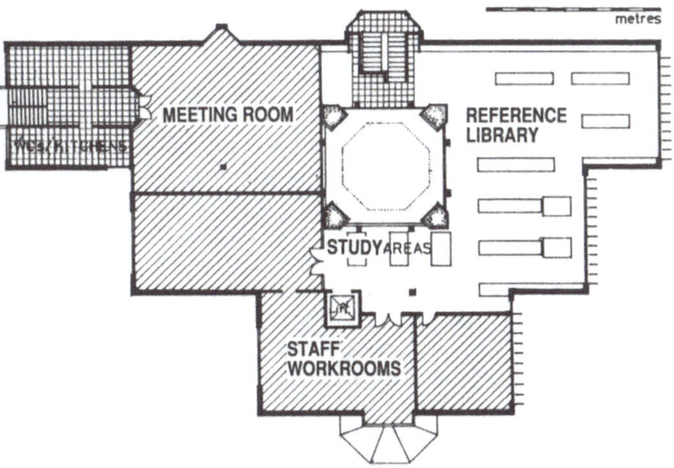

Figure 4. First floor plan

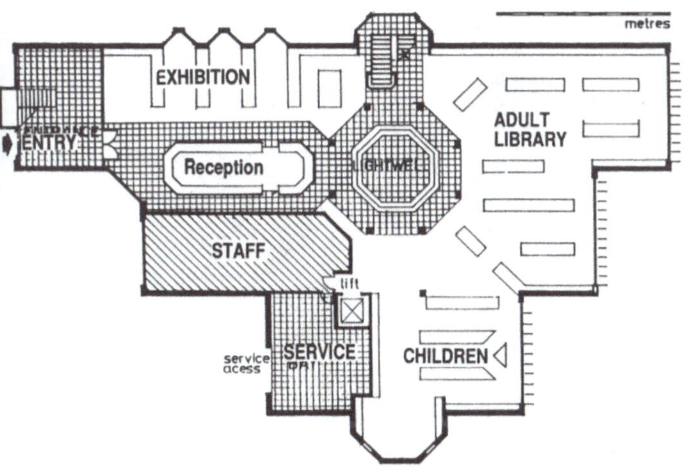

Figure 5. Ground floor plan

Construction Details

The building construction is typical of Berkshire with brick walls, clay roof tiles and timber trusses. U-values of some of the building elements are given in Table 2. The walls are of the insulated cavity type with an external layer of brick and an internal layer of plastered block and fairfaced brick. The ground floor consists of situ concrete. The suspended first floor is of pre-cast concrete beams with infill planks. A suspended ceiling void to the ground floor accommodates air handling ducts and services. The roof consists of clay roof tiles on insulated decking supported on exposed timber trusses. All windows are double glazed. Rooflights and windows which provide light for the study desks are additionally insulated with a low-emissivity coating.

To control winter ventilation and noise, the windows are sealed. An air distribution system supplies air to the perimeter zones and collects it at the ridge of the central lightwell. Zone controls allow for the contribution of internal and solar gains.

Floor	0.25
Walls	0.45
Glazing	3.0 and 1.8
Roof	0.35

Table 2. U-values of some building elements in W/m^2 K

DESCRIPTION OF PASSIVE SOLAR FEATURES/COMPONENTS

Introduction

In the UK, it is common for public libraries to be electrically lit for long periods. In deep-plan buildings, this results in high electricity consumption. In addition, lighting fitments generate heat which is expensive and may even produce an air conditioning load. Furthermore, it encourages the use of high ventilation rates and loss of useful energy. Savings can, therefore, result from using daylighting systems. Electricity consumption is reduced and the proportion of solar gains which can make a useful contribution to heating requirements is increased. In addition, there is a potential for making energy savings through ventilation control and preheat: in recent years, the increase in fabric insulation standards in libraries has caused the proportion of heat lost through ventilation in conventional buildings to rise above 40% - more than that lost through the building fabric.

In introducing passive solar systems into public libraries, certain standards of comfort must, however, be maintained (see Table 3). Discomforting direct solar gains and associated glare are unacceptable in the reading areas and bookstacks. Lighting levels by bookstacks must be better than 250 lux. Reading areas require at least 600 lux. Librarians' work stations need 600 lux and well-lit views of the main bookshelves. Heating and ventilating controls must respond to local zone heat gains. Lighting controls must enable automatic or user initiated response to daylight levels. Autonomous lighting at reading desks is desirable.

COMFORT TARGETS	Book Stacks	Reading & Reference.	Foyer	Exhibition
Environmental temperature.	21 deg C	21 deg C	19 Deg C	19 deg C
Natural Air Change per Hour.	0.5	0.75	1.0	1.0
Average Daylight Factor.	5%	5%	2%	5%
Average DF (electric supp.)	2%	2%	1%	2%
Limiting Daylight Glare Index	23	23	24	21
Electric Illuminance-Lux.	300	500/600	300	300/500
Limiting Lighting Glare Index	19	19	21	16/19
Occupancy period 9am-6pm.				

Table 3. Comfort targets

Daylighting

To achieve distributed daylighting with minimum glare and glazing area, several forms of rooflight and windows were studied. Figure 6 shows a central lightwell serving ground and first floors designed to increase the area of luminous surface seen from deep in the plan, diffuse sky and solar glare and relieve the contrast gradient across the space. Sloped ceiling planes surrounding the roof light reflect light deep into the upper floor. Reflectors on the lightwell balustrade increase the vertical component of daylight reaching the ground floor. This is distributed over the ground floor by the inter-reflection between lightwell base and ceiling. Diffusing lay-lights suspended beneath the rooflight increase the apparent surface area of internal luminous sources and improve daylight penetration. Reading tables surround the lightwell to optimize daylight use. Tables on the south-facing edge have rotating screens above for local solar glare control. Vertical glazing is concentrated into east and west gables to maintain a simple architectural form. Clerestorey and view windows in the gables plus a lightwell offer balanced daylight distribution. Window desks have local blind control. Louvers and lightshelves on upper windows prevent peak solar overheating and aid distribution of daylighting. All other shading is movable to give maximum diffuse daylight at all times.

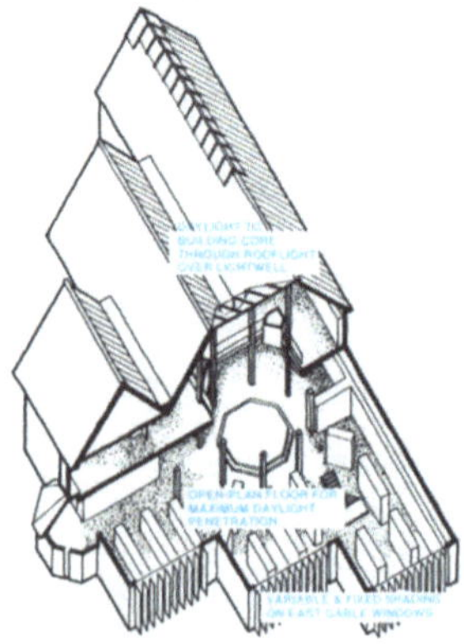

Figure 6. Central lightwell serving ground and first floors

Ventilation Control

The windows are sealed and during the heating season mechanical ventilation is used to maintain exchange rates of up to 12 l/sec per person. CO_2 sensors ensure air purity. Fresh air is ducted to the perimeter zones and exhausted at the roof over the lightwell core. Intake air is drawn through solar-heated zones to be preheated before delivery. Cavity rooflights act as solar ventilation preheaters. Summer-time ventilation needs are met by natural, stack-driven airflow. Venting from the ridge of the south-facing cavity rooflights and the lightwell induces cross ventilation. Fresh air inlets are positioned on the north and west building elevations, away from traffic noise. They are controlled by security dampers to allow a continuous flow. The direction of venting air flow is consistent with the needs of smoke extract.

Solar Thermal Gains

So that they can serve as both daylighting and passive solar heating components, the rooflights incorporate movable blinds in an airstream cavity. Solar glare, solar thermal transfer and glazing insulation are all controlled by the blinds. Direct solar gain is used preferentially unless it causes discomfort. When the rooflights are shaded, intake air through the cavity carries the radiant heat gains received on the blinds to the ventilation plant for delivery to the north zones of the building.

Thus, in winter solar preheating of intake air offsets ventilation losses. In summer, natural ventilation is encouraged by stack-driven exhaust flow along the rooflight cavity. Air is drawn into the base of the rooflights from the building and vented at the ridge of the rooflight (see Figure 7).

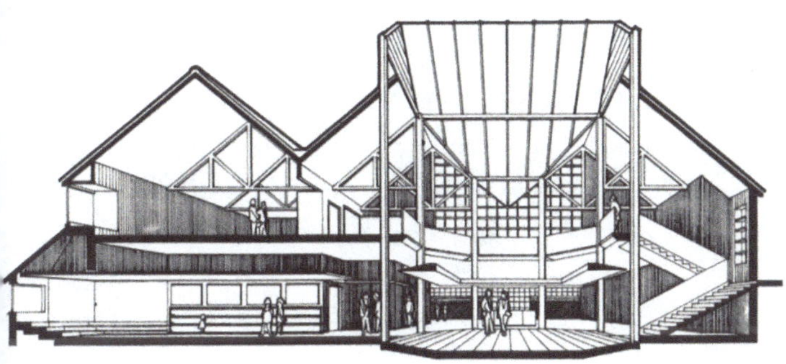

Figure 7. Cross section through building showing lightwell

Electric Lighting

The electric lighting is controlled according to the daylight received. There are automatic, photo-sensor controls in the public areas and timed switch-off in the staff areas. Automatic control is by progressive switching of banks of luminaires zoned according to distance from the windows. A central building energy management system predicts the daylighting distribution from photo-cell readings. Switching signals are distributed to luminaires on a control loop. Motorized blinds on rooflights are controlled under the same system. Manual over-ride is available at the librarians' control panel. Task lighting over reading desks is user-controlled. High-efficiency fluorescent and mini-fluorescent lamps are used with high frequency electronic ballast. Luminaires are mounted between bookstacks and local to reading areas. There is extra lighting from wall washers at positions remote from windows.

ENERGY CALCULATIONS PERFORMED AND DESIGN TOOLS USED

Introduction

The daylighting and passive solar heating features were designed with the help of a number of design tools. Comparisons with similar buildings were used for the initial energy analysis. Graphical and computer methods and physical models were used for daylighting performance analyses. The SERI-RES computer model was used for detailed assessment of the lighting energy consumption and the thermal performance of the building.

Preliminary Energy Analysis

A broad breakdown of energy use in a building of this type with no passive solar or light-switching features was obtained from energy data for existing similar libraries (see Figure 8). It was found that electricity costs exceed gas costs. Artificial lighting is the major user of electricity.

In the proposed design, insulation levels are high (average U-values are 0.57, including glazing) and therefore natural ventilation is the principal route by which heat is lost. It was calculated that by using daylight-responsive light switching systems, 35-50% electricity reductions are possible. In addition, passive solar gains and ventilation control may cut heating fuel use by 55%. Total annual fuel savings were calculated to be worth #1,900 (ECU 2,880). The approximate capital cost of extra passive solar measures could be up to #11,000 (ECU 16,670) to achieve a seven year payback.

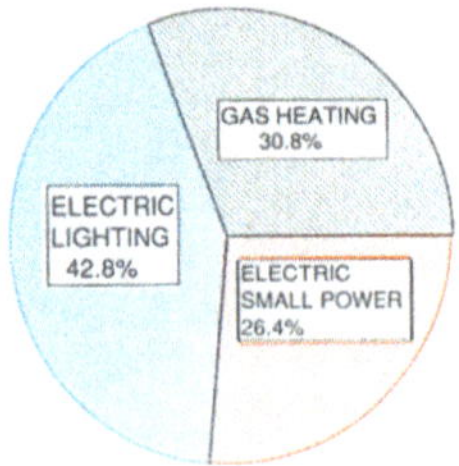

Figure 8. Breakdown of energy use for library with no solar features or light-switching system. (Total annual fuel consumption is #5,030 (ECU 7,620).)

Preliminary Estimates of Daylight Levels

In designing this building, the aim has been to achieve distributed daylighting with controlled electric lighting. The average daylight factor formula was used as a first step to estimate the window areas needed to achieve the target average daylight factors and the resulting electrical energy savings. It was found that average daylight factors between 2% and 5% offer energy savings with reduced use of electrical lighting. To judge whether the proposed glazing provides satisfactory daylight the designer needs to know the illuminance distribution. A sharp contrast between front and back of the room is unacceptable even if levels are above minimum requirements. In addition, if electric light switching is to be controlled by daylight levels, a representative daylight factor is required for each switching zone. Therefore the detailed analysis of daylight in the library was undertaken by defining ten switching zones. Figure 9 shows these zones on the first floor of the library with their average daylight factor and switching point daylight factor. Each zone represents a distinct activity space.

Each zone is differentiated by a particular daylight source and may be controlled under one switching strategy. Daylight factors were calculated using the BRE protractors (long-established graphic aids) and the results compared with those obtained by computer plotting. The positions chosen for switching control represent daylight factors close to the minimum. Commonly, the internally reflected component exceeds the sky component for these daylight factors and therefore detailed calculations of surface reflectances were conducted to establish accuracy. These daylight factors are particularly dependent on the reflectance values of interior finishes and furnishing. For daylight factors between 1.4% and 3.8% supplementary electric lighting will be required. The lightwell provides a level of daylight deep into the ground floor public space which would ensure a more uniform distribution throughout this area. Additional clerestory glazing significantly improves the sky component deeper into the room and is to be recommended. The meeting room (zone 10) is not intended to be dependent on daylight. However, calculations show that useful lighting may be gained there by means of a glazed screen to the lightwell.

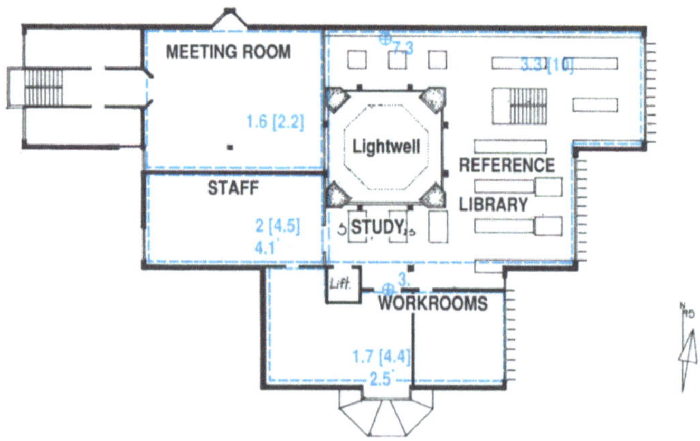

Key:

Figure 9. Floor plan showing average daylight factor and switching daylight factor for each lighting zone

Computer Simulation of Lighting Energy Consumption

The SERI-RES building energy simulation program was used to determine the effect of window size and position on daylight factors. The effects of window orientation, various electric lighting design choices, switching modes, lamp efficacy and room reflectances were also studied. The energy consequences of a number of lighting design options were examined for average (300 lux) and high (500 lux) target illuminances. The installed electric lighting power was calculated at 19 and 23 W/m^2 respectively, which was consistent with the existing libraries surveyed.

The energy savings achieved by the following switching systems were examined:

- continuous dimming (variable control) to all lamps in all zones in response to daylight levels, assuming high frequency dimmable fluorescent ballast under photo-sensor control;

- continuous dimming in the public areas of the library and manual switching in the staff and meeting rooms. Under manual switching the lights may be switched on by the staff in the morning if the daylighting is inadequate. Lights are automatically switched off at 12.00 GMT, but may be manually switched on. This offers the simplest strategy and may be preferred because it offers control to users.

- 4-step switching in the public areas and manual switching in the staff and meeting rooms. Stepped switching may be implemented with standard lighting gear and is therefore likely to have to lowest installed cost.

The energy used annually for electric lighting for ten zones, 1301 m^2 floor area, in use for 50 hours per week is shown in Figure 10. The reduction percentages indicate the potential for daylight substitution for each switching strategy. Without daylight substitution 44 kWh/m^2 (high) and 29 kWh/m^2 (average) of electrical energy per year are required to light the ten zones. The maximum use of daylight results from continuously variable electric lighting in response to internal daylight levels. Under this strategy 110 m^2 of window, 14% of the building wall plus 44 m^2 of rooflight over lightwell reduces electric lighting output by 56% (high) and 63% (average) of full consumption. This is worth £ 1,739 (ECU 2/m^2) and £ 1,305 (ECU 1.5/m^2) per year. The manual +4-step switching strategy reduces the lighting consumption to 53% of the full consumption for a 500 lux target level. This is 9% greater than the energy use under the continuously variable control strategy. This is worth £ 273 per annum. The initial capital cost difference between the two strategies should, therefore, not exceed £ 1580 (ECU 2394) to achieve savings after 7 years.

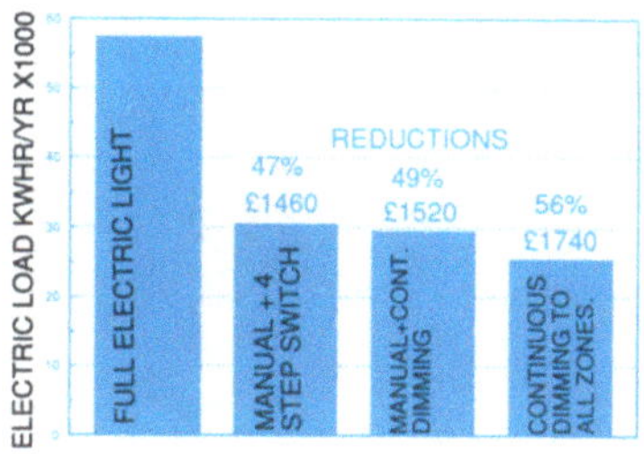

Figure 10. Annual electric lighting use with daylight controls up to 500 lux

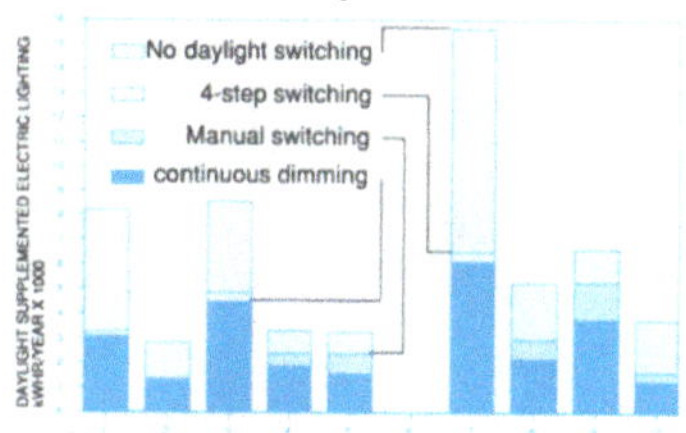

Figure 11. Electrical consumption for lighting for each switching zone. 500 lux target illuminance.

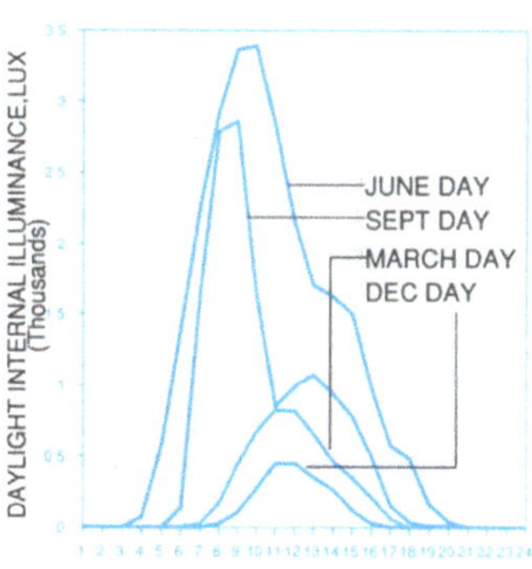

Figure 12. Daylight illuminance at zone switching point from lightwell and east-facing window

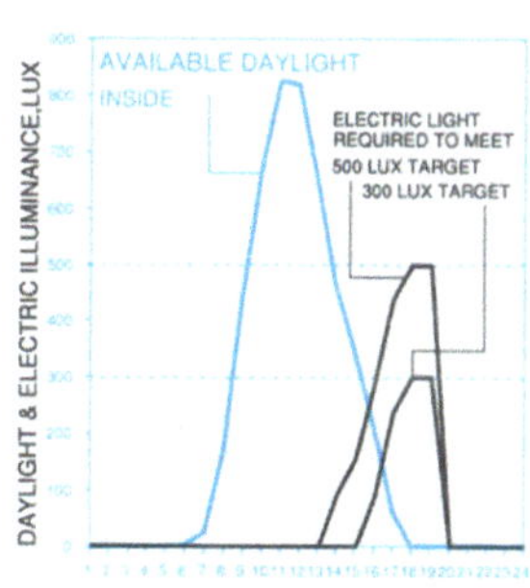

Figure 13. Electric light output under continuous dimming control

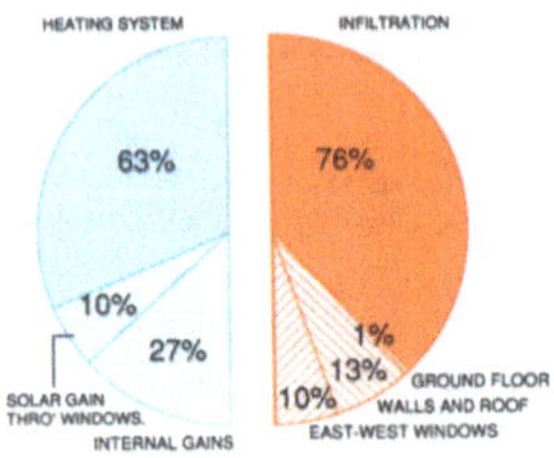

Figure 14. Thermal balance during the heating season when there are solar gains through windows only

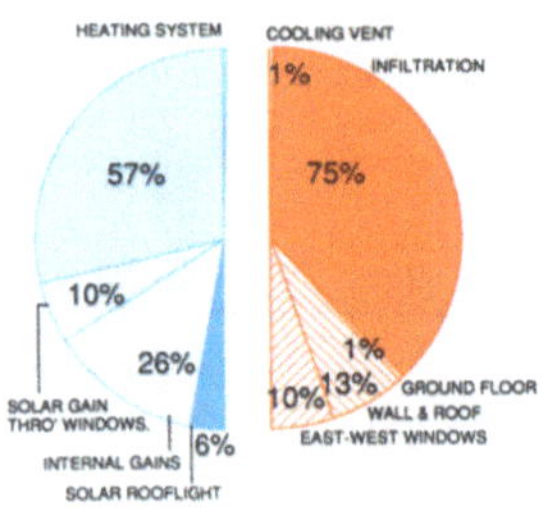

Figure 15. Thermal balance during heating season when there are solar gains through auxiliary rooflights and windows

Computer Simulation of Daylight Distribution

Computer calculations were used to obtain an idea of the daylight distribution in terms of daylight factors or illuminances for a grid of points within interiors with different windows, orientations, room shapes, obstructions and reflectances. They showed that a 3% daylight factor on the first floor can be maintained for more than 77% of the floorspace. Figure 16 gives the daylight factors across the central portion of the plan. The 41 m^2 rooflight to the lightwell adequately illuminates 193 m^2 of floorspace. This is 2.1 times the rooflight glass area. The average daylight factor across the open lightwell is 28%.

Thermal Performance Evaluations

SERI-RES was also used to evaluate the contribution of solar gains through gable glazing to the heating requirements of the building. A proposal for an airstream rooflight over the lightwell was also tested. This rooflight would allow solar pre-heated air to be delivered to the north zone.

Over the heating season, the east-west windows (with a U-value of 3.0 W/m^2 K) had thermal losses slightly exceeding solar gains. Use of shading devices results in a larger negative balance. Reducing the U-value to 1.9 by means of a low-emissivity coating re-establishes the thermal balance. For 80 m^2 of rooflights (with a U-value 1.9) over the lightwell, gains exceed losses by 13,500 kWh (equivalent to 10% of the heating requirement). This may only be fully utilized with a direct gain/ventilation gain strategy where absorbing/shading blinds are located in an airstream cavity. Controls on the rooflight allow direct solar gain unless it becomes uncomfortably hot in the south zone. Then, operation of the blinds and venting dampers allow the thermal gains to be transferred to the north zones. The ventilation mode is less efficient than direct gain but ensures a degree of ventilation preheat. An air supply rate of up to 1.2 building volumes per hour is required to transfer up to 90% of solar thermal gains. Excessive flow rates are not required to optimize the benefits.

Calculations on cooling needs were also conducted. During peak summer conditions up to 45 kW of unwanted solar gains on the rooflights must be vented away. Applying the standard stack formula an airflow of 1.8 m^3/sec is required for 80 m of rooflight. An airflow rate of 0.75 m^3/sec through a 0.3 m deep cavity would cope with this peak.

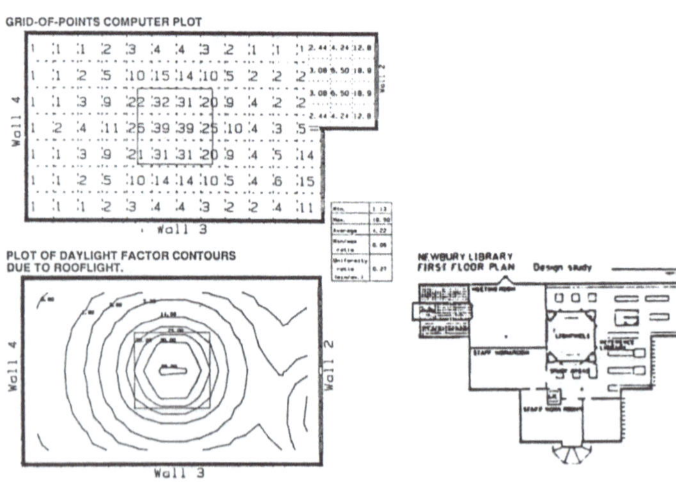

Figure 16. Daylight distribution study of first floor daylit by rooflight and two glazed gables in the clerestories

Daylighting Studies Using Physical Models

The daylight distribution was also tested using a physical model under a standard overcast artificial sky in the Polytechnic of Central London (see Figure 18). This allows easy examination of complex daylighting systems. The results are more precise than those obtained by manual calculation techniques or the currently-available simple computer calculation programs.

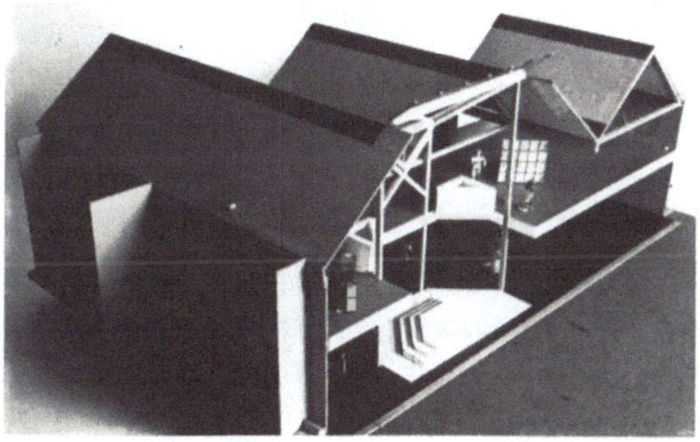

Figure 17. Model used in daylighting studies

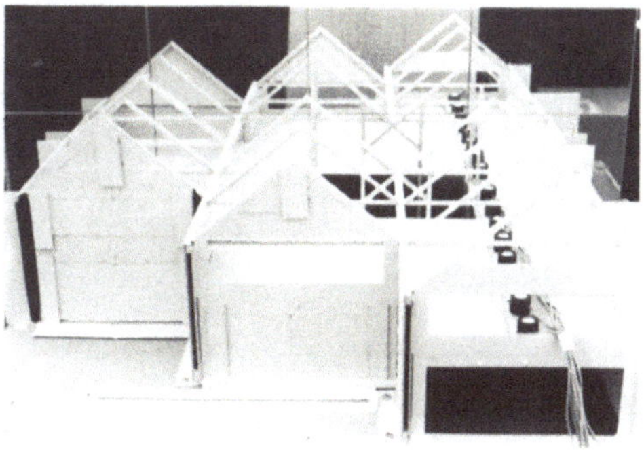

Figure 18. Model under test in artificial sky

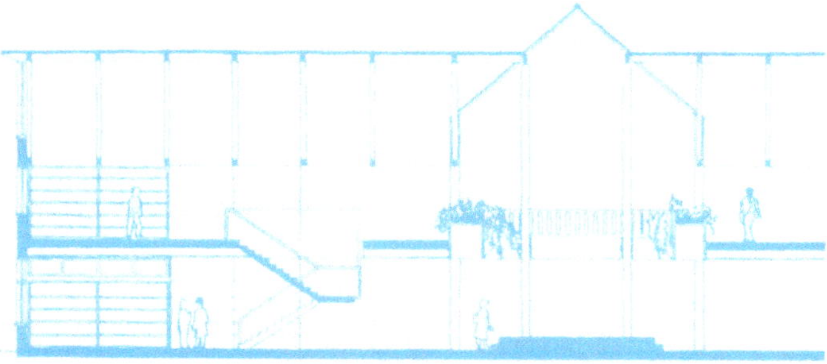

Figure 19. Daylight gradient produced by gable glazing and rooflight

The advantage of introducing daylight into the building core as well as having perimeter windows is illustrated in Figure 19. Gable windows, clerestories and a lightwell with reflective ceiling were all tested. Individually, the sources provide adequate local daylight but give a rapid diminution away from the glazing. With a lightwell by itself, for example, the predominance of the vertical component of light from the lightwell is apparent. When the lightwell is used in conjunction with the windows, however, the illuminance gradient is less sharp and the horizontal daylight factor is maintained above 3% along this line and the uniformity ratio (i.e. minimum daylight factor/average daylight factor) is 0.42. Without bilateral daylighting, permanent supplementary electric lighting would be required for 50% of the floorspace. Daylight values close to the glazing are sufficiently high to allow the use of reflectors at the gable glazing and beneath the rooflight. This would reduce the daylight factor close to the glass and raise the level of reflected daylight deeper into the space.

On the ground floor, the maximum value of the daylight factor within 3 metres of the lightwell base is 6.6% (see Figure 21) providing significant extra light to this core space. However the distribution needs improvement. Beyond the no-sky line of the rooflight, daylight is by reflection only. The initial low reflectance values of the first model tested produced a rapid diminution in light. Addition of diffusing high reflectance flooring to the lightwell base and a white ceiling will overcome the problem, improving daylight values by more than 40%.

The model studies were also used to define switching zones. A core floor space of 36% of this section has daylight factors below 3% and supplementary lighting will be required for long periods. A further 26% has daylight factors between 3% and 6% and may be switched as an independent zone. The remaining 38% is adequately daylit for most of the working year.

GENERAL DESIGN GUIDELINES/POINTS OF INTEREST RESULTING FROM THIS PROJECT

Savings in Lighting Energy Consumption

The aim of the studies described in this brochure was to explore the use of simple design tools to obtain information on daylight distribution and energy in the library design. They proved that, by introducing daylighting systems into the building, it was possible to use the longer daylight hours and higher diffuse sky illumination found in summer to reduce the artificial lighting energy consumption to 20-30% of what it would be with full electric lighting (see Figure 22).

Integration of Daylighting and Passive Solar Heating Systems

The studies also showed that, while in this building daylighting is the most cost-effective way of using solar energy, modest but useful savings in heating energy are also achievable which can offset heat losses through ventilation. In order to optimize the capital investment, in this building the daylighting and direct solar gains are produced by the same components. Thus, the windows contain adjustable shades to prevent glare and allow intake of ventilation air.

Usefulness of Design Tools

Manual formulae, graphic aids, computer plots and physical modelling all gave similar results for the daylighting performance of the building. Average daylight factors obtained from the formulae were well-matched to the measured and computer-predicted values. It was concluded that the formulae provided a reliable preliminary method. Photocell measurements in a model in an artificial sky were appropriate for testing complex window designs and reflectors. The ability to set targets for additional design refinements and to perform sunlight and photographic visualization studies are additional advantages of physical modelling. Computer plots provided a rapid and accurate assessment of daylighting. As programs become more sophisticated, computer plotting is likely to be the first choice of the designer.

Energy analysis was conducted using the SERI-RES dynamic simulation model. Fabric and ventilation heat flow analyses may be combined with daylight/electric light energy data to provide a highly flexible design tool. In the future, links to graphic software will offer designers fully integrated computer-aided design.

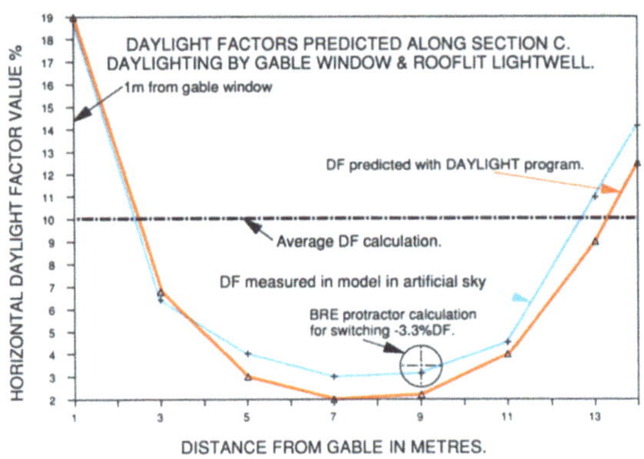

Figure 20. Comparison of computer calculations with physical model measurements for first floor of Newbury library

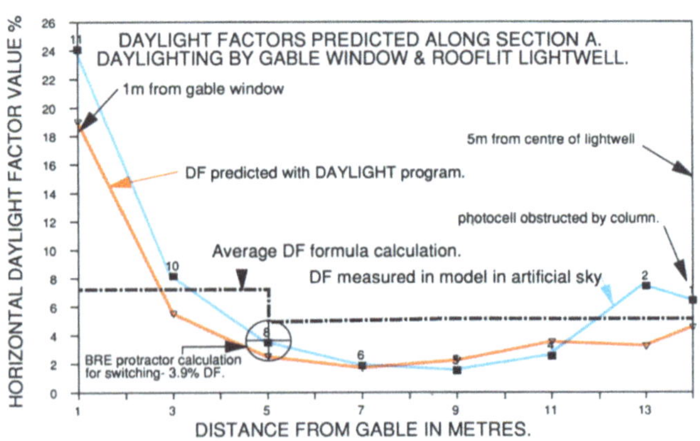

Figure 21. Comparison of computer calculations with physical model measurements for ground floor of Newbury library

Window Design and Position

Through the modelling work, a strategy was developed for relating glazing position and size to the required distribution across the floorspace. Figure 23 shows that windows, clerestories and rooflight serve distinct floor areas.

A doubling of the daylight factor in the mid-zone results from using clerestories and only increasing the wall glazing area by 1.5. Core daylighting from rooflight and lightwell is beneficial only when integrated with distributed windows. In open plan areas this produces bilateral daylighting, a system likely to be most effective in producing satisfactory daylight gradients. Perceived lighting adequacy depends on brightness contrast across the zone and across interior surfaces, including windows. Therefore shallow daylight gradients will minimize the desire for electric supplementary lighting.

Window Shading

For buildings under northern skies glazing is sized for overcast conditions. Shades provided to overcome glare or overheating must, therefore, be movable to avoid the unnecessary reduction of daylighting on overcast days. At times sunlight penetration is welcome but building occupants must have control. Simulations have shown that under usual lighting conditions, e.g. with daylight factors above 1% and service illuminances below 500 lux, variations in internal illuminance occur above the switching illuminance. Sunlight and movable sun shading have only a marginal effect on electric lighting energy consumption. In contrast, the use of tinted glazing or permanent shading reduces lighting energy savings markedly by obstructing the incoming diffuse illuminance at all times.

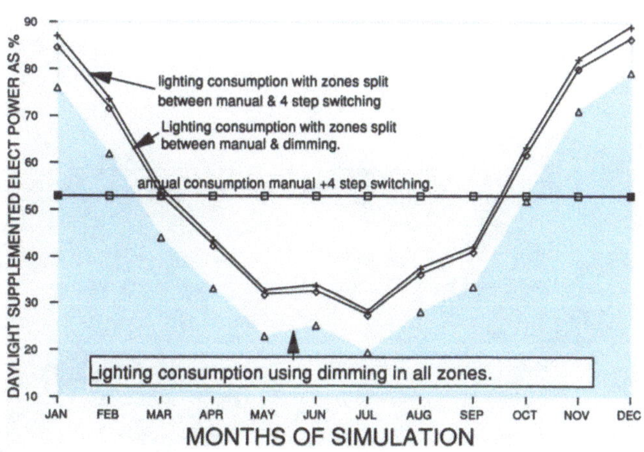

Figure 22. Electricity use for switching modes as percentage of full electrical usage

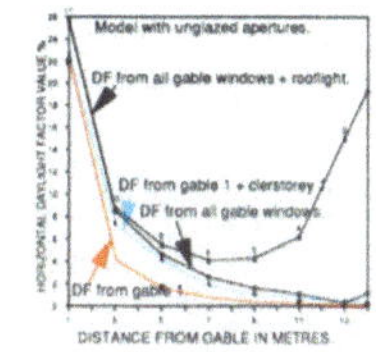

Figure 23. Daylight profile through first floor. Daylight factors measured along section C.

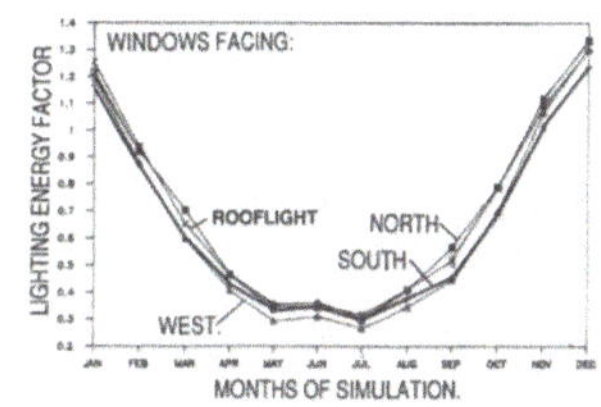

Figure 24. Orientation effect of window on lighting energy consumption for room. (Calculated using SERI-RES.)

Energy Saving Through Daylighting

Electrical savings are only realizable by controlling light switching. A balance between occupant satisfaction, capital cost and long-term savings must be struck. In addition, a division between controlled background electric lighting and local task lighting must be established. For the library, putting all background lighting under continuous dimming optimized savings (56% overall). Electronic dimmable ballast is expensive but more widespread use in future should bring cost reductions. Manual switching produces lower savings (up to 38%) but its lower cost and staff acceptability (through user control) make it suitable for staff rooms. For the public areas, stepped switching using conventional electronic ballast saves 47% of lighting consumption at lower cost than dimming. An energy management system progressively extinguishes bands of luminaires according to the daylight gradient. Local use of window blinds is monitored and their shading effect allowed for by the system.

BUILDING 2000

Building 2000 brochures are published by Directorate General XII of the Commission of the European Communities to show how design studies can help architects and other building designers use passive solar principles to the best effect to produce attractive energy-efficient buildings. Each brochure describes studies carried out with the support of the Commission during the design phase of one of thirty-six non-domestic buildings in the EC Member States. The studies were on such topics as daylighting, heating, cooling, ventilation, comfort, control systems and urban design. They were carried out with the help of acknowledged European experts in these fields and drew heavily on lessons learned and techniques developed through the Commision's research and development programme on solar energ applications to buildings.

Commission of the European Communities/Directorate-General for Science, Research and Developmen

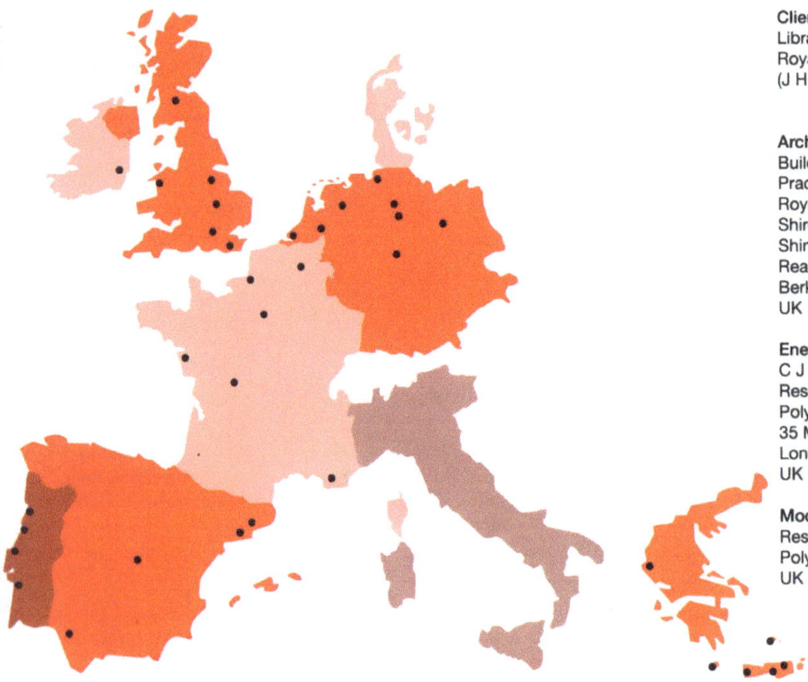

List of Design Team Participants and Advisers

Client
Library Services Committee
Royal County of Berkshire
(J Hicks, Chief Librarian Designate)

Architect
Building Design and Maintenance Practice
Royal County of Berkshire
Shire Hall
Shinfield Park
Reading
Berkshire
UK

Energy Consultants
C J Hancock and J Littler
Research in Building
Polytechnic of Central London
35 Marylebone Road
London NW1 5LS
UK

Modelling and Analysis
Research in Building Group
Polytechnic of Central London
UK

This set of **Building 2000** brochures illustrates how architects and other building designers can successfully apply passive solar principles to produce energy-efficient buildings.

BUILDING 2000 Participants

Project Director
Theo C. Steemers

Coordinator
Cees den Ouden

Technical Steering Committee
Dean Hawkes
Nick Baker
Alex Lohr
Jean P. Lepoivre

Regional Liaison Agents
(D) Jörn Behnsen
(E) Vicente Sifre
(F) Michel Raoust
(GB) Alan Hildon
(GR) Matheus Santamouris
(P) Eduardo Maldonado

Further information or copies of the brochures can be obtained from prof. ir. Cees den Ouden, EGM Engineering BV, P.O. Box 1042, 3300 BA Dordrecht, The Netherlands.

BUILDING 2000

Commission of the European Communities

Conversion of old building into centre for ecological community.

Glazed roof gives good daylighting and enables solar gains to contribute to space and domestic hot water heating.

Frostproof lean-to greenhouses provide direct solar gains, preheat ventilation air and provide buffer space and an area for growing plants

COMMUNITY AND EDUCATIONAL CENTRE

STEYERBERG/ FEDERAL REPUBLIC OF GERMANY

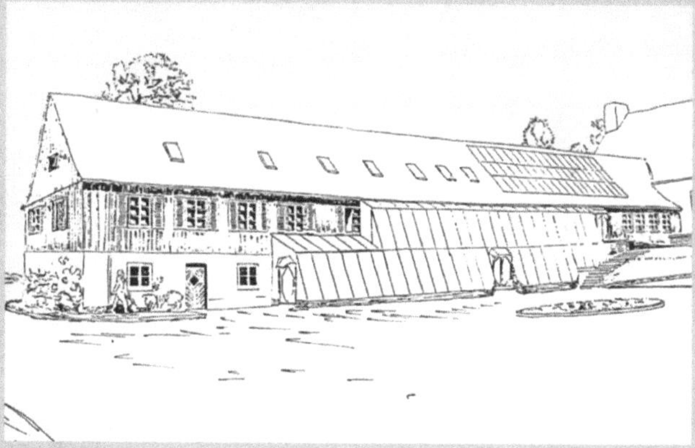

Building 2000 is a series of design studies illustrating passive solar architecture in buildings in the European Community.

ISSUE

P 4

FEB 1991

PROJECT DESCRIPTION

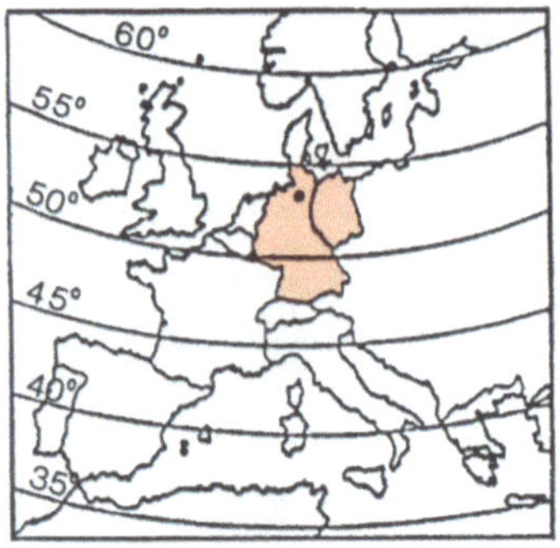

Figure 1. Location of Steyerberg

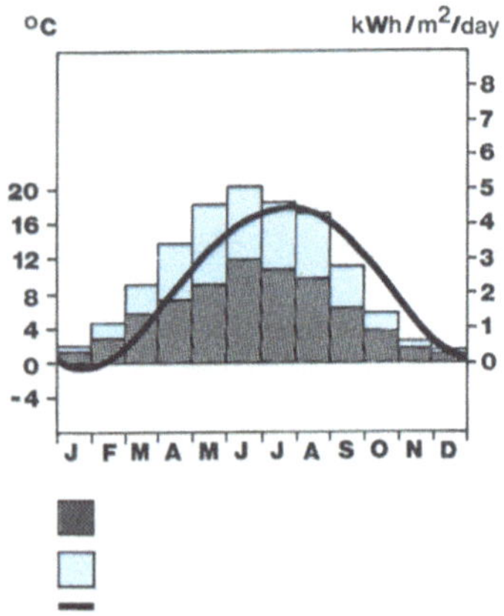

Figure 2. Mean solar irradiation and temperature data

Degree days (base 20 ° C)	3800
Sunshine hours a year	1600
Mean global solar irradiation per year	950 kWh/m^2

Table 1. Some key climate data

Building Type

This project concerns the conversion of the former administration and service building of a second world war munition workers' camp into a centre for Lebensgarten Steyerberg community. The original building consisted of a central hall capable of holding 600 people and east and west wings. The finished building will contain a kindergarten, offices, kitchen, bakery, dining rooms, food stores, lecture rooms and related facilities. The conversion is being carried out in stages and this brochure describes the first phase, the renovation of the west wing (see Figure 3).

Location

The site is 80 m above sea level in the hills overlooking the Weser valley in Steyerberg (latitude 52 ° 3 ' N, longitude 9 ° 3 ' E). This is between Hannover and Bremen in Lower Saxony in the north of the Federal Republic of Germany (see Figure 1). The building is in the centre of the site and sheltered to a certain extent by neighbouring residential buildings to the west and south and by pine trees to the north and west. The south side of the building, however, opens onto the camp's central square and is virtually unobstructed.

Site Microclimate

The climate is typical of the coastal region of north-west Germany. The site is fairly windy because of the presence of coastal winds. The prevailing wind direction is west to south-west. Mean solar irradiation and temperature data are given in Figure 2. Other key climate data are given in Table 1.

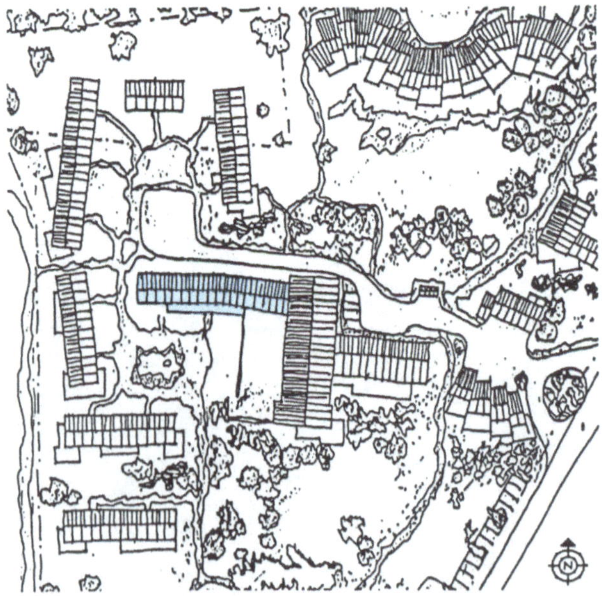

Figure 3. Site plan

Design and Construction Details

Introduction

The Lebensgarten Steyerberg community consists of 100 or so people who are exploring new approaches to solving today's ecological and social problems. They call themselves an ecological and spiritual community. They offer a wide-ranging seminar programme and attract increasing numbers of visitors and seminar guests; the current visitors' rate is 2000-3000 a year. The community's aim in developing this centre is to demonstrate how ecological and energy-saving techniques can be used in the renovation of an old building with a construction type typical of the region.

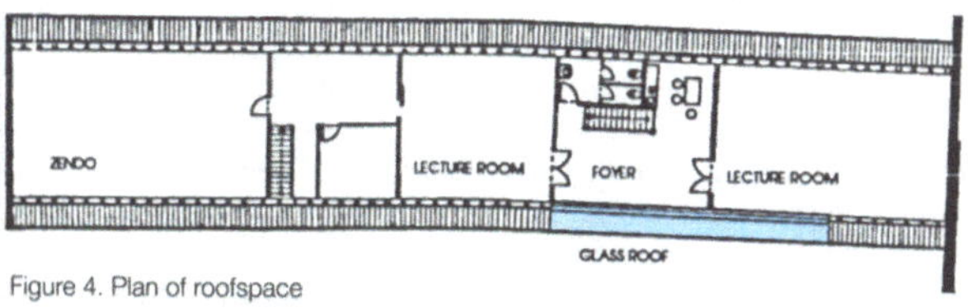

Figure 4. Plan of roofspace

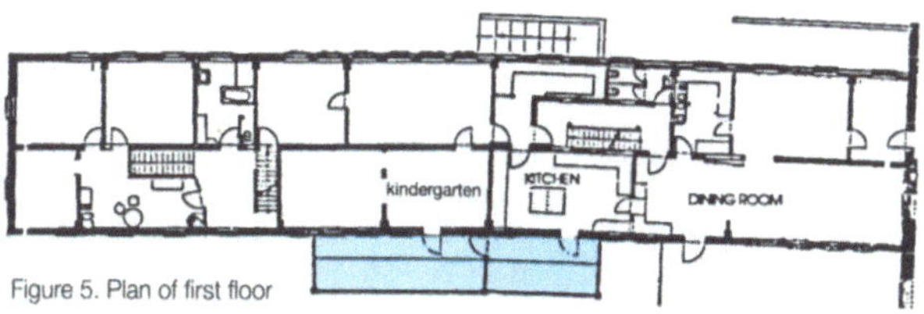

Figure 5. Plan of first floor

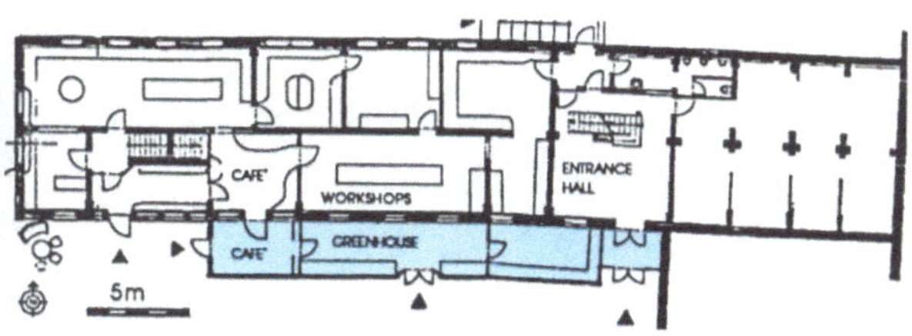

Figure 6. Plan of ground floor

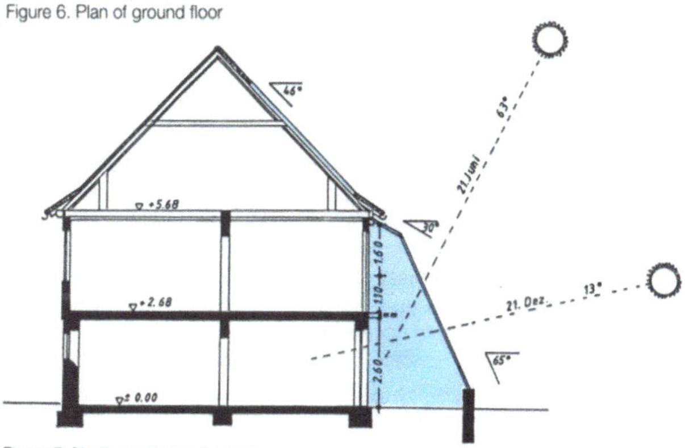

Figure 7. North-south cross section

Design Details

The west wing covers 460 m^2 of land and contains three storeys including the roofspace. Plans of the three floors are shown in Figures 4-6. A north-south cross section is given in Figure 7.

Lean-to greenhouses (most of which are two storeys high) cover half of the south facade. In order to optimize their thermal buffer effect, they are placed against those parts of the building which need to be kept at a relatively high temperature (such as the kindergarten and cafe) or have a high level of internal gains (such as the kitchen and bakery).

Each room of the parent building has at least one window which opens directly onto the outside. In addition, a 60 m^2 area of glazing has been introduced into the south-facing slope of the roof over a 62 m^2 lecture room and its 20 m^2 foyer. These measures ensure good daylighting and effective ventilation in summer.

Construction Details

The building has a traditional brick and wood structure. The external walls are 360 mm thick on the ground floor and 240 mm on the first floor. Those on the northern facade have additional insulation. The roof has been insulated with 120 mm processed recycled paper plus wood panelling. All windows are double glazed. The glazed roof and lean-to greenhouses have double glazing and wooden frames. The greenhouses have a frostproof concrete foundation. U-values of some building elements are given in Table 2.

Roof	0.3
External walls	
ground floor	0.64
first floor	0.56

Table 2. U-values (W/m^2 K) of some key building elements

DESCRIPTION OF PASSIVE SOLAR FEATURES/COMPONENTS

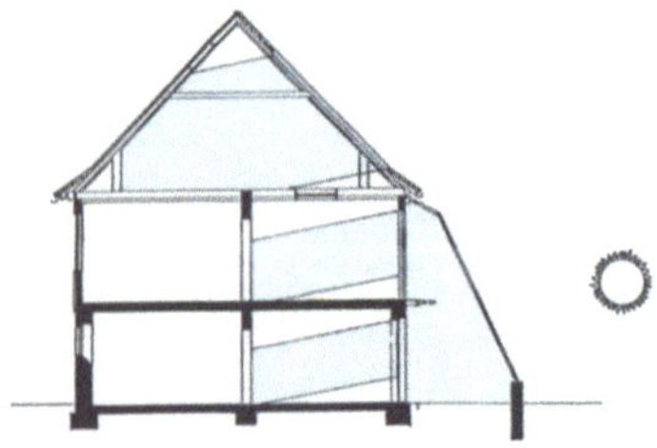

Figure 8. Direct gain on a winter's day with the shading devices open

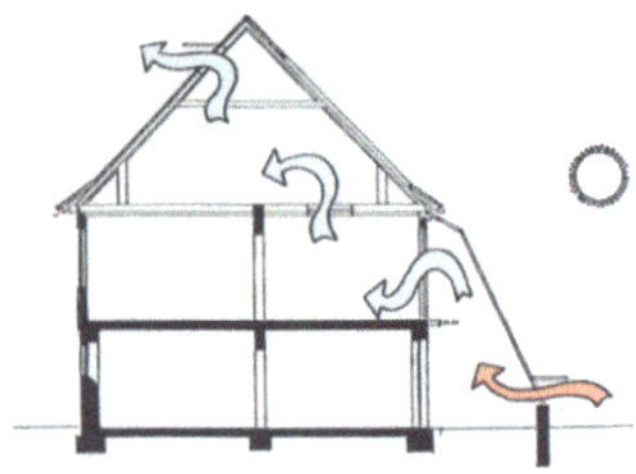

Figure 9. Solar preheated ventilation on a spring/autumn day with the shading devices open

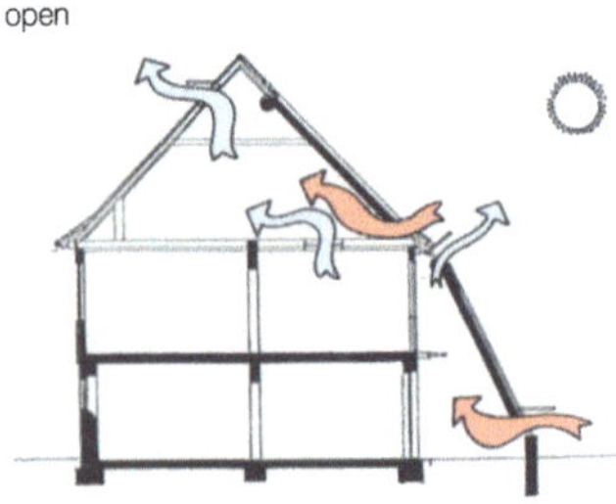

Figure 10. Direct ventilation on a summer's day with the shading devices closed

Auxiliary Heating

Auxiliary heating is provided by a gas-fired co-generation system. The heating system is controlled by room thermostats connected to the radiators. In winter, the heating demand determines the mode of operation of the co-generation system; surplus electricity is distributed to neighbouring residential buildings. From May to October, hot water is supplied by the solar water heating system; power is provided by wind turbines and the public supply grid.

Introduction

In developing the design, the architects began with the aim of optimizing solar gains and making use of the large thermal mass of the traditional brick and wood construction of the original building. This led to using components such as lean-to greenhouses which collected solar energy quickly and systems such as preheated ventilation which were compatible with the building type. Each passive solar feature was designed to contribute to at least three of the following: direct gain for space heating; domestic hot water supply; preheating of ventilation air; space for growing plants; additional living space in spring and autumn. The features had to have a construction and mode of operation easily understood by visitors to the centre and be capable of being applied cheaply to the visitors' own homes.

Greenhouses

The lean-to greenhouses have a 0.8 m concrete frostproof foundation, wooden frames and double glazing put together from two single panes 60 mm apart. The total glazed area is 97 m^2 and the total space in the greenhouses 254m^3. Fresh air enters through doors and inlets at the bottom of the glass front. Outlets at the top of the greenhouses allow pre-heated ventilation air to enter the kindergarten and kitchen or be vented outdoors. Overheating is prevented by light-coloured fabric blinds. Most of the greenhouse space is two storeys high; the ceiling between the ground and first floors consists of a metal grid to minimize shading of the ground floor.

Glazed Roof

Construction of the frame and glazing for the 60 m^2 glazed roof is similar to that for the greenhouses. The frame is mounted on the original rafters. Solar water heaters are installed behind the panes in the top third of the glazed roof. Fresh air enters the roofspace through openings at the bottom of the glazed area. Exhaust air leaves via windows in the northern slope of the roof and an opening in the loft of the adjacent hall. There is an innovative shading system in the lecture room under the glazed part of the roof which provides effective temporary insulation at night and can act as an air collector, contributing solar gains to the hot water supply. The glazing in the roof of the foyer to this lecture room is shaded by light- coloured fabric blinds.

Mode of Operation of the Solar Systems

Because the local climate changes rapidly, it is not possible to define steady-state operating modes for the passive solar systems for each season. It is necessary on cool days to capitalize on direct gains when they occur even if they are of a short duration by having quick-reacting control systems, making efficient use of the thermal mass of the building and minimizing heat losses. The modes of operation for some typical winter, spring/autumn and summer days, however, are illustrated in Figures 8-10. On a sunny winter's day, direct solar gains can make a contribution to space heating and hot water supply. Heat loss is reduced by using preheated air from the greenhouses to ventilate the rooms, by double glazing and by using shading systems as additional insulation. The total solar contribution is most significant in spring and autumn. In summer, overheating of the greenhouses and glazed roofspace is prevented by shading systems and direct ventilation. The building's orientation is such that the prevailing west/south west winds aid cross ventilation through the greenhouses, first floor rooms and roofspace. Control of the ventilation and shading systems is automatic in greenhouse areas used for growing plants. There is manual control in the glazed roof rooms and the greenhouse spaces open to the public. When the glazed roof lecture room is unoccupied the shading system is controlled automatically to optimize energy savings.

ENERGY CALCULATIONS PERFORMED AND DESIGN TOOLS USED

Introduction

Design of the greenhouses and rooms under the glazed roof was carried out with the help of Building 2000 experts. In the course of the discussions a detailed evaluation of the lecture room under the glazed roof was conducted using a computer simulation model.

Computer Study of Thermal Behaviour of Lecture Room under Glazed Roof

The Calculations

The thermal behaviour of the room was predicted for given ventilation conditions, comparing different strategies for reducing space heating energy consumption and estimating the need for shading during sunny periods. All the calculations, which were carried out by Peter Voit of Stuttgart, assumed that the room had dimensions and ventilation rate, etc., as shown in Table 3. The weather data were the long-term averages for Bremen.

Three simulations were carried out. In the first (case 1), a constant room thermostat setting of 20 ° C was assumed. In the second run (case 2), it was assumed that the temperature dropped at night (i.e. from 10 pm to 8 am) to 10 ° C because of the low thermal mass of the room. In the third run (case 3), it was assumed that the temperature dropped at night and the south- facing roof glazing was triple-glazed.

Prediction of Annual Heating Energy Consumption

The predicted annual heating energy consumption for the three cases is shown in Figure 11. Case 1 required 78.3 kWh/m^2, case 2 52.1 kWh/m^2 and case 3 45.4 kWh/m^2. Thus, allowing the room temperature to drop at night gave a saving of over 33% compared with case 1. A further saving is achieved by installing a third pane of glass, but this is not cost-effective.

It was found, however, that the peak heating demand in case 2 is almost twice that of case 1. Further, when the temperature is allowed to drop at night it is sometimes necessary to have some auxiliary heating, even in summer, to maintain comfort conditions.

Prediction of Room Temperatures

The computer simulations indicated that there would be some fluctuation in room temperature throughout the year. The January figures for case 2 are shown in Figure 12. They indicate that, despite the fact that ambient temperatures may be 0 ° C or below, the room temperature may rise to above 25 ° C because of solar gains through the south-facing roof glazing.

The given air change rate of 40 volumes per hour means that the room temperature will never exceed 39 ° C even without shading devices. However, in July it will never be below 20 ° C and the daytime levels will always be uncomfortably high. It was found that this problem can be solved by installing a shading system. Calculations showed that, when there is appropriate shading, the room will only be 3 ° or 4 ° C above that outside when there are twenty-five people in the room and a slide projector is in use.

Use of Shading System as Temporary Insulation to Reduce Space Heating Demand

Further simulations showed that closing the shading system at night and during sunless winter days when the room is not in use prevents heat loss and helps reduce space heating energy demand by a considerable amount.

Room volume	195 m^3
Roof glazing	
net area	25.3 m^2
orientation	due south
inclination	44 °
Ventilation rate	44 vols/hour
Shading	nil

Table 3. Dimensions, etc., of lecture room undergoing thermal evaluation

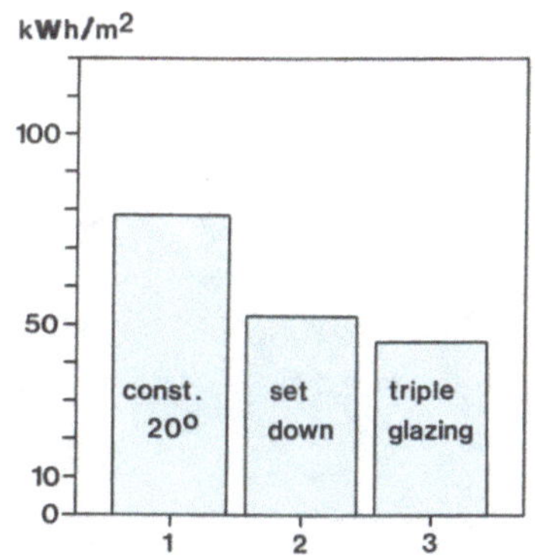

Note. The diagram shows the energy consumption for the following three cases:
Case 1. Constant room temperature of 20° C
Case 2. Permitting room temperature to drop at night
Case 3. As case 2 but with triple- glazing in roof

Figure 11. Annual space heating requirements for lecture room under glazed roof

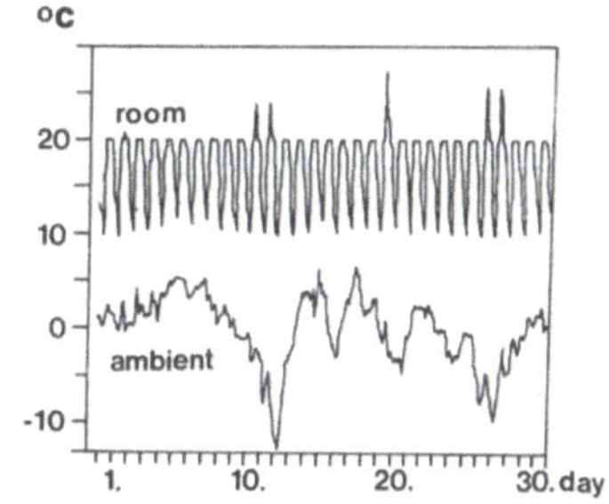

Figure 12. Predicted temperatures in the lecture room under the glazed roof in January with night- time temperature drop

Energy Savings

Introduction

Because the conversion of the building is still underway and its occupancy is increasing, it is difficult to predict accurately the energy balance of the building without resorting to precise monitoring. Energy consumption data for earlier years cannot be used for comparative purposes to predict the energy savings due to the passive solar and other energy- saving features.

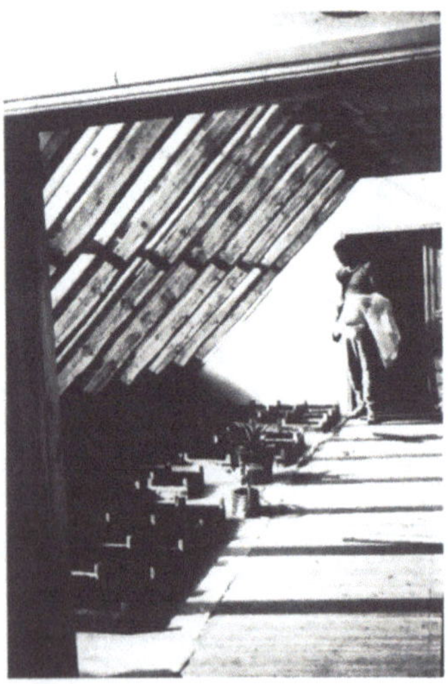

Figure 13. The foyer under the glazed roof in use as a plant nursery

Solar Gains Through Glazed Roof

The calculations outlined above suggest that during the heating season the solar gains in the room under the glazed roof are slightly higher than the space heating energy demand. However, because these rooms do not have much thermal mass, some of the solar gains are not usable but will be lost through ventilation. It is estimated that, if the energy savings due to the temporary insulation referred to above are not taken into account, then the solar contribution to space heating the rooms under the glazed roof is around 50-60%. The energy balance is much more positive if the savings from the temporary insulation are included.

Solar Water Heating

A further major contribution to the overall energy requirements of the building is made by the solar water heating system mounted behind the upper third of the glazed roof. If this system has an efficiency of about 60%, it can be expected to provide some 2,600 kWh a year of energy for heating water.

Daylighting

Thanks to the excellent daylighting conditions under the glazed roof, no electric lighting is necessary there from sunrise to sunset even on cloudy winter days. The electricity savings are estimated to be about 2,000 kWh a year.

Preheating of Ventilation Air in Greenhouses

At the moment, the conventional energy savings likely to accrue from the presence of the greenhouses cannot be predicted. This is because their principal contribution is expected to be via solar preheating of ventilation air and the structure of the main building has not yet been modified to allow the best use to be made of this.

Overall Savings

Overall, the passive solar and other energy-saving features are expected to reduce the space heating fuel bills by 25-30%.

Cost Effectiveness

The total construction cost of the glazed roof system, including the shading systems and the solar water heaters, amounts to about 15,000 ECU - some 12,000 ECU more than the cost of providing standard insulation to the same roof area. The contribution which the glazed roof will make to the conventional energy demand for space heating, hot water supply and artificial lighting are estimated to be about 500 ECU a year. In addition, some 700 ECU-worth of seedlings, herbs and other plants will be grown in the foyer (see Figure 13).

The overall cost of constructing the lean-to greenhouses is around 25,000- 30,000 ECU.

GENERAL DESIGN GUIDELINES/POINTS OF INTEREST RESULTING FROM THIS PROJECT

Human Aspects and Users' Response

The lecture room and foyer under the glazed roof were put to use as soon as they were completed, even though they were in an unfinished state. They have now been seen and used by some hundreds of seminar guests. Despite the fact that they were subjected to ongoing construction work, nearly all the visitors were highly complimentary about the good daylighting and the general ambience of the rooms. They reported that they experienced improvement not only in subjective comfort but also in ability to concentrate on the course work. This resulted in the visitors being willing to operate the manual control systems.

Further, when part of the foyer began to be used as a plant nursery in spring 1989, the daily encounters with the growing of their own food had an unquantifiable but definite beneficial effect on seminar guests and teachers. A similar response is expected when the lean-to greenhouses are built. Working conditions in the kitchen will undoubtedly be helped by the improved ventilation in autumn, winter and spring and there will be added pleasure from having easy access to the herbs and spices growing in the greenhouses.

The lecture room and foyer in the glazed roofspace and the cafe in the greenhouses will allow visitors to gain direct insight into the advantages of passive solar architecture. Hopefully, this will encourage some of them to consider using passive solar components when they redesign, build or renovate their own homes.

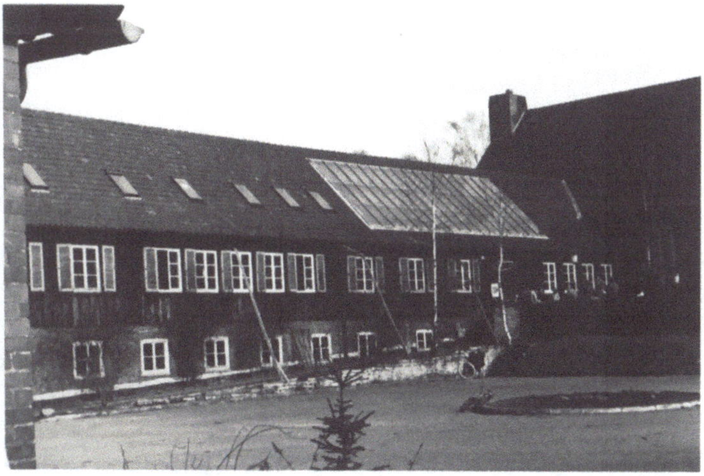

Figure 14. View of the glazed roof.

Conclusions on Design

The experiences and calculations carried out in the course of this study, together with comments made by the Building 2000 experts, lead to the conclusion that:

- passive solar features combine well with traditional buildings with large thermal mass;

- care should be taken over the location and geometry of the passive solar components. Glazing should face due south wherever possible and be inclined at an angle which optimizes the collection of diffuse radiation;

- the design should take into account the fact that solar preheating of ventilation air is the principal mechanism for transferring heat;

- provided there is sufficient thermal mass in the original building, the effective use of thermal storage mass should be regarded as being of prime importance in the design of sunspaces/greenhouses;

- overheating can be controlled by good ventilation combined with an effective shading system;

- both lean-to greenhouses and glazed roof areas can be cost- effective multi-functional measures for improving energy use in an old building.

BUILDING 2000 Building 2000 brochures are published by Directorate General XII of the Commission of the European Communities to show how design studies can help architects and other building designers use passive solar principles to the best effect to produce attractive energy-efficient buildings. Each brochure describes studies carried out with the support of the Commission during the design phase of one of thirty-six non-domestic buildings in the EC Member States. The studies were on such topics as daylighting, heating, cooling, ventilation, comfort, control systems and urban design. They were carried out with the help of acknowledged European experts in these fields and drew heavily on lessons learned and techniques developed through the Commision's research and development programme on solar energy applications to buildings.

Commission of the European Communities/Directorate-General for Science, Research and Development

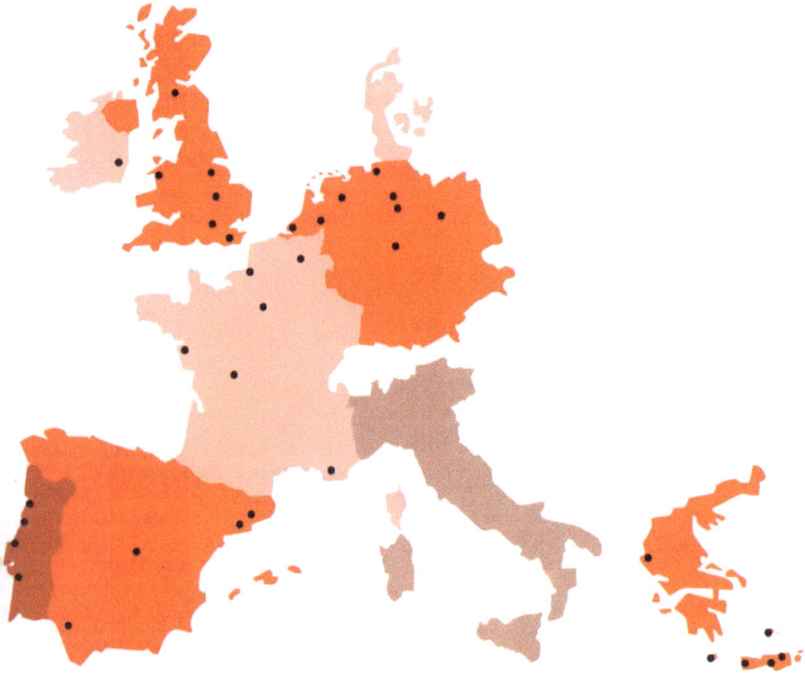

List of Design Team Participants and Advisers

Client
Lebensgarten Steyerberg eV
Akazienhain 4
D-3074 Steyerberg
Federal Republic of Germany

Architect
Caspar Harbecke
Gilsatalstr. 12
D-3585 Neuental-Gilsa
Federal Republic of Germany

Consultants
Planungsburo Kennedy
Ginsterweg 4-5
D-3074 Steyerberg
Federal Republic of Germany

Dr Roland Kuhne
Rosenanger 3
D-3074 Steyerberg
Federal Republic of Germany

Energy Consultant
Achim Kreuzer
Akazienhain 2
D-3074 Steyerberg
Federal Republic of Germany

Funding Support
Schweisfurth-Stiftung
Barerstrasse 9
D-8000 Munchen 2
Federal Republic of Germany

This set of Building 2000 brochures illustrates how architects and other building designers can successfully apply passive solar principles to produce energy-efficient buildings.

BUILDING 2000 Participants

Project Director
Theo C. Steemers

Coordinator
Cees den Ouden

Technical Steering Committee
Dean Hawkes
Nick Baker
Alex Lohr
Jean P. Lepoivre

Regional Liaison Agents
(D) Jörn Behnsen
(E) Vicente Sifre
(F) Michel Raoust
(GB) Alan Hildon
(GR) Matheus Santamouris
(P) Eduardo Maldonado

Further information or copies of the brochures can be obtained from prof. ir. Cees den Ouden, EGM Engineering BV, P.O. Box 1042, 3300 BA Dordrecht, The Netherlands.

BUILDING 2000

Commission of the European Communities

Civic building containing town hall, public registry and notary office and offices for electricity and water supply services

Direct solar gain in winter

Shading from solar radiation in summer

Thermal insulation of building envelope

Combination of natural and mechanical ventilation

Zoned auxiliary heating with control according to occupancy and indoor and outside temperatures

Preheating of ventilation fresh air by heat recovery of exhaust air

Daylighting from overhead in deep areas

Automated switching of electric lighting

TOWN HALL

ILHAVO/PORTUGAL

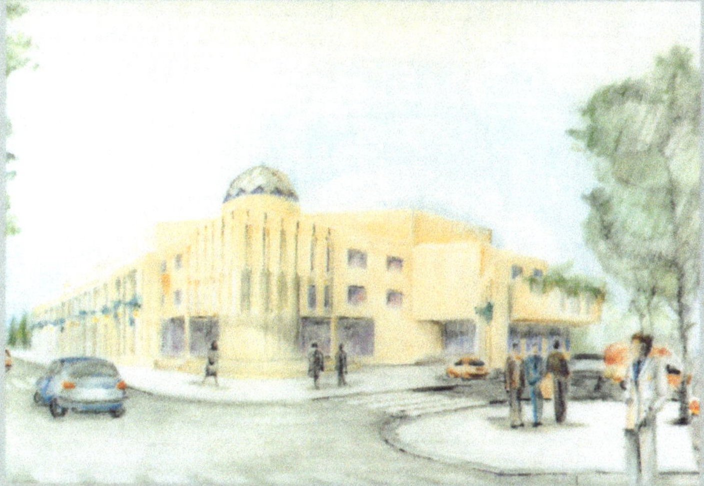

Building 2000 is a series of design studies illustrating passive solar architecture in buildings in the European Community.

ISSUE

P 5

FEB 1991

2 PROJECT DESCRIPTION

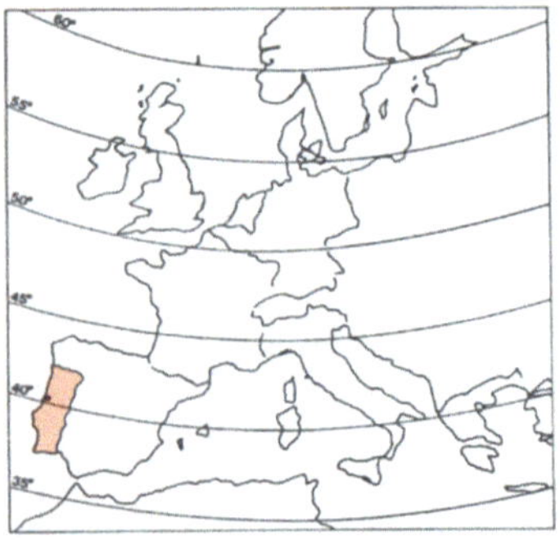

Figure 1. Location

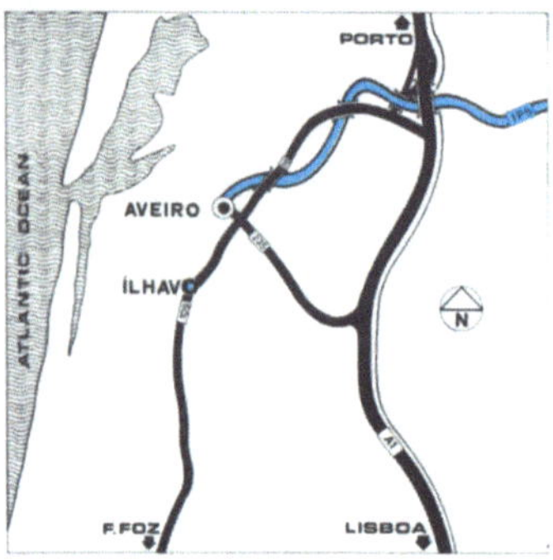

Figure 2. Location of Ilhavo

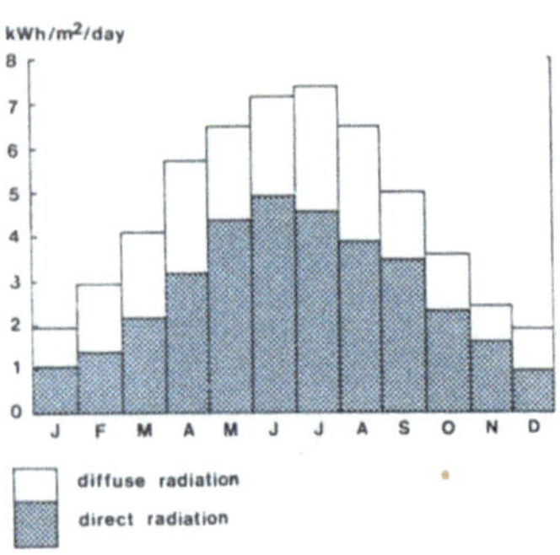

Figure 3. Mean solar irradiation data

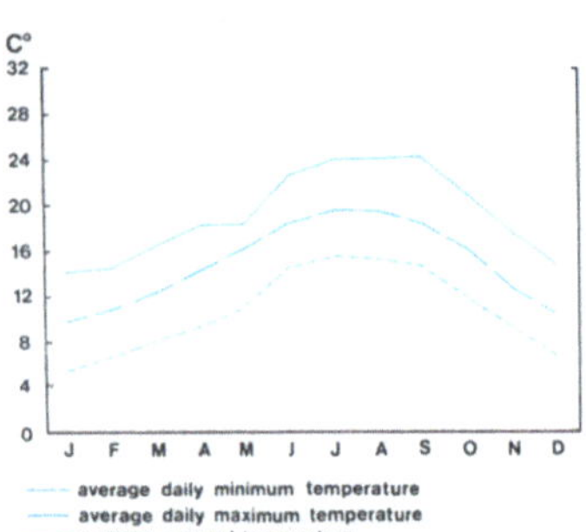

Figure 4. Mean daily temperatures

Building Type

This project concerns a civic building, completed in 1990, which contains a town hall with its executive and administrative offices and cultural services, a public registry and notary office and offices for the public electricity and water supply services.

Location

The building is in Ilhavo (latitude 40 ° 40 ' N), about 5 km from the town of Aveira near the Atlantic coast of the northern part of central Portugal (see Figures 1 and 2). Ilhavo is located in a marshy area named the Ria de Aveira. The area contains a lagoon, canals, salt-pans and mud-flats.

The site is in a square at the south end of Ilhavo's main avenue in the town centre (see Figure 5). It is flat and 9 m above sea level.

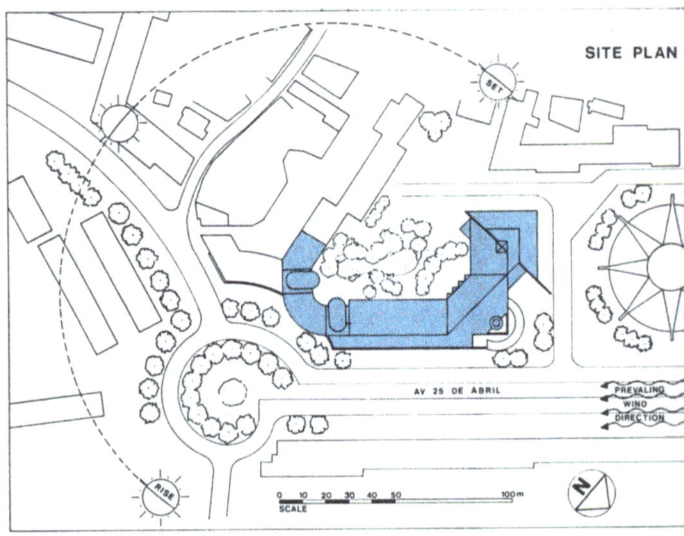

Figure 5. Site plan

Site Microclimate

The climate is of the temperate maritime type with moderate variations in daily air temperatures. The air humidity is above the national average because of the presence of the marsh. The average daily minimum temperature is 5.2 ° C (in January) and the average daily maximum temperature is 24.3 ° C (in September) - see Figure 3. The summers are particularly mild (see Table 1 and Figure 4). On many days, a fresh northern wind blows.

Design and Construction Details

Aim

The aim was to create an important civic building with surrounding spaces which forms a fitting end to the town's central avenue and makes a definite contribution to the town centre.

Design Details

The building is in three blocks which house the town hall, the public registry and notary and the public utilities' offices. The three parts of the building form a continuous architectural whole and are wrapped around a public garden, open to the south, which provides an area of transition between the building and an old, more residential quarter of the town (see Figure 7).

The building is on three storeys and has a total floor area of 5,800 m^2 and a volume of 19,000 m^3. The town hall has two main entrances so that access to the executive offices is independent of the public entrance. The ground floor of the town hall (see Figure 6) houses the municipality's cultural services (including the library), some of its executive offices and the accesses to the upper floors. The first floor (see Figure 11) contains the executive, administrative and planning offices and the conference hall. The offices used by the public are in the central part of this floor; this is a deep area, lit from above by a light well. The second floor (Figure 12) houses the archives and the offices for private and municipal works. Most of the ventilating and heating plant is installed in the roofspace (Figure 13).
On the side of the building which adjoins the avenue, there is a covered gallery at street level. On the other side, there is the public garden. This contains a pond and a small open-air amphitheatre - part of the municipality's cultural facilities.

Most of the offices are located along the south facade of the building (see Figure 8), overlooking the garden, and have large windows. Spaces with low daylighting requirements or which have to have artificial lighting (such as circulation areas, archives, building services rooms and large drawing offices needing high lighting levels) are placed on the north or north-east sides, where the windows are small.

Figure 6. Ground floor plan

Global irradiation on horizontal plane	
Oct-Apr in kWh/m^2	993
(Oct-Apr in MJ/m^2	3575)
Jan-Dec in kWh/m^2	1668
(Jan-Dec in kWh/m^2	6005)
Sunshine hours per year	2500
Heating degree days (Oct-May)	
base 18 ° C	1295
Cooling degree days (Jun-Sep)	
base 20 ° C	73
base 22 ° C	29

Table 1. Some climate data
(Source: Instituto Nacional de Meteorologia e Geofisica)

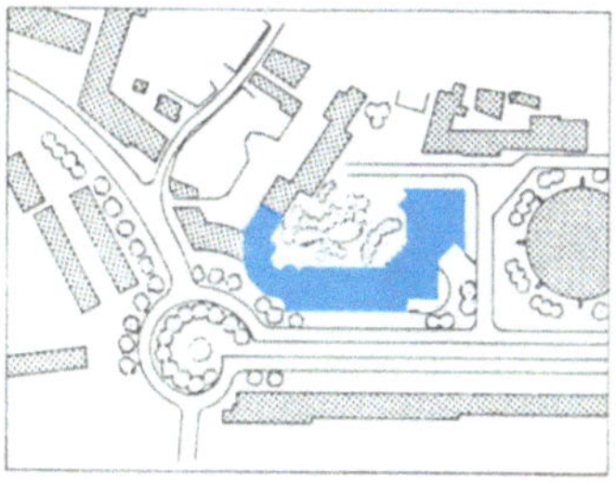

Figure 7. Building form

Figure 8. Orientation

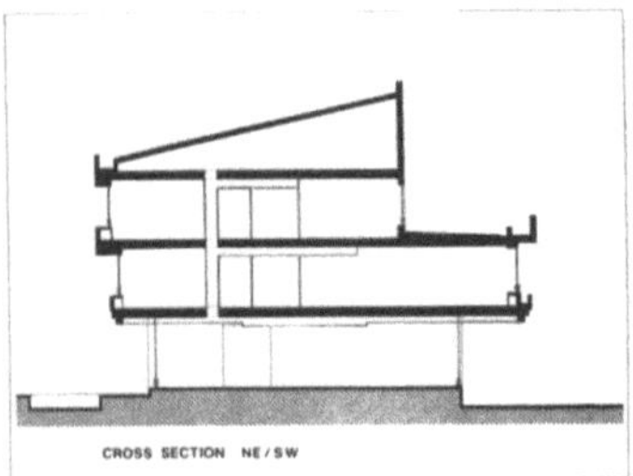

Figure 9. NE-SW cross section

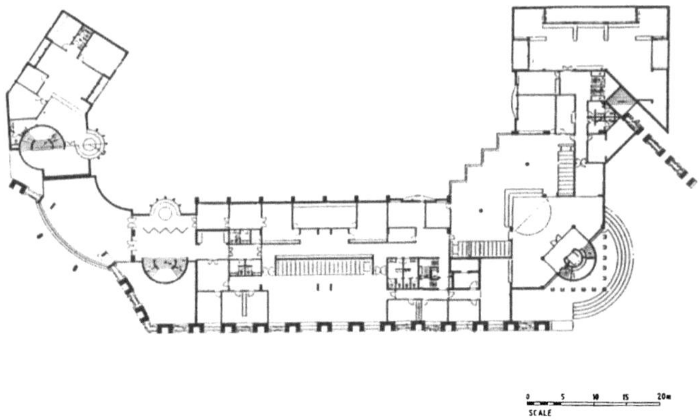

Figure 11. First floor plan

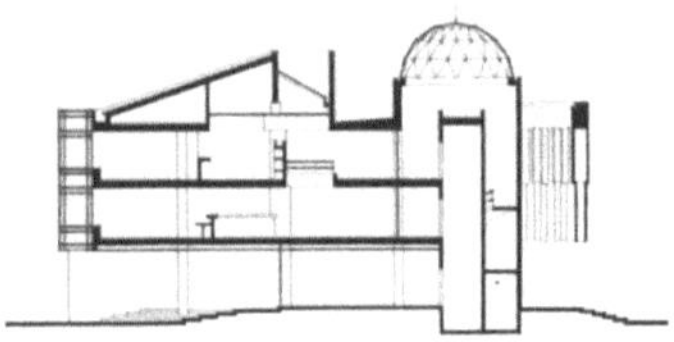

Figure 10. N-S cross section of entrance hall

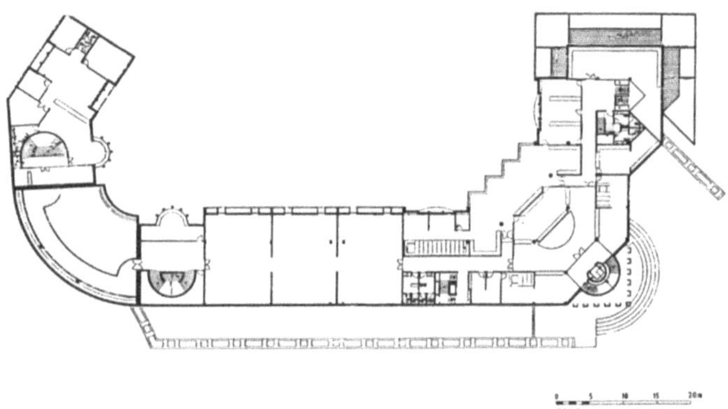

Figure 12. Second floor plan

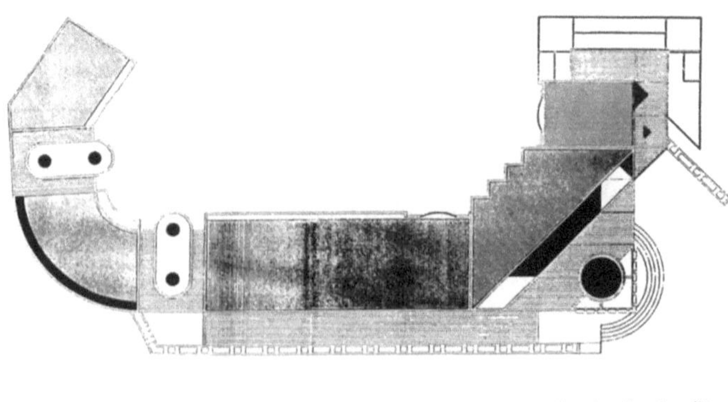

Figure 13. Roof plan

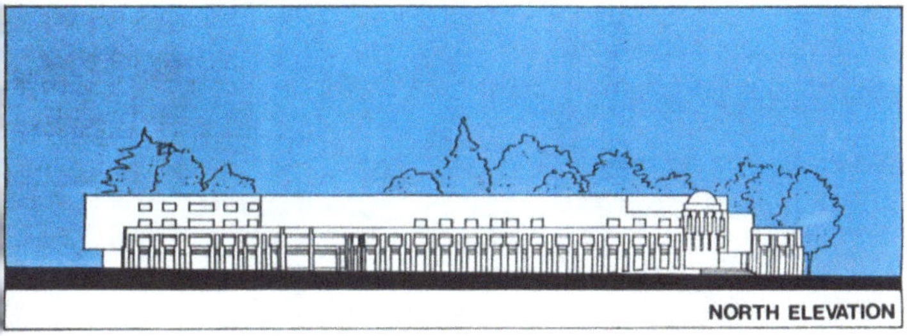

Figure 14. North elevation

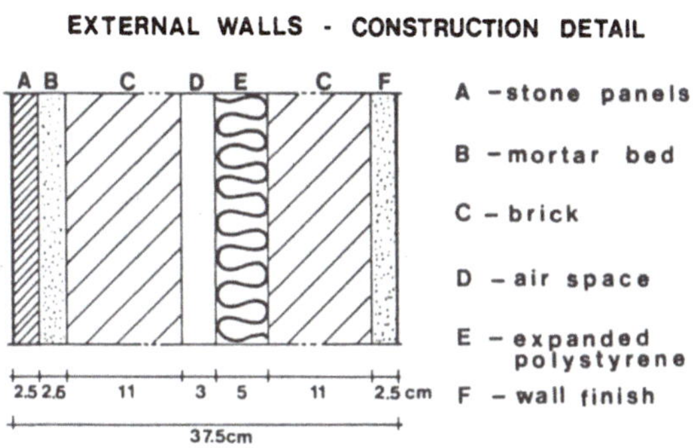

Figure 15. External walls - construction detail

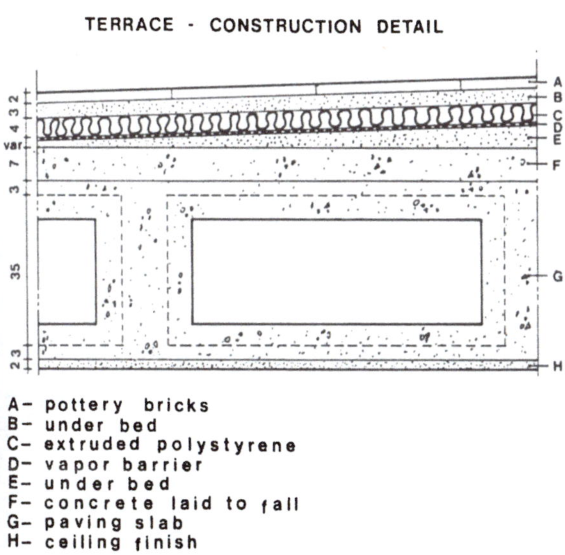

Figure 16. Terrace - construction detail

Construction Details

The building (which cost ECU 3,350,000) has a precast concrete structure and, because the soil is alluvial, sits on a foundation of concrete stakes.

The external walls (Figure 15) are made of a double layer of bricks with squared voids. Between the layers there is a ventilated air space and a 0.05 m thick slab of insulating material. The external facade of the walls are covered with polished white stone.

The roof is made of a light concrete slab and bricks and is covered with traditional gutter tiles. The top slab is insulated with extruded polystyrene. The roofspace is naturally ventilated to prevent heat accumulated in the roof from entering the building.

The pavement slabs have rectangular voids (0.35 m in height) and are topped with a layer of concrete (0.07 m thick). The terraces (Figure 16) are constructed with the same slabs and covered with pottery bricks (0.02 m thick) and thermally insulated with a 0.05 m layer of extruded polystyrene.

The windows have lacquered aluminium frames and are double glazed with tinted glass in working areas and single glazed in other areas.

The U-values of some of the building elements are given in Table 2.

External walls	0.4
Floor slab in roofspace	0.4
Terrace	0.4
Windows	
double glazed	3.4
single glazed	5.6

Table 2. U-values of some building elements in $W/m^2 K$

DESCRIPTION OF PASSIVE SOLAR FEATURES/COMPONENTS

Introduction

In the design of this building one of the objectives was to provide a comfortable indoor environment with the lowest possible consumption of fuel throughout the year. A number of passive solar features were introduced to help warm the building in winter, keep it cool in summer and provide daylighting.
The modes of operation of the passive solar heating and cooling systems in winter and summer are shown in Figures 17 and 18.

Direct solar gains help to warm the building in winter. Because the orientation of the building could not be altered, these gains are achieved by including large windows on the south and south-west facades and minimizing the size of the north-facing windows.

To avoid excessive heat losses in winter and provide a thermally-stable building, high levels of thermal insulation were incorporated in the external walls and roof. High-quality aluminium window frames and double glazing were used in areas where thermal comfort is particularly needed.

To prevent excessive amounts of solar radiation from reaching the face of the building in summer, large deciduous trees are planted close to the building to provide shade. In addition, overhangs are used so that the windows are set back and the sun's rays cannot enter. The polished white stone surface on the facade also contributes to minimizing solar gains through the building envelope. The pond in the garden creates evaporative cooling which makes the atmosphere more agreeable during hot weather.

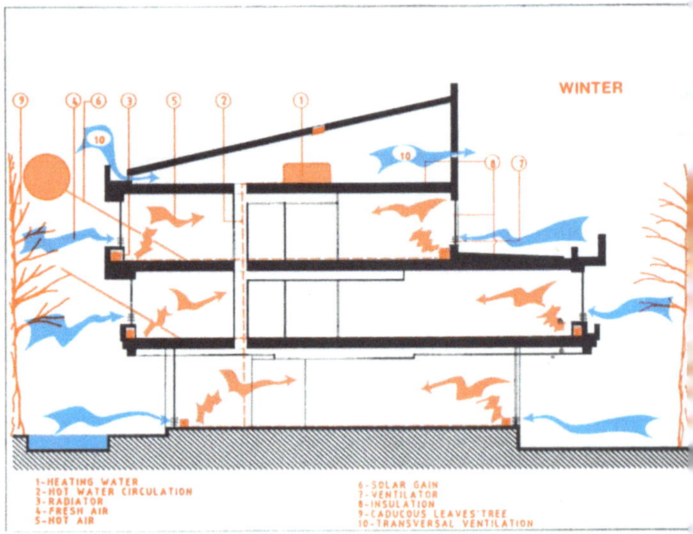

Figure 17. Operation of passive solar systems in winter

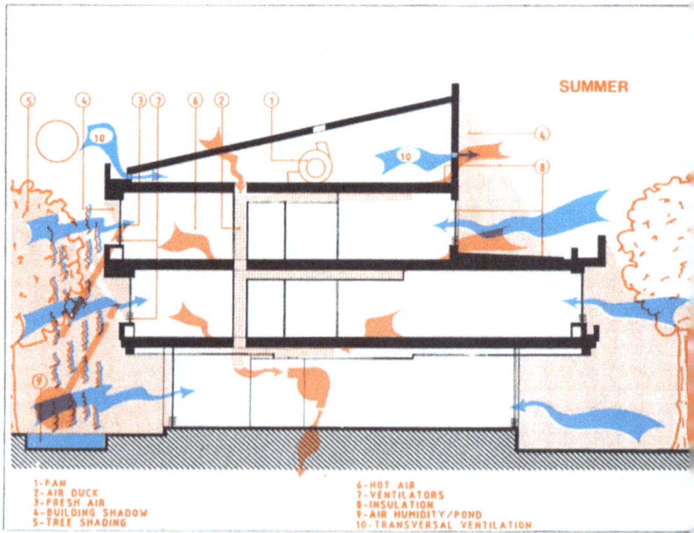

Figure 18. Operation of passive solar systems in summer

Air Ventilation and Auxiliary Heating Systems

Design and Energy Conservation

Because different spaces in the building are used for different types of activities, a number of strategies were adopted for heating and ventilating the building in winter.

In the offices and most of the small rooms, fresh air enters through ventilators in the windows and exhaust air leaves by mechanical means using central extraction ducts. Heating is provided by large area convective radiators (through which hot water circulates) placed beneath the windows. The heat from the radiators balances part of the heat lost through the ventilation air changes and the building envelope.

Additional heating for large central areas of the building such as the entrance hall is provided by a warm air system. The air is preheated in a cross-flow heat recovery unit by means of hot water and by heat recovered from the exhaust air in an exchanger before it is rejected. The ducts through which the exhaust air flows are insulated.

The conference room is heated and ventilated by means of a mixed system involving radiators and supply and exhaust of heated air. A heat recovery unit saves energy. A similar system with another heat recovery unit is used in the archive room on the second floor.

The library on the ground floor is heated by an independent system based on a fan-coil unit which supplies the room with a mixture of outdoor and recirculated air. The air mixture is heated to the desired temperature by hot water.

To allow good temperature control and efficient use of energy, the hot water distribution system consists of a considerable number of circuits, each directed to a different zone of the building. The zones were chosen according to their orientation and pattern of use. The hot water is produced in a propane-burning boiler which has six burners and is operated on a modular basis.

In summer, the building is kept cool by natural ventilation (involving opening the windows) and mechanical extraction. An air conditioning system is not required because of the building type, dimensions and pattern of use.

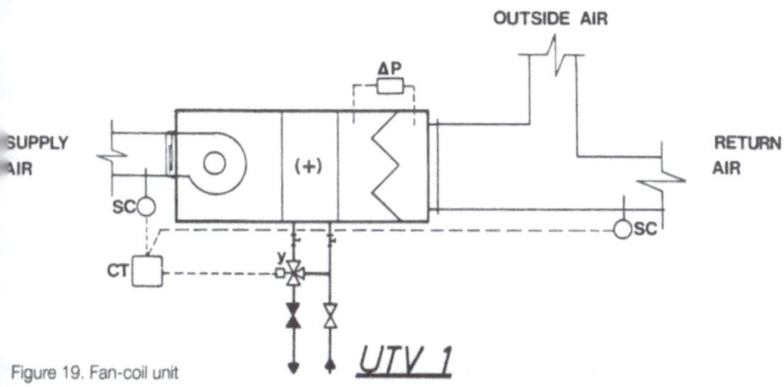

Figure 19. Fan-coil unit

Controls

Both the fan-coil and cross-flow heat recovery units are provided with control systems which regulate the rate of flow of hot water through the units so that the air is heated to the desired temperature (see Figures 19 and 21).

The control of the flow of hot water to the radiators through each of the independent distribution circuits is based on a master control unit and several slave control units, one for each of the distribution circuits (Figure 20). This control system allows the heating periods to be programmed daily and throughout the year; it also permits fast heat-up during start-up periods, the setting of minimum and maximum temperatures for the water supplied and the cut- off of the supply of hot water if the exterior temperature is too high. Each slave control unit controls the flow of hot water to the radiators by actuating the circulation pump and a 3-way valve. Sensors for the temperature of the ingoing hot water, for the interior temperature (placed in a typical room) and for the exterior temperature provide input signals for this control unit. The exterior temperature sensor is conveniently placed outside the building and gives a direct input to the master control unit. In addition to these controls, each radiator is provided with a thermostatic valve.

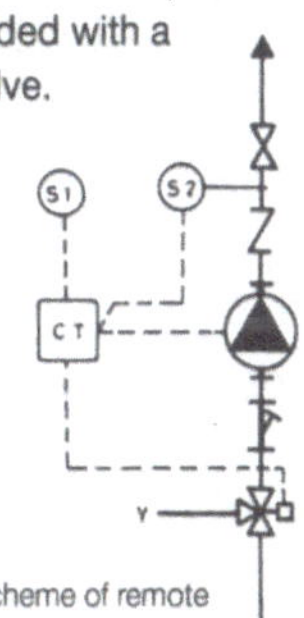

Figure 20. Control scheme of remote stations

Lighting

In designing the building, the apertures were sized to give the maximum natural light penetration possible without glare or summer- time overheating. Overhead lighting through a light well is used for the public entrance hall in the centre of the building. Light wells are also used in various staircases and provide partial lighting for the conference hall.

These daylighting systems limit the amount of electric lighting required. To reduce lighting energy consumption still further, task (as opposed to general) lighting is provided in offices and energy- efficient luminaires and lamps are used. In addition, an electric lighting control system has been installed in which the communication with the control units is by means of infrared receivers and transmitters, making the traditional vertical wiring unnecessary. It incorporates a control system which automatically switches off lamps as soon as the outdoor illuminance exceeds a given level. However, the occupant is able to switch the light on if he wishes. In addition, the system can also be used to ensure that only a limited amount of lighting is used during periods when the building is being cleaned.

The daylighting and electric lighting control systems are expected to save some 25% of the energy used for lighting a similar conventional building.

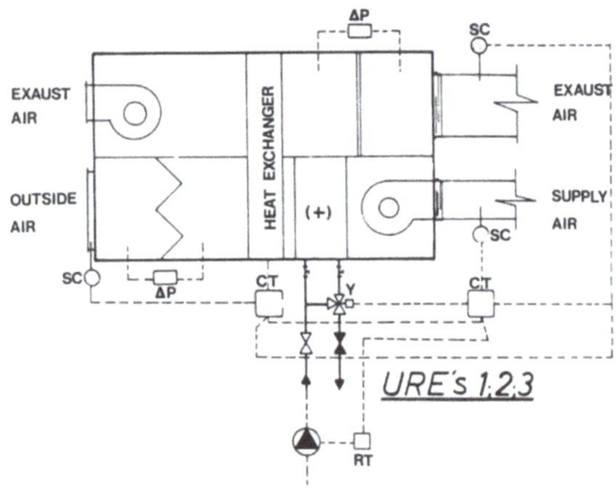

Figure 21. Heat recovery unit

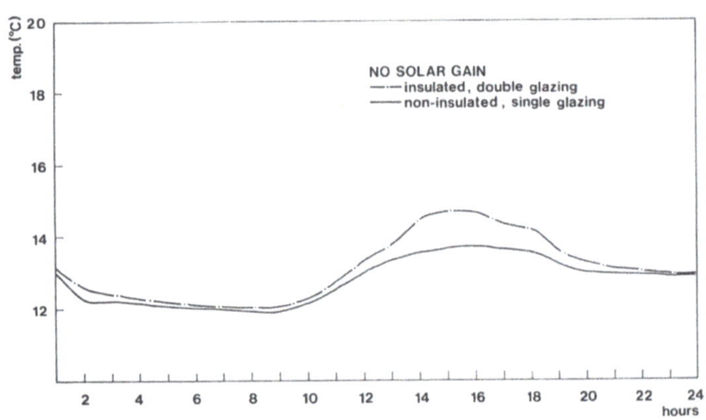

Figure 22. Temperature profiles for ground floor library on typical Janaury day with no heating and no solar gain

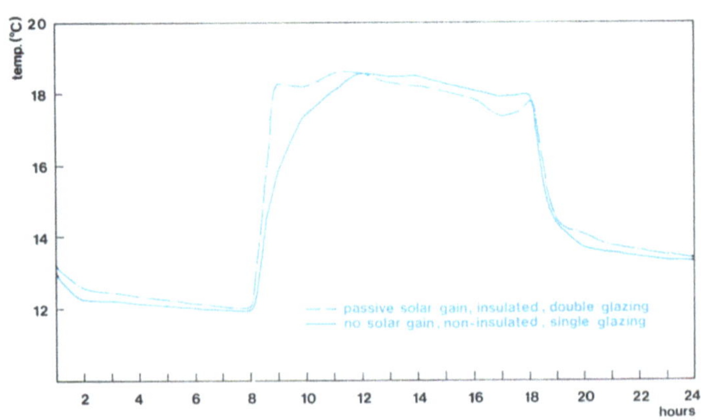

Figure 23. Temperature profiles for ground floor library with heating on a typical January day

ENERGY CALCULATIONS PERFORMED AND DESIGN TOOLS USED

Introduction

The building was designed to be heated during the winter seaon and mechanically ventilated (but not air conditioned) during the summer. The mild climate in summer coupled with the passive solar features means that air conditioning is not needed during the warm season. This, in itself, produces a large energy saving. In addition, no domestic hot water is provided and there are only two (six-person) lifts. Under these circumstances, the highest energy-user in the building will probably be the lighting and the second highest the heating system.

The thermal performance of the building was studied using the HVACSIM+ computer simulation program developed at the National Institute of Standards and Technology in Portugal. This uses a hierarchical and modular approach to perform dynamic simulations of the interactions between a building envelope and internal spaces, its heating, ventilating and air conditioning systems and the building controls.

The daylighting performance was studied using the GENELUX computer program developed at the Laboratoire Sciences de l'Habitat (ENTPE) in France.

Thermal Performance Calculations

Computer simulations were carried out for each of the three floors of the building. These were divided into zones chosen primarily according to their function, usage and orientation, which gave variations in their energy requirements. The computations were carried out for two basic situations: one in which the building received passive solar gains and one where it did not. For each situation, two cases were considered: one where no insulation was included and the windows were single-glazed (as in a normal office building in the area) and one (which corresponded to the actual design) where the walls were insulated and the windows double-glazed. In all cases it was assumed that there were 1.0 air changes an hour - which is what would be expected in the winter period.

Computations were carried out of the temperature profile in a ground-floor room (the library) for a typical January day with no heating and solar gains (see Figure 22). This is a large room covering the entire depth of the building with windows facing south-west and north-east. The maximum temperature reached was about 13.7 ° C if there was no insulation in the building and about 14.7 ° C with the insulation and double glazing.

Computations were also carried out for a typical day for each of the five months from November to March if a heating system was in operation from 8 am to 6 pm. The supply of heat was controlled to give a temperature of 18 ° C. For the library, it was assumed that the heating system had a power of 5 kW. The results showed that, if the building were insulated and there were solar gains, the temperature reached 18 ° C within a one-hour heating period (see Figure 23). When there were no solar gains or insulation, it took about 3 hours to reach this temperature.

The average energy consumptions for the various zones when the building was heated were also calculated. The results show that, for a typical January day, passive solar gains give energy savings on the ground, first and second floors of an insulated, double-glazed building of between 11% and 25% (see Figure 24)

(cont.)

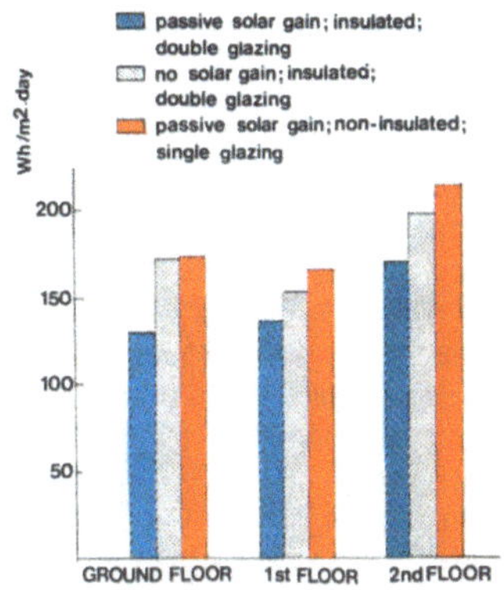

Figure 24. Heating energy consumption per day for the various floors on a typical January day

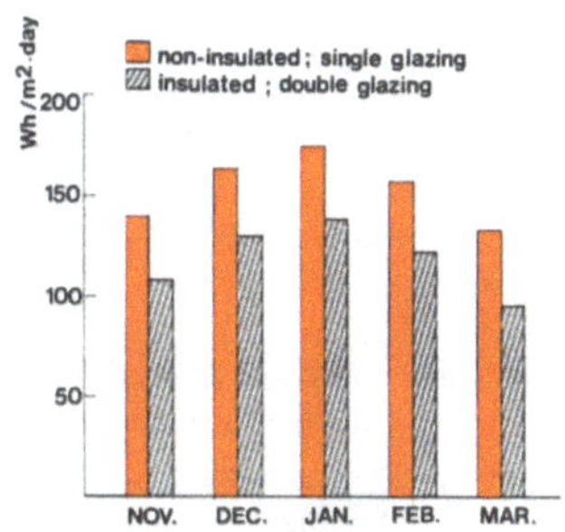

Figure 25. Average daily heating energy consumption for building with solar gain, November to March

	non-insulated single glazing	insulated double glazing
Groud Floor	23	16
1st Floor	22	18
2nd Floor	30	24

Table 3. Annual heating energy consumption of passive solar building in kWh per square metre heated floor area

Key: control of solar gain
no control of solar gain

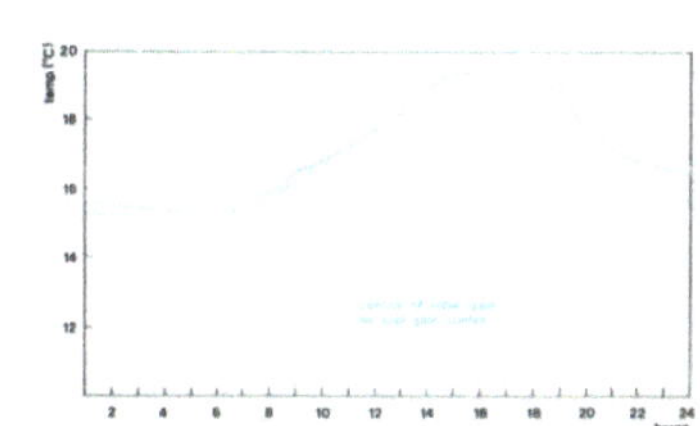

Figure 26. Temperature profiles of first floor office facing south- west on typical July day

(cont.)
For each of the five months from November to March, calculations were made of the average energy consumption with heating for the whole building per square metre of heated floor taking into account the solar gains (see Figure 25). Adding insulation and double glazing reduced the overall energy consumption by 19% to 29%. The greatest savings were made in March and November.

The annual energy consumption with heating per square metre of heated floor area were made for each of the three floors. The results are shown in Table 3. The second floor has the highest energy consumption because of the smaller area of glazing and the proximity of the roof.

To evaluate the behaviour of the building in summer without air conditioning, computations were carried out of the temperatures inside the building assuming two air changes per hour. The results (see Figure 26) show that in a first floor office facing south- west, the temperature reaches only 17.6 ° C on a typical July day with a maximum outdoor temperature of 22.5 ° C if the solar gains in the building are controlled. Even if there is no control of solar gains the temperature in the office only reaches 19.6 ° C.

Daylighting Calculations

Light Well in Entrance Hall

The public entrance hall in the central part of the building where the two main blocks join together receives light from a light well. A detailed analysis was carried out using the GENELUX program to evaluate the light distribution pattern in the entrance hall and determine how deeply the natural light penetrates the lower floor and how far away from the aperture the natural light will be useful. The results are shown in Figure 27. The light pentrating the well will generate high illuminance levels on the horizontal plane. On the second floor daylight penetrates efficiently more than five metres away from the sides of the cavity. On the first floor, natural light is still abundant under the aperture but the illuminance decreases more sharply with distance from the centre. Typically, daylight factors drop to below 1% within a few meters. A light well system can, however, usefully balance the luminous environment when it is employed in combination with a perimeter system. In such a case, double side lighting can be achieved efficiently provided that the distance between the facade and the cavity is 6 to 8 metres.

Conference Hall

Light can enter the conference hall through vertical windows behind the audience and through slots at the back of the room and behind the speaker initially directed towards the ceiling. The resulting illuminances will be even and low. Simulations (see Figure 28) show that theoretically the daylight factor on the seats varies between 0.2% and 1%. If the absorption due to the window frames and furniture is taken into account the value is predicted to be between 0.1% and 0.5%. Such levels are sufficient to allow the audience to read and write during transparency projections under natural light conditions but are a little too high for comfortable viewing of the screen. The level can be reduced with dark paintings and shading the aperture behind the screen. Illuminances from a slide projector are of about 1 to 60 lux on the screen and, if a disturbing light throws 5 lux or more on the screen, the vision of the slide is diminished, particularly in the darkest zones. Therefore furniture and wall coverings in this room should be of dark and warm colours. This is particularly true of the wall facing the screen, which should have a reflectance of about 5-10%. This decreases the daylight factor to the 0.05% to 0.25% range, which is acceptable. Between conferences, a better luminous environment is achieved by drawing back to their fullest extent the black curtains on the vertical windows behind the audience.

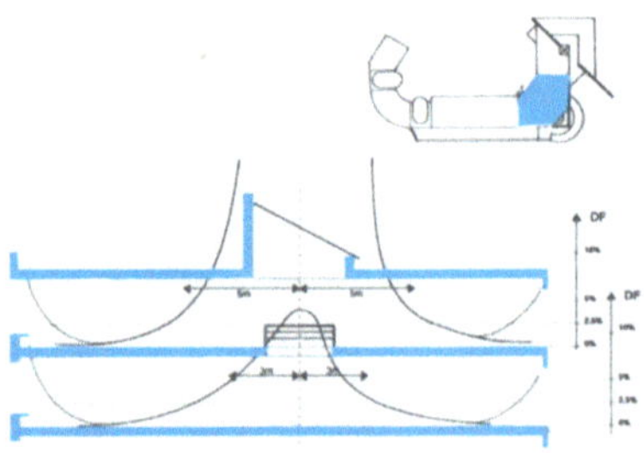

Figure 27. Daylight factor distributions in entrance hall

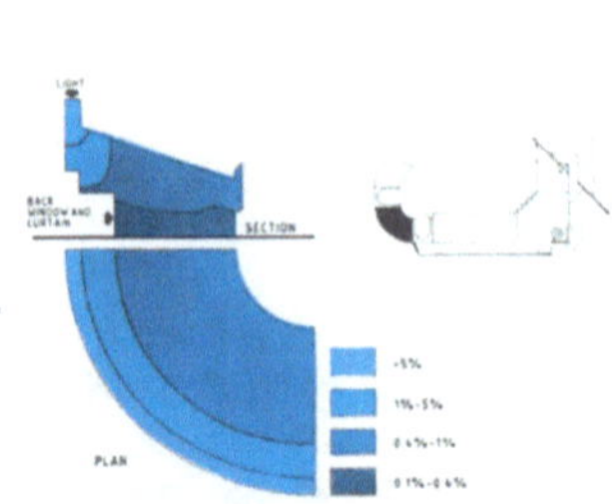

Figure 28. Daylight factor distributions in conference hall

GENERAL DESIGN GUIDELINES/POINTS OF INTEREST RESULTING FROM THIS PROJECT

General

This building is a good example of a classic, low-temperature energy- efficient design which is applicable to the Portuguese climate and to temperate maritime climates in general.

It shows how the use of passive solar features to control solar gains by the simplest means (namely planting and overhangs) can reduce substantially the energy requirements during the summer season. No air conditioning is required to obtain indoor comfort conditions, the building being naturally ventilated with mechanical support. Expensive and high energy consuming systems are, in this manner, avoided.

The design optimizes the daylighting of the spaces and reduces the use of electric light. The thermal properties of the building were designed to limit the period over which the building has to be heated and the auxiliary heating system, with its own energy conservation features, is simple and straightforward.

With such a design, the energy requirements of the building are reduced to about 50% of the calculated current requirements of a conventional building with no air conditioning.

The Role of the Design Advice

The idea of using passive solar features in this project was introduced in the early stages of the design process. Although orientation was not a factor which could be changed to improve the passive behaviour of the building, care was taken with other aspects of the design to provide indoor spaces which are comfortable throughout the year.

The objective was to combine the highest possible direct solar gain in the cold season with a high level of insulation to reduce the use of auxiliary heating and, at the same time, control the solar gains in the warm season so that temperatures inside the rooms do not exceed a comfortable level. In addition, good daylighting (from overhead, in some deep areas) would help create visual comfort conditions inside the building and reduce energy requirements. The installation of a lighting control system would further reduce consumption of electricity for lighting.

These objectives were strengthened and the solutions developed during various phases of the design process in which Building 2000 support made it possible to profit from the advice of European experts. This advice covered various aspects of the passive solar features, energy conservation, the auxiliary heating systems and the lighting systems.

Advice on the Thermal Aspects of the Design

The advice received on the following is particularly worth mentioning: the use of ventilators in the windows to allow air to enter the rooms when the windows are shut during the winter; the level of insulation; the ventilation of the roofspace.

An early idea of including Trombe walls in the design to heat the building indirectly was abandoned because of the system's low efficiency and the fact that the building is not used at night. The use of solar panels for domestic hot water was also rejected because the consumption of hot water in the building is very low; basically such a system would not be cost- effective unless a 50% grant was available. These decisions are in line with the rule (which should be stressed) that the best ways of obtaining indoor comfort are the simplest ones.

Advice on Lighting Systems

The advice received on lighting helped define the characteristics of the lighting system (i.e. that it should supply task lighting rather than general lighting) and the important features of the lighting control system. The simulations carried out by the expert using GENELUX to determine daylight factors under the light well in the entrance hall were of importance in determining how far away from the aperture the natural light will be useful on the first and second floors and where it ceases to be efficient so that the provision of additional electric lighting is necessary. The daylight factor simulations for the conference room enabled the wall reflectance to be defined which would enable transparencies to be seen easily on a screen. This study was of special interest in identifying the particular lighting problems which occur in a conference hall.

BUILDING 2000

Building 2000 brochures are published by Directorate General XII of the Commission of the European Communities to show how design studies can help architects and other building designers use passive solar principles to the best effect to produce attractive energy-efficient buildings. Each brochure describes studies carried out with the support of the Commission during the design phase of one of thirty-six non-domestic buildings in the EC Member States. The studies were on such topics as daylighting, heating, cooling, ventilation, comfort, control systems and urban design. They were carried out with the help of acknowledged European experts in these fields and drew heavily on lessons learned and techniques developed through the Commision's research and development programme on solar energy applications to buildings.

Commission of the European Communities/Directorate-General for Science, Research and Development

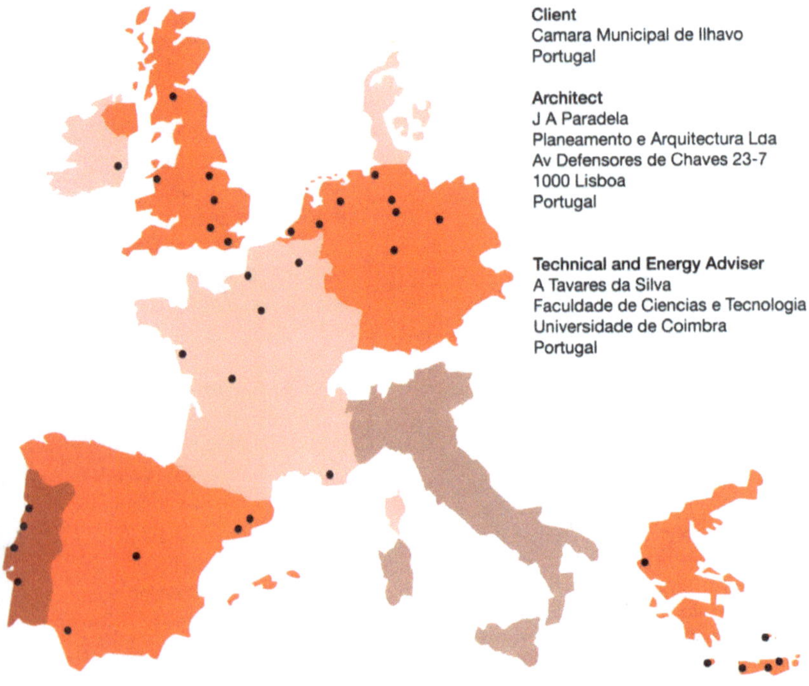

List of Design Team Participants and Advisers

Client
Camara Municipal de Ilhavo
Portugal

Architect
J A Paradela
Planeamento e Arquitectura Lda
Av Defensores de Chaves 23-7
1000 Lisboa
Portugal

Technical and Energy Adviser
A Tavares da Silva
Faculdade de Ciencias e Tecnologia
Universidade de Coimbra
Portugal

Passive Solar and Energy Conservation Adviser
P O'Sullivan
Bartlett School of Architecture and Planning
University College, London
UK

Daylighting Adviser
Marc Fontoynont
LASH
Ecole Nationale des T.P.E.
69120 Vaux-en-Velin
France

Engineering - Building Systems
Fernando Baptista
R. Cons. Jose Silvestre Ribeiro 8-3D
1600 Lisboa
Portugal

Engineering - Power and Lighting
J Nave Ferreira
Av Antonio Augusto de Aguiar 5
1000 Lisboa
Portugal

This set of Building 2000 brochures illustrates how architects and other building designers can successfully apply passive solar principles to produce energy-efficient buildings.

BUILDING 2000
Participants

Project Director
Theo C. Steemers

Coordinator
Cees den Ouden

Technical Steering Committee
Dean Hawkes
Nick Baker
Alex Lohr
Jean P. Lepoivre

Regional Liaison Agents
(D) Jörn Behnsen
(E) Vicente Sifre
(F) Michel Raoust
(GB) Alan Hildon
(GR) Matheus Santamouris
(P) Eduardo Maldonado

Further information or copies of the brochures can be obtained from prof. ir. Cees den Ouden, EGM Engineering BV, P.O. Box 1042, 3300 BA Dordrecht, The Netherlands.

Commission of the European Communities

BUILDING 2000

HOTELS
AND
HOLIDAY
COMPLEXES

H1
Tourist Complex, Agios Nicolas, Crete, Greece
H2
Hotel and Conference and Service Centre, Barcelona, Spain
H3
Tourist Hotel, Bethune, France
H4
Holiday Apartments, Gournes, Crete, Greece
H5
Tourist Complex, Chania, Crete, Greece
H6
Tourist Complex, Drios, Paros, Greece

Building 2000 is a series of design studies illustrating passive solar architecture in buildings in the European Community.

VOLUME 2 / **KLUWER ACADEMIC PUBLISHERS**

BUILDING 2000

Commission of the European Communities

- Luxury tourist complex with approximately 550 beds, reception, administration offices, multi-purpose hall, restaurants, lounge, shops, theatre, swimming pools, gym and saunas
- Complex is built around secluded bay in series of small-scale buildings, some of which are partly buried in the slope of the land
- In summer, outdoor living is the norm and extensive landscaping has been used to create a comfortable microclimate
- Summer-time overheating has been prevented by cross and night ventilation, shading, evaporative cooling and air-to-ground heat exchangers
- Daylighting is used carefully, for practical reasons and to enhance the architecture
- The annual contribution of passive solar systems to the heating requirements of the complex is 82%. Their contribution to the cooling load is 52%.

TOURIST COMPLEX

AGIOS NICOLAS/CRETE/GREECE

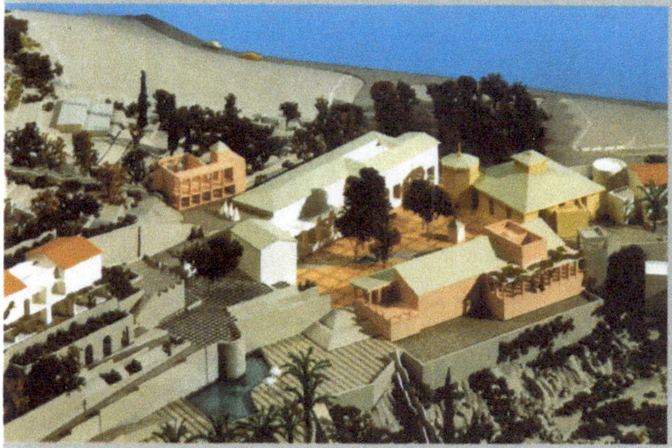

Building 2000 is a series of design studies illustrating passive solar architecture in buildings in the European Community.

ISSUE

H 1

FEB 1991

PROJECT DESCRIPTION

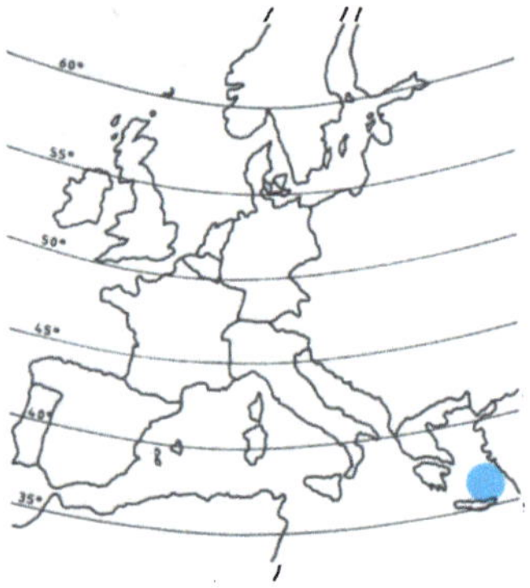

Figure 1. Location of Crete

Sunshine hours	1131
Global irradiation on a horizontal plane	
Jan-Dec	1743 kWh/m²
(Jan-Dec	6275 MJ/m²)
Oct-Apr	713 kWh/m²
(Oct-Apr	2567 MJ/m²)

Table 1. Some key climate data. (Source: Solar radiation atlas of Greece.)

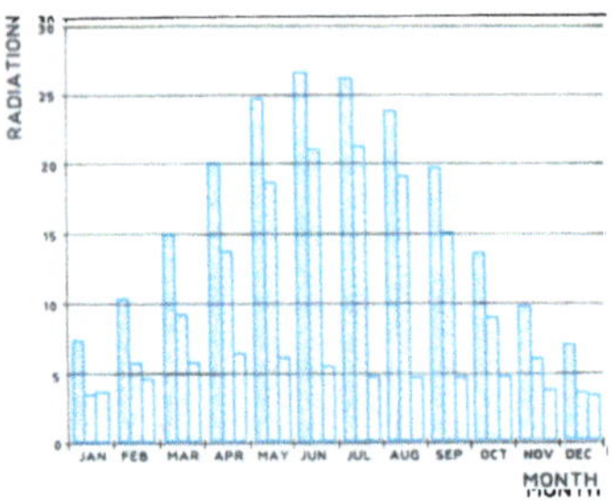

Figure 2. Mean daily global, direct and diffuse solar irradiation per month (MJ/m²)

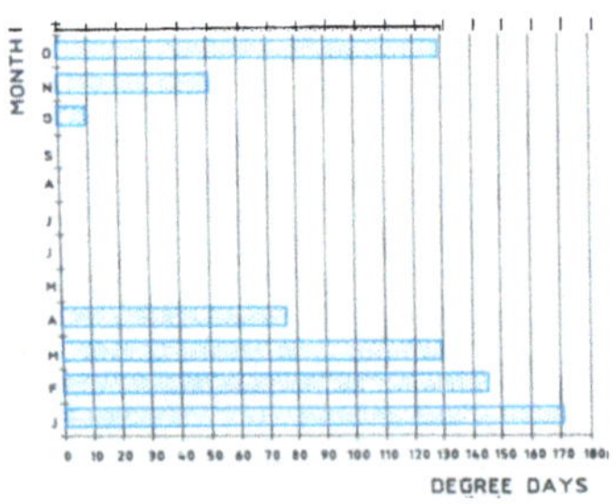

Figure 3. Degree days per month

Building Type

This scheme, known as Hellas Holiday Hotels, is a luxury tourist complex containing 237 double rooms and 28 suites, reception and administration offices, a multi-purpose hall for conferences and exhibitions, restaurants, a lounge, shops, a theatre, a health centre with gym and saunas and indoor and outdoor swimming pools.

Location

The complex is located in Lasithi County (latitude 35° N), south of the town of Agios Nicolaos in Crete (see Figure 1). It is adjacent to the bay, which runs east-west (see Figure 4). The site is approximately 180,000 m² in area and has a 20-30% slope to the north. There are views to the east and to the open sea to the north.

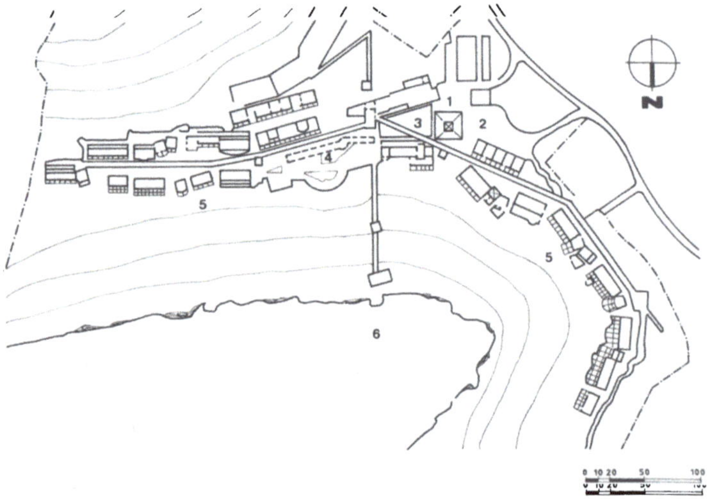

Key:

1. main entrance
2. service yard
3. main square and public facilities
4. swimming pools
5. guest rooms
6. sea

Figure 4. Site plan

Site Microclimate

Details of the site climate are given in Table 1 and Figures 2 and 3.

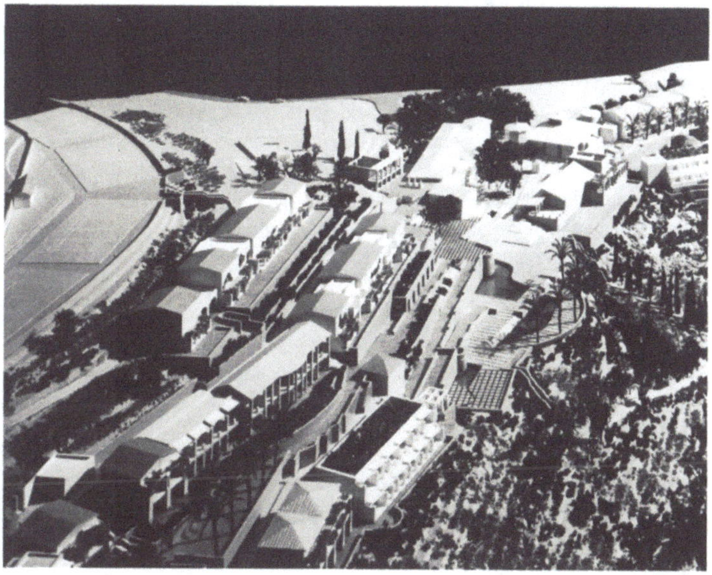

Figure 5. General view of model of site showing the linear building layout following the shape of the bay

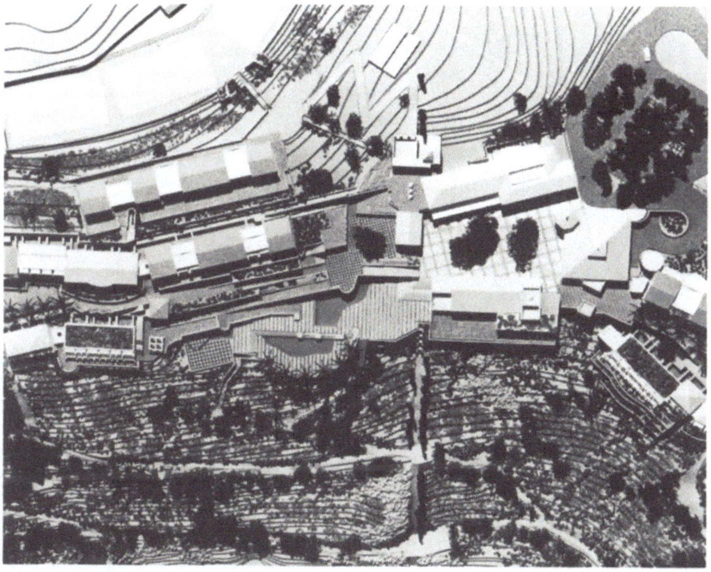

Figure 6. General view of the central part of the model

Design and Construction Details

General Layout

The complex has been developed along the lines of a traditional village and individual buildings are based on forms which have been close to the hearts of people in Crete for generations.

Indoor and outdoor communal facilities are at the site centre and the guest rooms are on the edges so they are quiet and secluded. The layout makes full use of the views and exploits the slope of the land.

The individual buildings are positioned to follow the shape of the bay (Figure 5). Between them they have a floor area of 21,859 m^2 and a volume of 100,514 m^3 . Most face north with their main axis running east-west so that in summer they receive the sun in the late evening. The U-values of some key building elements are given in Table 2.

There are two entrances to the site at the uncovered parking lot and service yard. Vehicles are not allowed beyond this point. The site is designed to encourage outdoor living and the main pedestrian pathway, which runs parallel to the seashore, and main square form an open-air promenade with sitting and recreation areas.

roof	0.32-0.34
floor	0.82-0.83
external walls	0.33-0.56

Table 2. U-values of some building elements (W/m^2 K)

Main Square

The main square is in the centre of the complex adjacent to the main entrance (see Figures 6 and 7). It provides a focus for activities in the day and, especially, evening.

The elevations of the public buildings around the main square are shown in Figure 8. On two sides there are arcades for sitting-out, paved with white marble strips and red brick to absorb heat in winter and shaded by large deciduous trees in summer. A water fountain provides evaporative cooling.

On the south side are the reception and administration offices. This is a long two-storey building with an east-west axis. The reception lobby and some administration offices are on the ground floor. More offices and top management accommodation is on the top floor. The building is cross ventilated and naturally lit from the south and north. The ground floor has a high ceiling both as an architectural feature and to provide cooling.

On the western side of the main square is a square building on the ground floor of which is located the multi-purpose hall. The latter can be divided and is used for conferences, exhibitions and games. Service facilities are in the basement. The building is daylit from all sides through deeply recessed doors-windows and from the top by a glazed turret on the wooden roof.

On the north side of the main square is a two-storey restaurant building. In the upper level there is a speciality restaurant which opens out into an open-air eating area under a covered porch at the edge of the square. On the north side of this restaurant is a terrace with a view of the sea. Crossventilation keeps the central area cool in summer. Daylight enters from both sides.

South-facing clerestory windows provide direct solar gain for heating the speciality restaurant in winter. The main entrance area to the restaurant building is covered with a cupola and is top lit. A stairway off this leads to the main restaurant, kitchen and laundry which are on the lower level. The main restaurant has a large terrace on the north side, shaded by deciduous creepers on trellis. Because it is single-sided, the main restaurant is cooled by overhead fans.

Secondary Square

To the east of the main square is a smaller, secondary square or piazza. The lounge, main bar, shopping mall and theatre are located around this. The lounge, which is for use on cool days, is enclosed and has a large fireplace. It is nearly square in plan and is cross lit and ventilated from the east and west. The main bar opens directly onto the square and provides an outdoor sitting area. The back of the bar is dug out of the slope of the land. In the shopping mall, small individual shops spill out onto the walkway. The indoor, multi-use three-sided stepped theatre with specially-designed top lighting is dug out of the hillside.

Swimming Pools and Health Centre

Below the secondary square, to the north, are grouped the swimming pools on terraces connected by waterfalls, a pool snack bar, the health centre with gym/sauna, an indoor pool and children's corner. This whole area is embedded in the hillside and, apart from the side which faces the sea, is provided with a variety of top lighting.

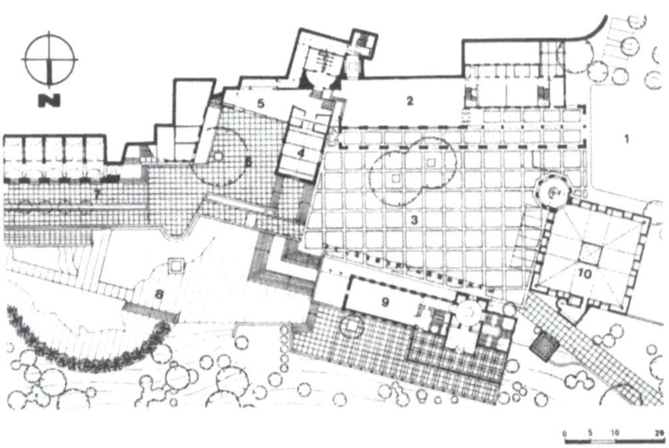

Key:
1. entrance
2. reception
3. main square
4. lounge
5. bar
6. secondary square
7. shops
8. swimming pools and terraces
9. restaurant with covered porch on one side and terraces on the other
10. multi-purpose hall

Figure 7. Plan of central squares and public facilities

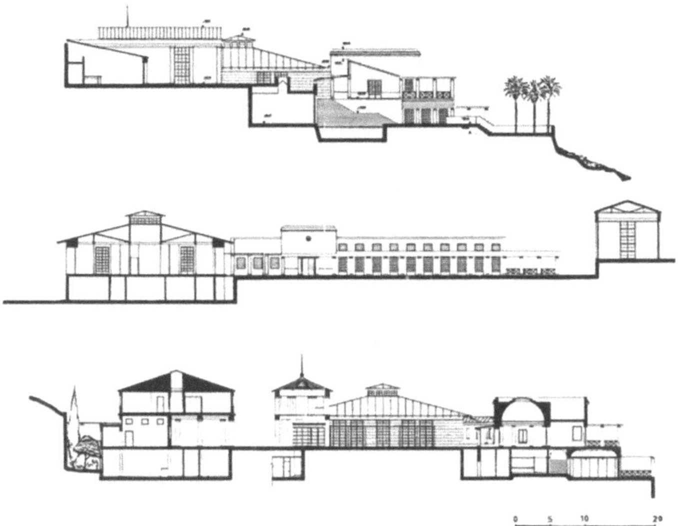

Figure 8. Elevations of the public buildings round the main square

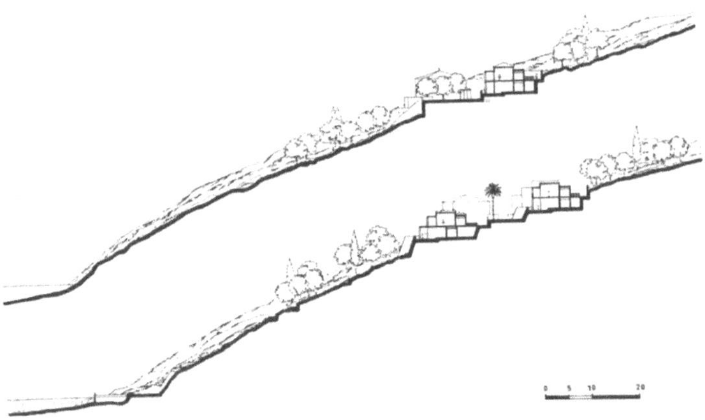

Figure 9. Typical sections through guest rooms

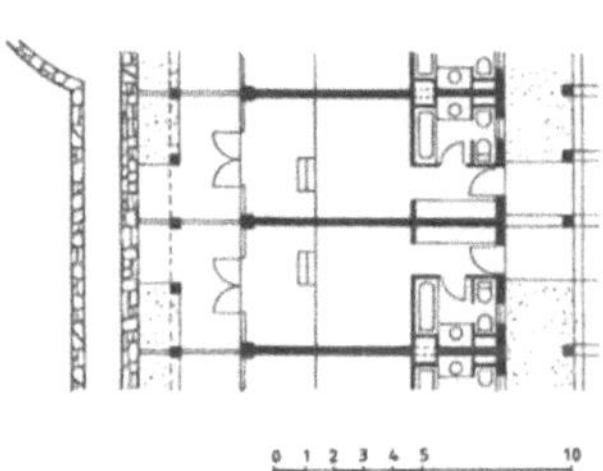

Figure 10. Plan of guest room module

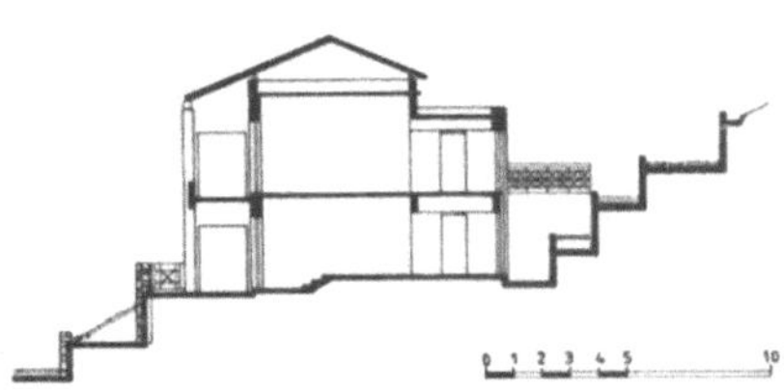

Figure 11. Section of guest room module

There are 237 double guest rooms of four basic types and 28 suites. They are located in small two-storey buildings around the bay along the natural contours of the land (see Figure 9). All have a linear plan with an east-west axis. A plan and section of a typical module are shown in Figures 10 and 11.

All the guest rooms have an entrance and bathroom on the south side and a spacious private veranda on the north, looking towards the sea. Access to the lower-storey rooms is via an exterior pathway on the lower level. Access to the upper rooms is through little wooden bridges built in pairs from the high ground behind the buildings. Thus, even though the site is steeply sloped, all the rooms have openings on front and back and are cross ventilated and cross lit. Some of the upper rooms have south-facing clerestories.

Pathways and walks covered with trellises and climbing plants interlink the buildings and give access to the bay from different parts of the complex. A few seats are arranged in secluded corners or points with a special view. Running water provides cooling and environmental features in the form of gullies, pools, fountains and waterfalls. Wooden gangways, jetties and sundecks run out from the paths along the seashore. Floating decks and gangways stretch out into the shallow bay providing areas (shaded by awnings) for use in daytime and at night, when they will be lit up.

DESCRIPTION OF PASSIVE SOLAR FEATURES/COMPONENTS

Introduction

The complex's main occupancy period is from April to October. Most of the passive solar and other energy-saving features, therefore, are aimed at cooling and daylighting. High levels of insulation have been incorporated in the buildings to improve their performance in summer and winter.

Cooling Systems

Ventilation

Nocturnal cross and fan-forced ventilation is used to reduce cooling loads at times when there is a large difference in interior and ambient temperatures. The cross ventilation systems in the multi-purpose hall, restaurants and guest rooms are shown in Figures 12-15.

Minimizing Direct Solar Gain

Small areas of glazing on the buildings' south side minimize solar gains while large areas of north-facing glazing aid heat loss as well as provide excellent views over the bay.

Shading devices - some fixed, some adjustable - cut out direct solar radiation at times when overheating in likely. Planting on pergolas and the use of deciduous trees also minimize the possibility of excess solar gain.

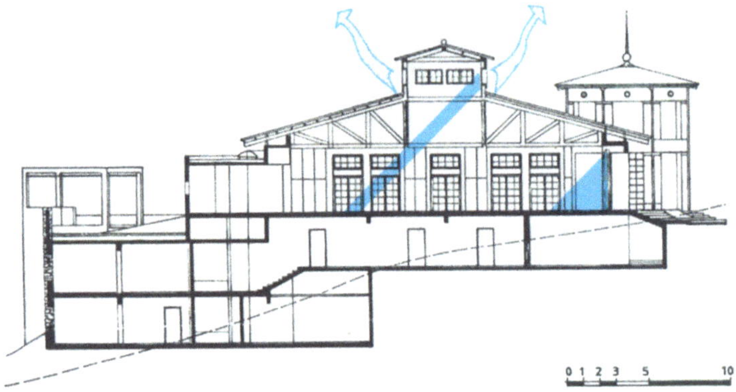

Figure 12. Section through multi-purpose hall showing cross ventilation and daylighting

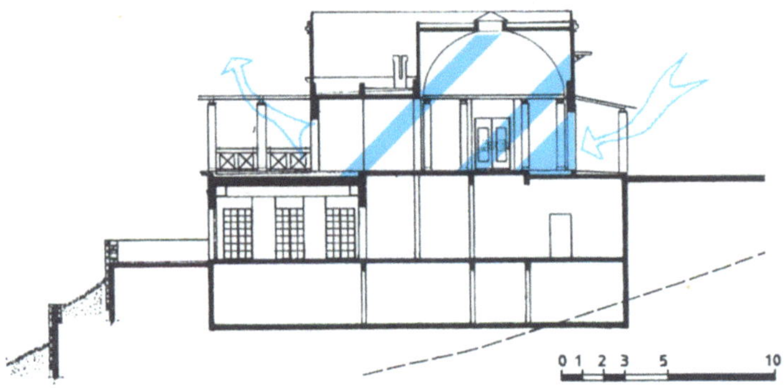

Figure 13. Section through restaurants showing cross ventilation and daylighting

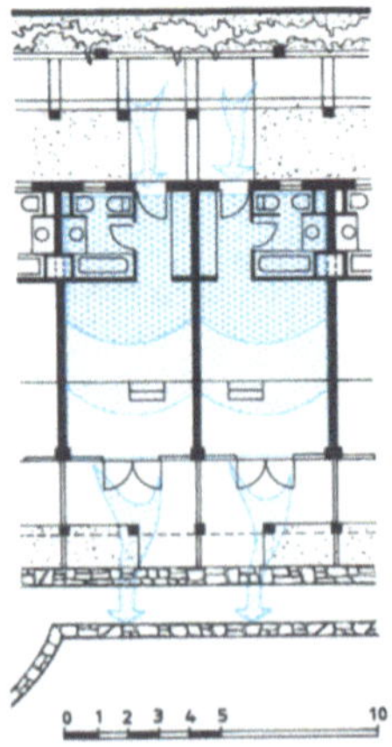

Figure 14. Plan of guest room module showing cross ventilation and daylighting

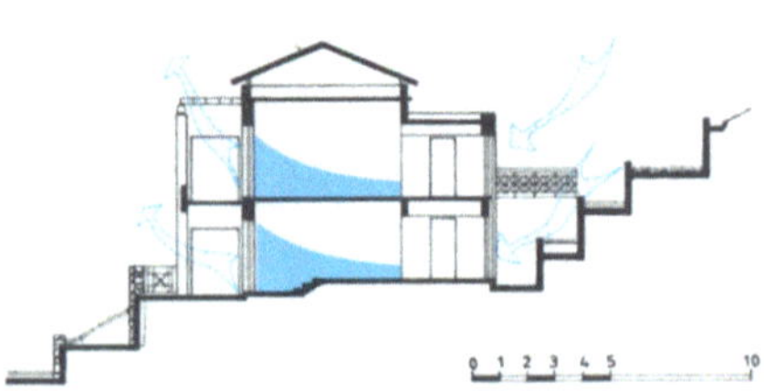

Figure 15. Section of guest room module showing cross ventilation and daylighting

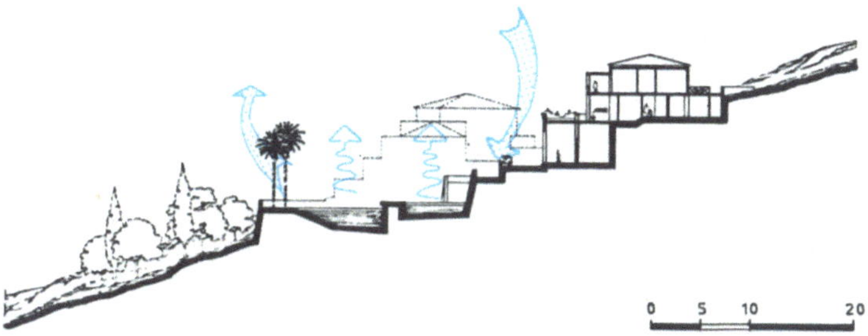

Figure 16. Section through swimming pool area showing evaporative cooling effect

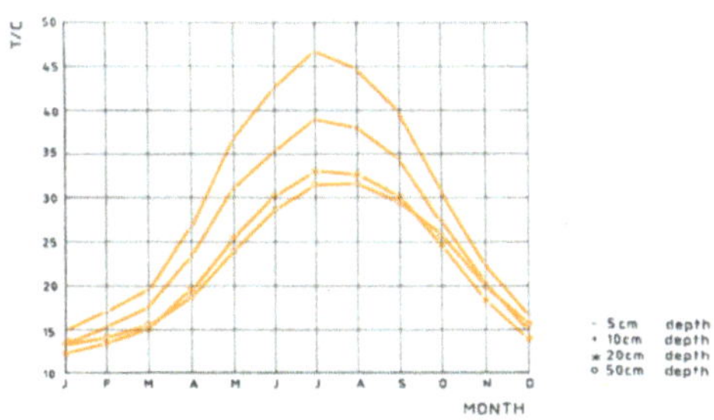

Figure 17. Mean soil temperatures

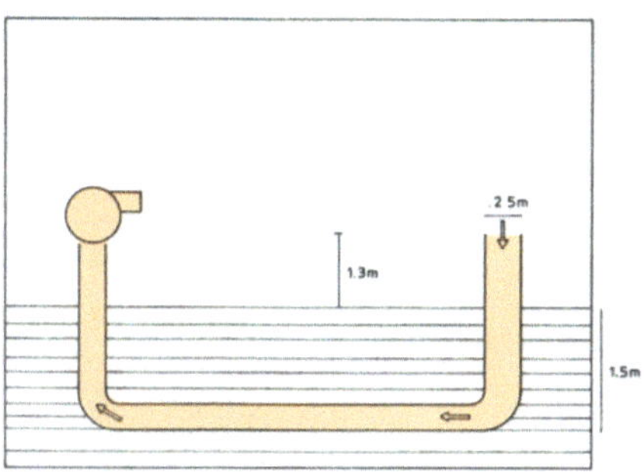

Figure 18. Buried pipes

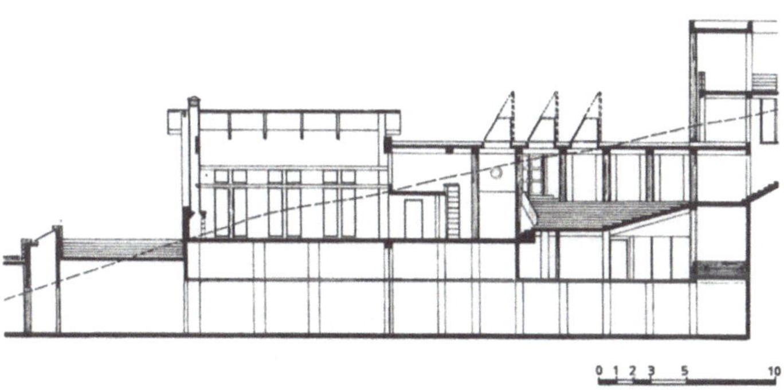

Figure 19. Section through the entrance reception and bar showing daylighting

Evaporative Cooling

Evaporative cooling results from the proximity of the sea and the swimming pools in the centre of the complex (see Figure 16). Fountains and streams of water near the buildings increase this effect.

Air-to-Ground Heat Exchangers

These are used for cooling and heating. The system is based on the difference in temperature between the ground and the air. Figure 17 shows the variation in temperature at different ground depths. The principle of the system is illustrated in Figure 18. Pipes are buried underground and air circulated through them. The air loses or gains heat according to the season.

Auxiliary Cooling and Heating Systems

Conventional air conditioning and fan coil units have been installed in the communal buildings. Auxiliary heating in the guest rooms is supplied by split electrical resistance units.

Daylighting

North daylighting provides uniform natural light in the living spaces (see Figures 14 and 15). Rooflights allow natural light to penetrate into deep and otherwise unlit spaces such as the centre of the multi-purpose hall, speciality restaurant, and the reception and bar areas (see Figures 12, 13 and 19).

ENERGY CALCULATIONS PERFORMED AND DESIGN TOOLS USED

The design of the passive solar and other energy-saving systems was carried out with the help of studies using a number of design tools. The thermal performances of the heated or cooled spaces were calculated using the admittance method (a simplified thermal network method) developed by Building Research Establishment in the UK. This calculates the heating and/or cooling loads, the contribution of the passive systems to heating or cooling and the daily temperature variations. The performance of the air-to-ground heat exchangers and the effect of night ventilation were determined using the PASCOOL computer program for simulating passive solar cooling components. Calculation of daylight factors inside all the spaces was carried out using MICROLIT 1.0, a computer program developed in the USA. Thermal comfort was evaluated using the COMFORT computer program developed at the Technical University of Denmark. This determines the predicted mean vote (PMV) which indicates how close to comfort conditions a space is, taking into account indoor air temperature, mean radiant temperature, relative air velocity, water vapour pressure, clothing, metabolic rate and external work.

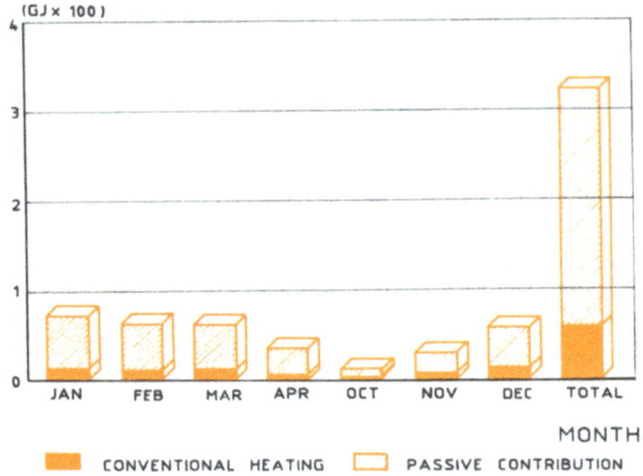

Figure 20. Heating load and contribution made by passive solar features

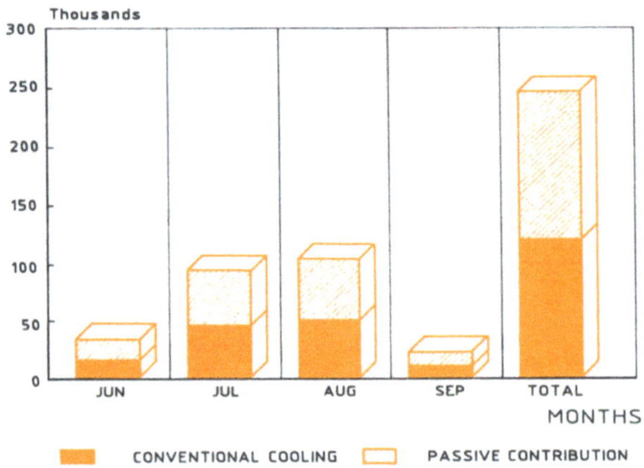

Figure 21. Cooling load and contribution made by passive solar features

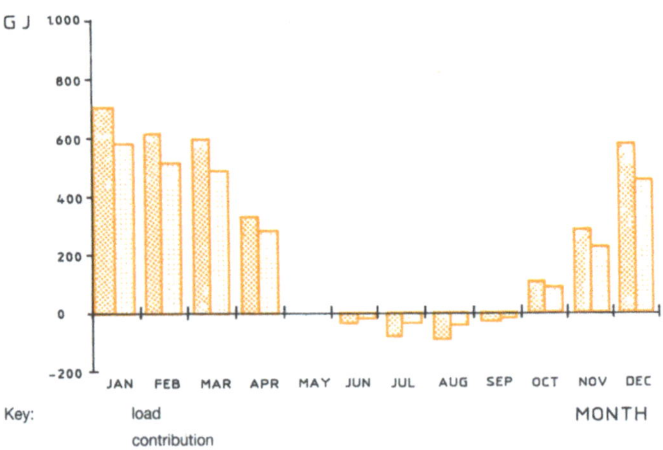

Figure 22. Monthly heating and cooling loads and contribution made by passive systems

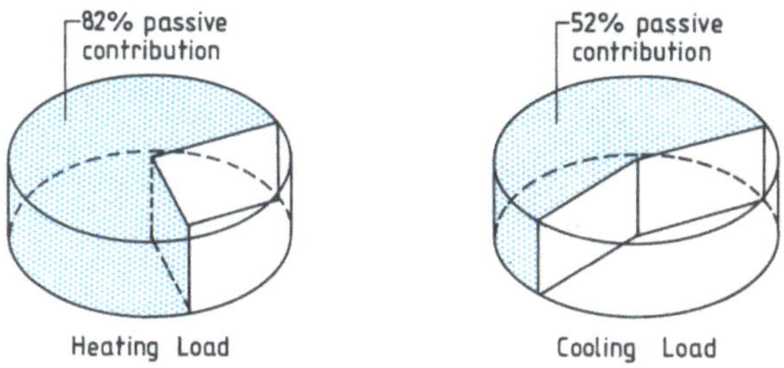

Figure 23. Annual contribution of passive solar systems to heating and cooling loads

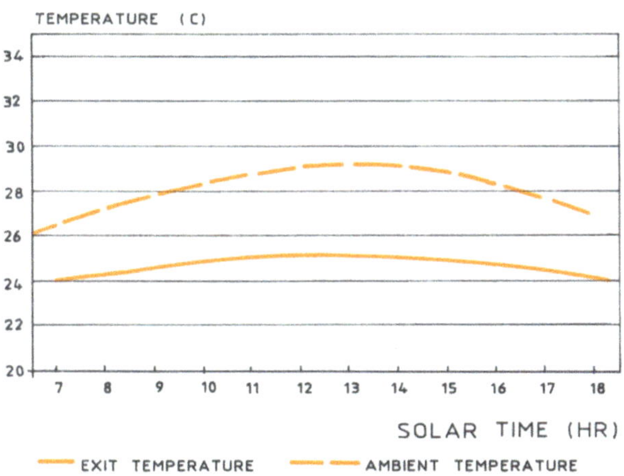

Figure 24. Performance of ground-to-air heat exchanger system in July

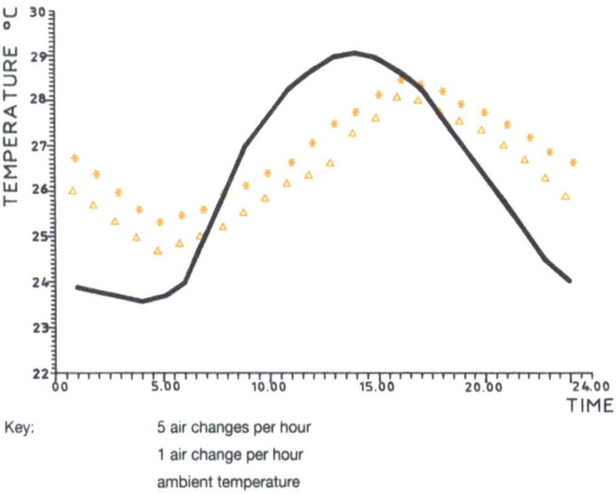

Figure 25. Indoor temperatures of a room in July with night ventilation

Thermal Performance Evaluations

The monthly heating and cooling loads and passive contribution to heating and cooling, together with the internal temperature, were calculated for all the buildings in the complex. Each room in the reception building was treated as a separate zone; the multi-purpose building was considered as one zone; the restaurant building was divided into three (each restaurant was treated separately); each room in the guest room buildings was considered separately. Auxiliary and non-heated spaces were not taken into account. Occupancy parameters for each building type were taken from ASHRAE data. U-values were assumed to be as follows (the units are W/m² K): single glazing 5.8; double glazing 3.7; external walls 0.58; floors 0.95; roofs 0.38-0.44. The results are given in Figures 20-23. The passive systems were found to contribute 82% of the total heating load and 52% of the total cooling load.

Performance of Ground-to-Air Heat Exchanger System

The outlet air temperatures likely to result from the tubes in July are given in Figure 24. The modelling showed that the air- to-ground heat exchanger system could provide 30% of the cooling load for the multi-purpose building in June, 37% in July and 43% in August.

Evaluation of Night Ventilation

The effect on room temperature of increasing ventilation rates at night was evaluated month-by-month. Figure 25 shows the results of increasing night ventilation in July from one to five air changes an hour. It can be seen that this technique would decrease the temperature throughout the 24 hours by nearly 1° C. The best contribution to cooling loads is achieved in June - a month when the performance of the ground cooling system is at its poorest. Thus night ventilation and ground cooling can be regarded as complementary techniques.

Daylighting Performance

The results of the daylighting studies at a height 1.5 m above the floor in a 9.5 m² ground floor bedroom with north-facing glazing at 12 noon on 21 December with an overcast sky are shown in Figures 26 and 27. Under these conditions daylighting is independent of orientation. It can be seen that the daylighting is uniform with daylight factors in the range 3% to 9%, except for the zone near the glazing where higher values are reached. Similar calculations for offices produced daylight factors in the range 5% to 13%. The studies showed that daylighting performance is good in bedrooms and offices alike.

Thermal Comfort Evaluation

The results of thermal comfort studies in a bedroom without auxiliary heating or cooling in January and July are given in Figures 28 and 29. In each case, the predicted mean vote (PMV) is perfectly satisfactory, being slightly below the optimum in January and slightly above it in July.

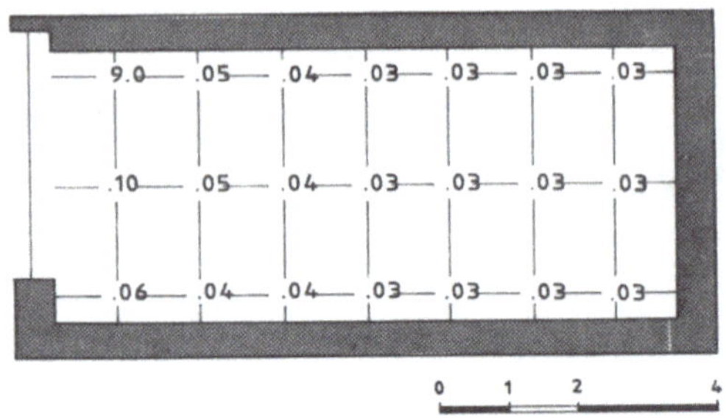

Figure 26. Daylight factor distribution in plan of typical guest room

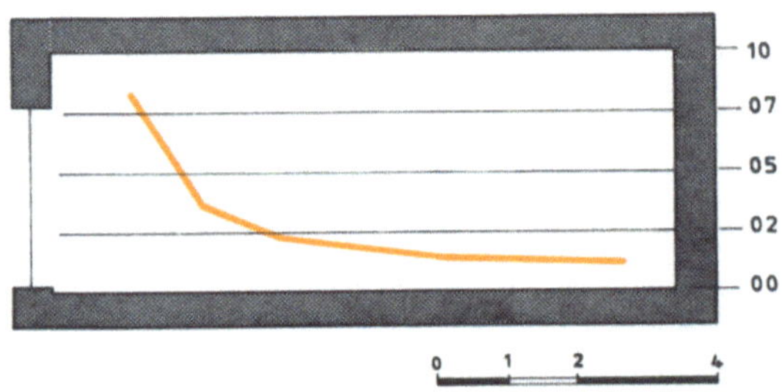

Figure 27. Daylight factor distribution in section of typical guest room

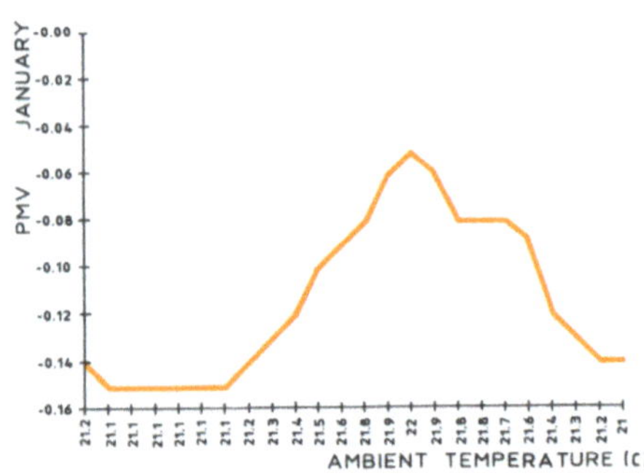

Figure 28. Values of PMV (predicted mean vote) in a typical bedroom in January

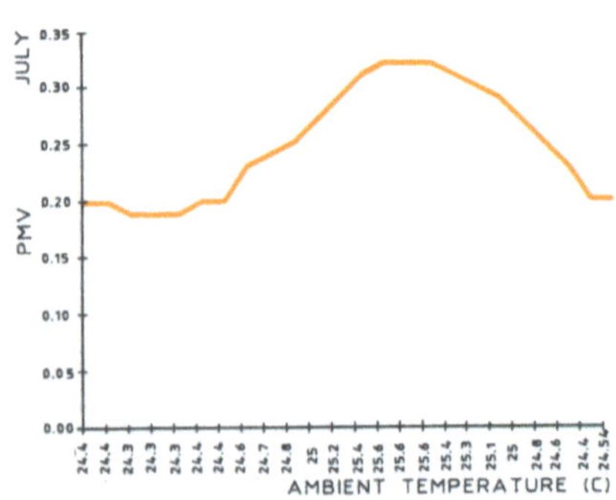

Figure 29. Values of PMV (predicted mean vote) in a typical bedroom in July

GENERAL DESIGN GUIDELINES/POINTS OF INTEREST RESULTING FROM THIS PROJECT

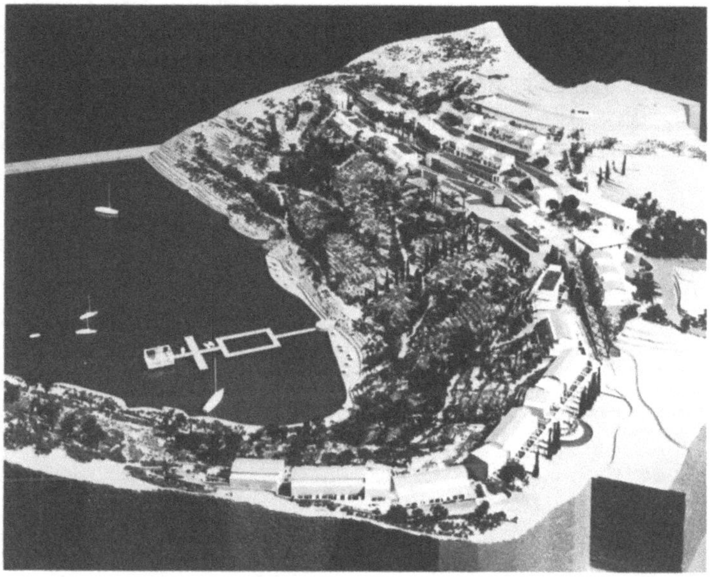

Figure 30. General view of the model of the site

This project showed that, in developing schemes of this sort, the following general guidelines can apply:

- provide a harmonious relationship between architecture, landscape, human intervention and nature, creating a dialogue between design and accident;

- improve the microclimate round each building by using evaporative cooling and shading;

- apply simple but innovative concepts for achieving cooling such as an air-to-ground heat exchange system;

- reduce the use of auxiliary heating, cooling and artificial lighting in comparison with similar conventional buildings;

- passive solar features, no matter how important they are to the scheme, need not dominate architectural design;

- encourage social contact and outdoor living in a hierarchy of squares, terraces, secluded corners and open spaces;

- provide links with the past by using classical elements in the site organization and building form.

BUILDING 2000 Building 2000 brochures are published by Directorate General XII of the Commission of the European Communities to show how design studies can help architects and other building designers use passive solar principles to the best effect to produce attractive energy efficient buildings. Each brochure describes studies carried out with the support of the Commission during the design phase of one of thirty-six non-domestic buildings in the EC Member States. The studies were on such topics as daylighting, heating, cooling, ventilation, comfort, control systems and urban design. They were carried out with the help of acknowledged European experts in these fields and drew heavily on lessons learned and techniques developed through the Commission's research and development programme on solar energy applications to buildings.

Commission of the European Communities/Directorate-General for Science, Research and Development

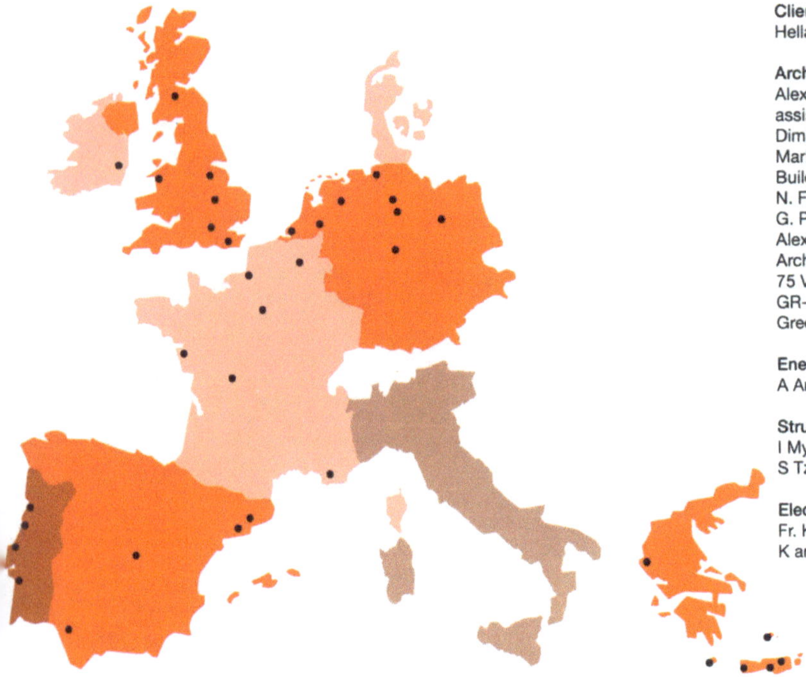

List of Design Team Participants and Advisers

Client
Hellas Holiday Hotels

Architect
Alexandros N Tombazis
assisted by
Dimitris Potiropoulos, architect
Maria Spyridaki, architect for the Building 2000 aspects of the project
N. Fletorides, L. Nella-Potiropoulou and G. Panetsos, design team architects
Alexandros N Tombazis and Associates, Architects
75 Vasilisis Sofia Avenue
GR-11521 Athens
Greece

Energy Consultant
A Argiriou

Structural Engineer
I Mylonas
S Tzivanakis

Electrical and Mechanical Engineer
Fr. Krispis
K and A Alexopoulos

This set of **Building 2000** brochures illustrates how architects and other building designers can successfully apply passive solar principles to produce energy-efficient buildings.

BUILDING 2000 Participants

Project Director
Theo C. Steemers

Coordinator
Cees den Ouden

Technical Steering Committee
Dean Hawkes
Nick Baker
Alex Lohr
Jean P. Lepoivre

Regional Liaison Agents
(D) Jörn Behnsen
(E) Vicente Sifre
(F) Michel Raoust
(GB) Alan Hildon
(GR) Matheus Santamouris
(P) Eduardo Maldonado

Further information or copies of the brochures can be obtained from prof. ir. Cees den Ouden, EGM Engineering BV, P.O. Box 1042, 3300 BA Dordrecht, The Netherlands.

BUILDING 2000

Commission of the European Communities

- Hotel with 46 bedrooms, conference rooms, a service centre, shops and parking facilities
- In summer, an open patio on the north side encourages loss of heat from adjacent parts of the building and incoming ventilation air is cooled by passage underground
- Direct solar gain in south-facing rooms coupled with thermocirculation of warm air to the cooler parts of the building helps keep the hotel warm in winter
- The building is daylit through windows and light ducts
- Overheating and glare are prevented by means of automatically controlled venetian blinds

HOTEL AND CONFERENCE AND SERVICE CENTRE

BARCELONA/SPAIN

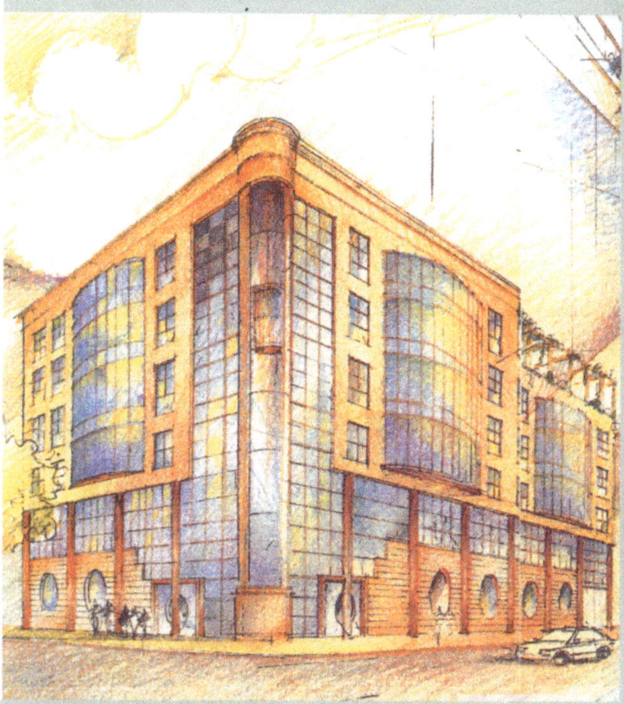

Building 2000 is a series of design studies illustrating passive solar architecture in buildings in the European Community.

ISSUE

H 2

FEB 1991

2

PROJECT DESCRIPTION

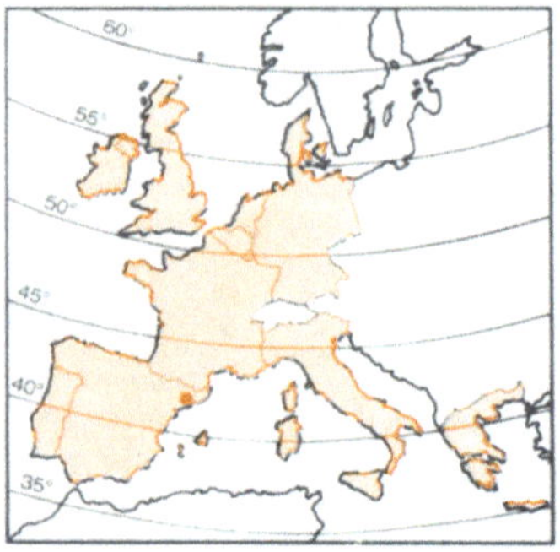

Figure 1. Location of Barcelona

Building Type

This project concerns a hotel containing 46 bedrooms, conference rooms, a service centre, restaurant, shops and parking facilities. It is due to be completed in time for the 1992 Olympic Games and will be used initially by visitors to the games and later by people attending international business fairs in the area.

Location

The hotel is located in Barcelona (latitude 41.2 ° N, longitude 0 ° W) in Catelonia, north east Spain (see Figure 1) at an altitude close to sea level. The site is protected from the wind and has minimal overshading. The site layout is shown in Figure 3.

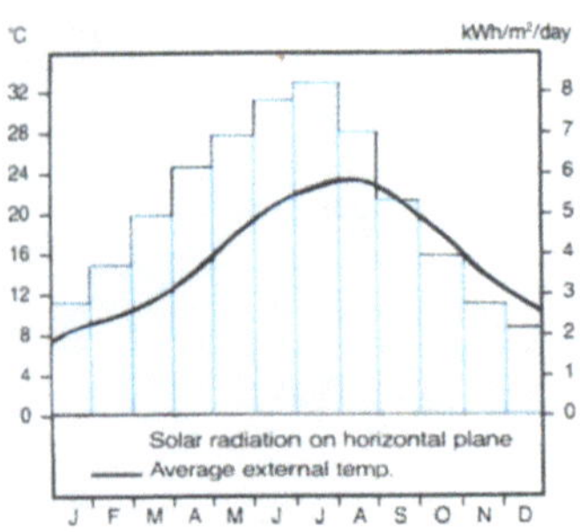

Figure 2. Mean solar irradiation and temperature data

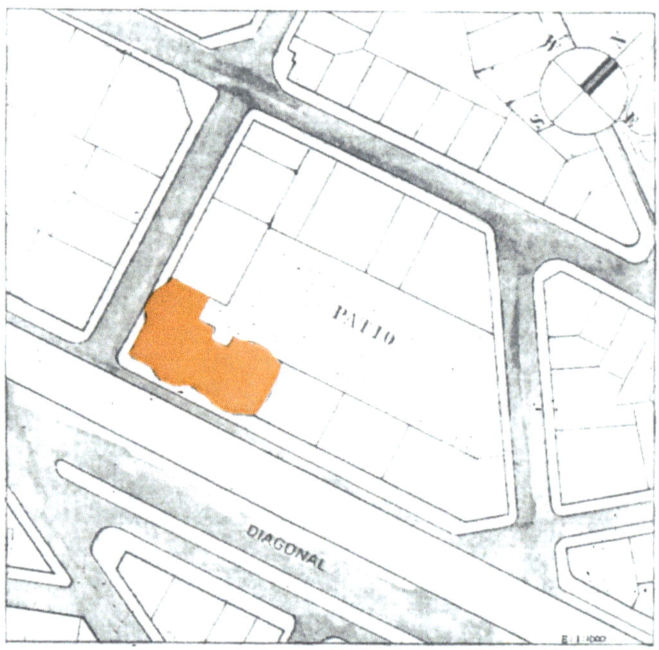

Figure 3. Site layout

Solar irradiation on horizontal plane	
in kWh/m² /year	1463
(in MJ/m² /year	5266)
Sunshine hours per year	2439
Mean daily temperatures	
Jan	8.4 ° C
Aug	23.6 ° C
Jan-Dec	15.9 ° C

Table 1. Some key climate data

Site Microclimate

The climate in Barcelona is of the Mediterranean type with wet and mildly cold winters and hot summers. Mean solar radiation and temperature data are shown in Figure 2. Additional climate data are given in Table 1. Sunpath diagrams are given in Figure 5.

Design and Construction Details

A facade and an axonometric drawing of the building are given in Figure 4. The net floor area is 2,850 m^2 and the volume is 7,838 m^3. The area underground is 1,840 m^2. Part of the building has three storeys and part has four. The main facade faces south-south east. The hotel entrance is in a corner lobby. Entrance to the shops, restaurant and parking area is independent of the hotel.

The building has a concrete structure. The walls are of a standard construction with a high thermal mass. U-values of some building elements are given in Table 2.

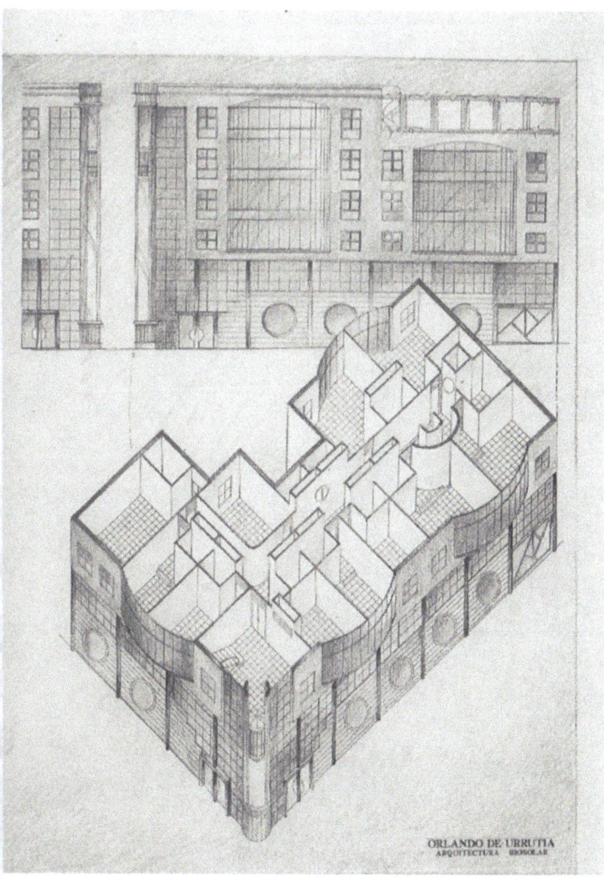

Figure 4. Facade and axonometric

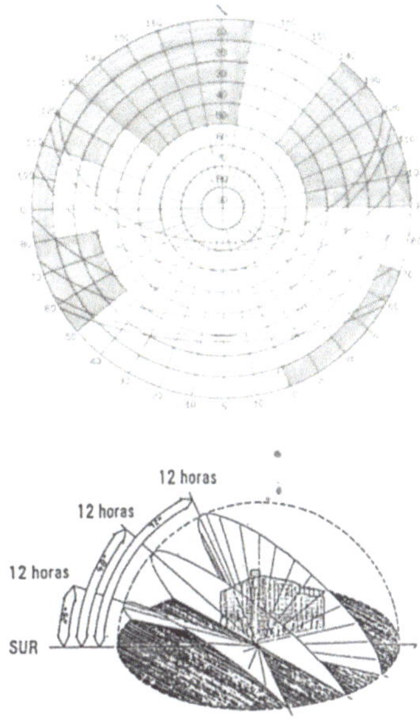

Figure 5. Sunpath diagrams

Roof	0.34
Floor	0.64
Walls	0.57
Windows	2.50

Table 2. U-values of some building elements in $W/m^2 K$

DESCRIPTION OF PASSIVE SOLAR FEATURES/COMPONENTS

Introduction

The building contains a number of passive solar features (see Figure 6) which help to keep the building cool in summer (particularly important in a Mediterranean climate), warm in winter and adequately daylit throughout the year. An open patio on the north side (see Figure 3) encourages excess heat to be lost from adjacent parts of the building and cools incoming ventilation air. Light shelves on the south side prevent direct entry of solar radiation into the building in summer but permit maximum solar gains in winter. Blinds prevent overheating of the sunspace in spring and autumn. In winter, the air warmed in the sunspace is circulated to the rest of the building by thermocirculation. Light ducts bring natural light into the guest room bathrooms throughout the year.

Solar collectors are used to preheat the domestic hot water throughout the year.

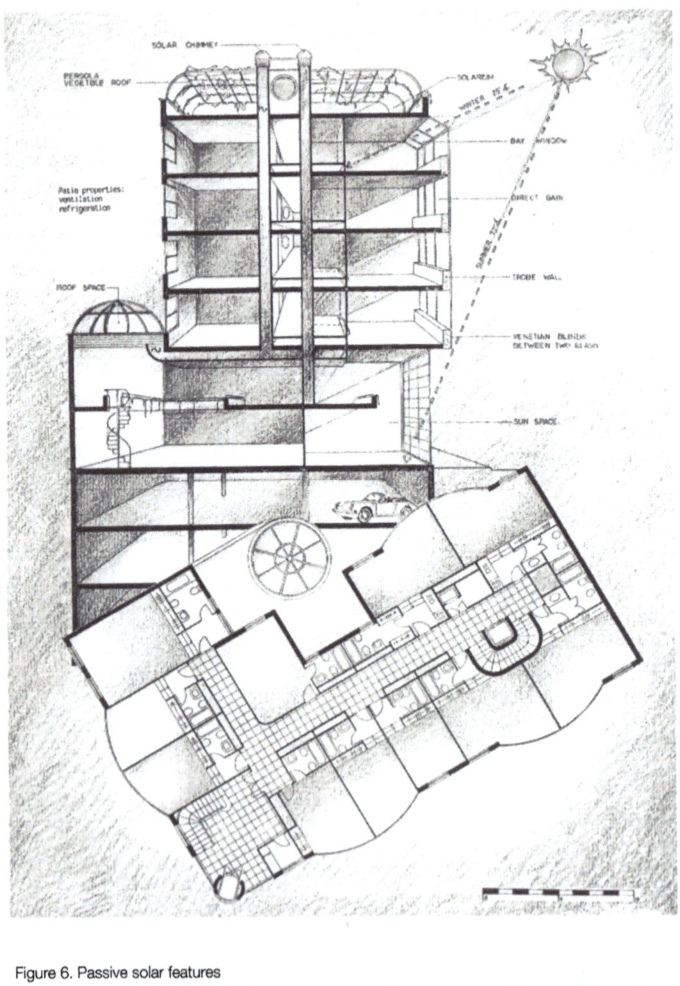

Figure 6. Passive solar features

Mode of Operation of Passive Solar Features

Summer

In summer, the south side of the building is protected against solar radiation by means of awnings and venetian blinds between double glazing. The circulation of fresh air in the rooms is increased. This air is collected in a ground floor gallery on the north side (where it is protected from the sun's rays) and passed through the building via a series of underground pipes (see Figure 7). The system is designed to provide enough cool air for the whole building during even exceptionally hot periods.

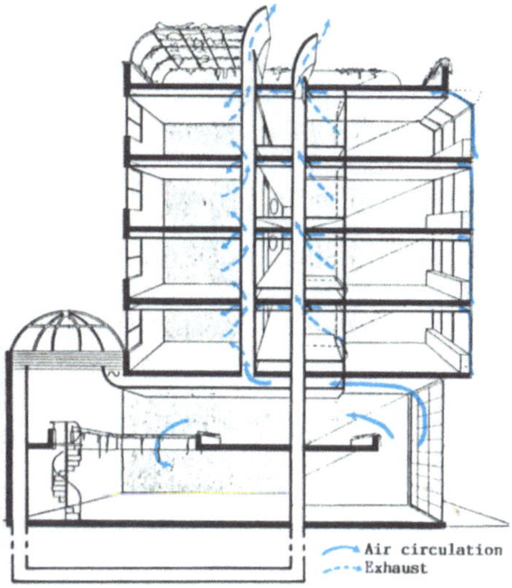

Figure 7. Natural ventilation on a summer's day

Winter

In winter, the air circulation is adjusted to allow preheated air from the south side to heat the north-facing rooms (see Figure 8). Redistribution of the solar gains reduces the internal temperature fluctuations in the south-facing rooms and minimizes the effect of the evening temperature drop, thus improving the average temperature. The auxiliary heating system is zoned and controlled according to occupancy.

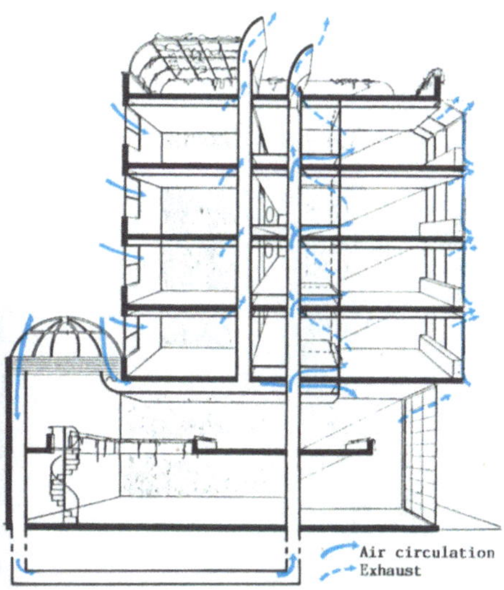

Figure 8. Natural ventilation on a winter's day

Lighting of Bathrooms Using Light Ducts

Design of Light Ducts

On all four floors, light ducts are used to bring daylight into the guest room bathrooms. The top opening of these ducts is illustrated in Figure 11. The height of the curve is 2.1 m, three times the duct width. The curved design was felt to be more effective than one with angled straight sides because it has a larger surface area (1.9 m x 2.1 m = 3.99 m^2 - say, 4.0 m^2) and allows incoming light to be dispersed after its first reflection so that a uniform light enters the room. The horizontal section of each duct is rectangular (the dimensions are 1.9 m x 0.7 m = 1.3 m) so that it can be positioned between the bathroom and the corridor and can be serviced from the corridor.

Mode of Operation

Light enters the bathrooms via the ceiling spaces (see Figure 12). The size of the opening at each ceiling space increases successively from the top to the ground floor of the hotel so that the same amount of light enters each room.

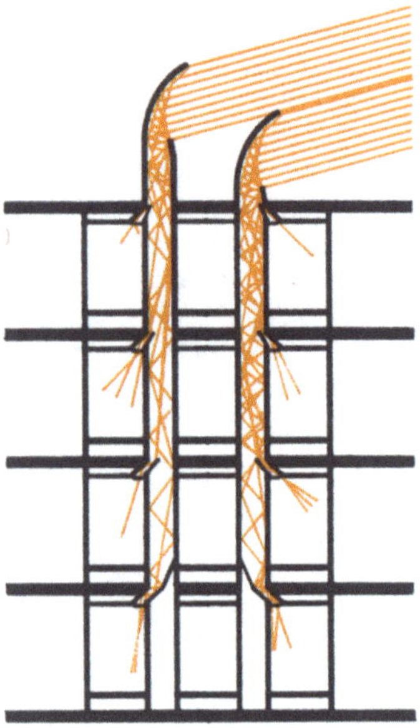

Figure 9. Entry of light through ducts when sun angle is 15 °

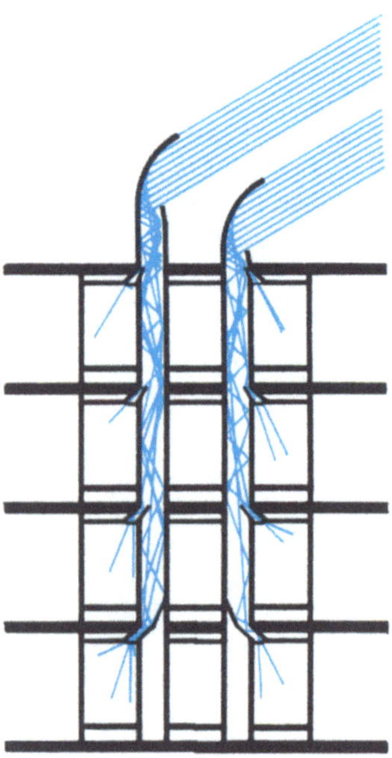

Figure 10. Entry of light through ducts when sun angle is 30 °

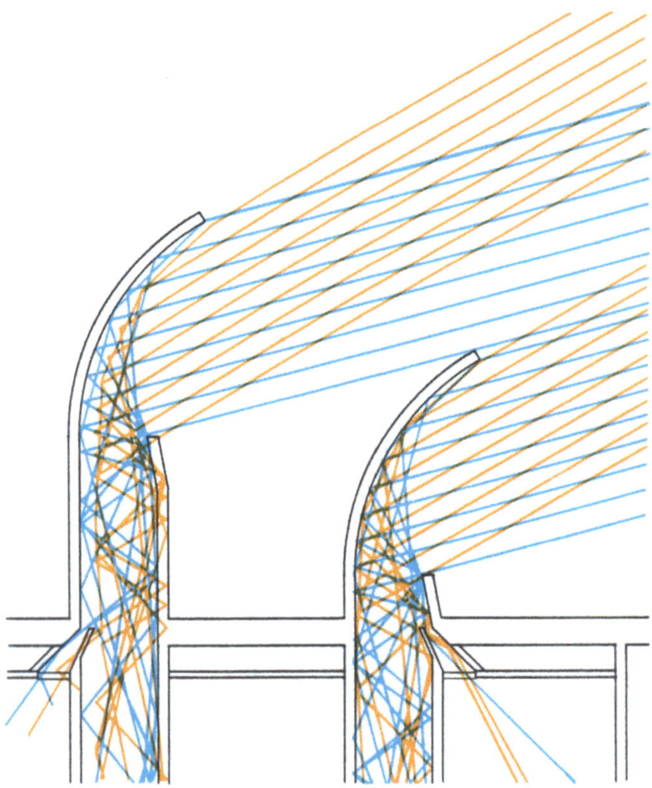

Figure 11. Top opening of light ducts

ENERGY CALCULATIONS PERFORMED AND DESIGN TOOLS USED

Introduction

Calculations were carried out to establish the amount of daylight entering the bathrooms from the light ducts. The lighting performance of the bathrooms were also simulated using a computer model. The daylighting performance in the other rooms was determined using a scale model of a module of three rooms and computer graphics. The thermal performance of the building was also assessed.

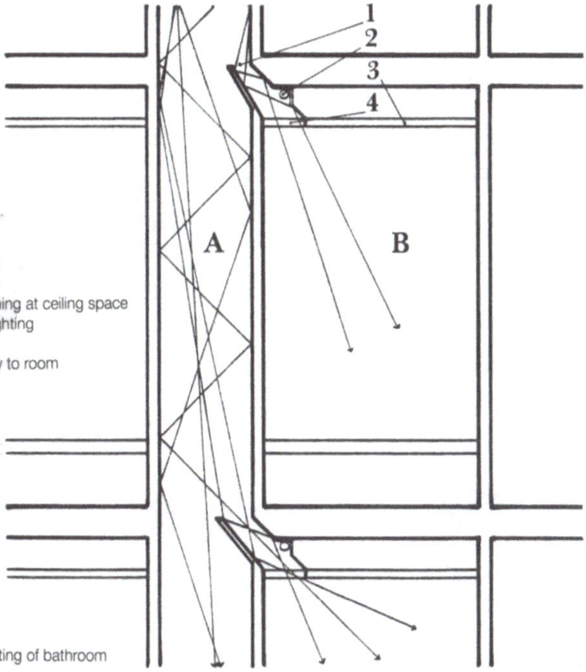

Key: A. Light duct
B. Bathroom
1. Duct opening at ceiling space
2. Artificial lighting
3. Ceiling
4. Light entry to room

Figure 12. Daylighting of bathroom

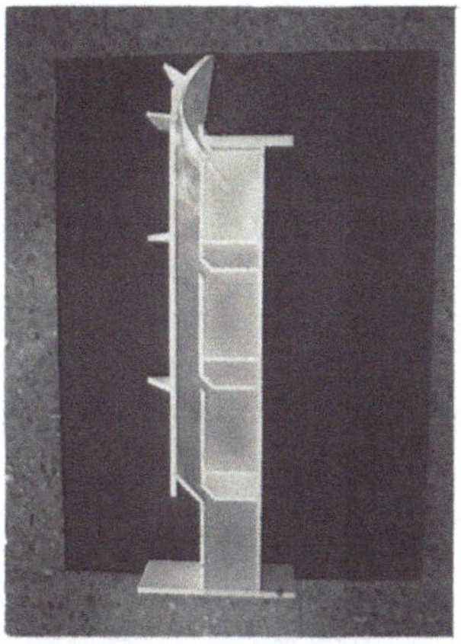

Figure 13. Model of light duct section used for simulation of daylighting

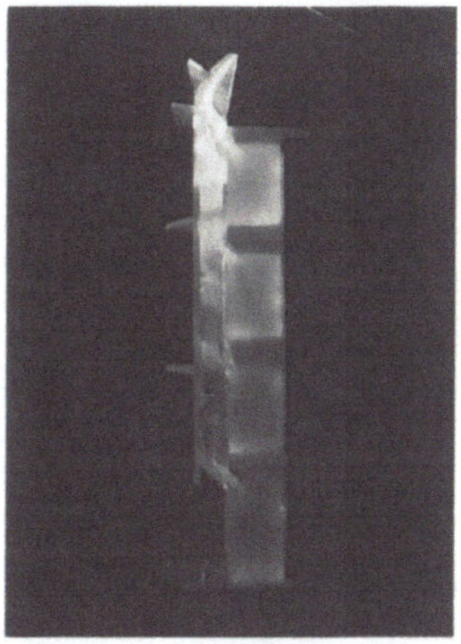

Figure 14. Simulation of the daylighting in rooms on the four floors

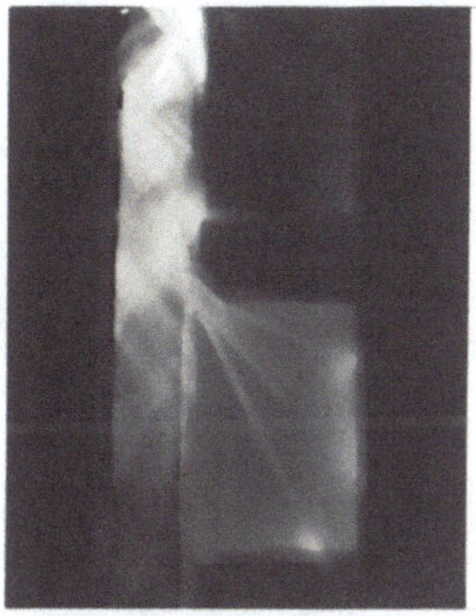

Figure 15. Light distribution in room

Daylighting Performance of Bathroom

The amount of light entering a bathroom from the light duct was calculated for two cases - when the sun is shining and when it is not shining.

For the case when the sun is shining it was assumed that 50,000 lux falls on the vertical plane of the duct entrance. The flux entering through the opening is therefore 50,000 x 4.0 = 200,000 lumens. Therefore, there are 10,000 lumens entering each bathroom. If the utilization factor is 0.5 and the bathroom floor area 3.75 m^2, the mean value of the lighting in each bathroom is 10,000 x 0.5 - 3.75 = 1333 lux.

When the sun is not shining, it was assumed that 5,000 lux falls vertically on the duct entrance. This gives 133 lux in each bathroom.

It was concluded that even on the worst winter's day the light ducts give sufficient illumination in the bathroom. When the sun is shining,

Daylighting Performance of Other Rooms

A scale model of a three-room module was used to assess the daylighting performance of the other rooms, qualitatively and quantitatively. The model is illustrated in Figures 16, 19 and 20. Computer graphics of the daylighting distribution are given in Figures 17, 18 and 21. The measurements were made under clear sky conditions at the times of the winter and summer solstice and spring and autumn equinox and under overcast sky conditions. The results of some of the measurements are shown in Figure 22. The modelling proved that the use of a control system (such as venetian blinds between double glazing) to reduce the level of illumination in the rooms can improve the quality of the daylighting.

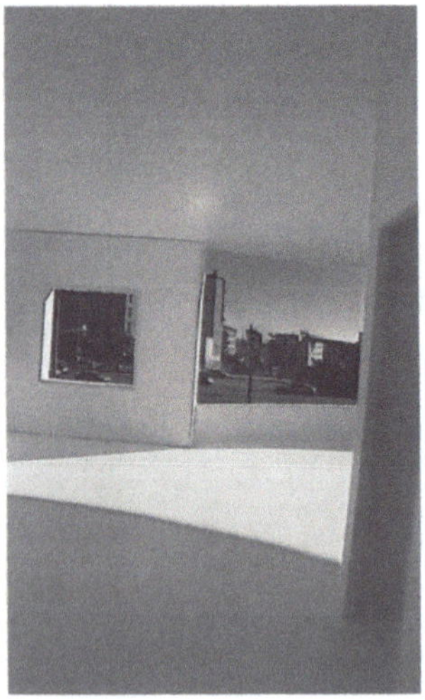

Figure 16. Scale model

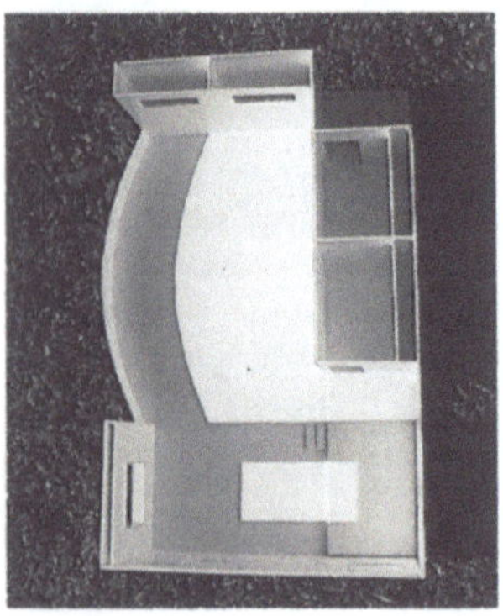

Figure 17. Computer graphics showing light distribution in interior of building

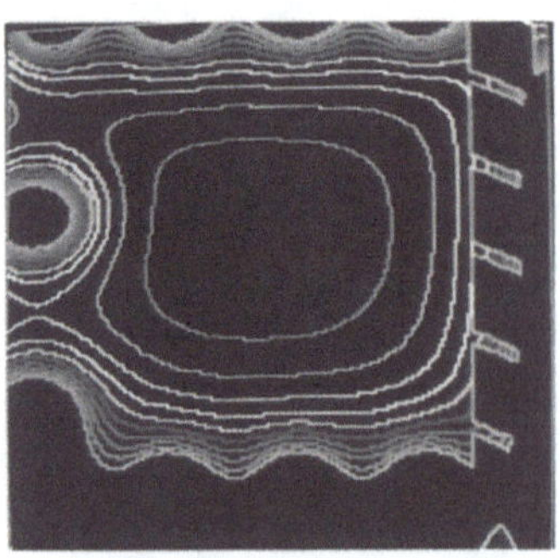

Figure 18. Isolux curves

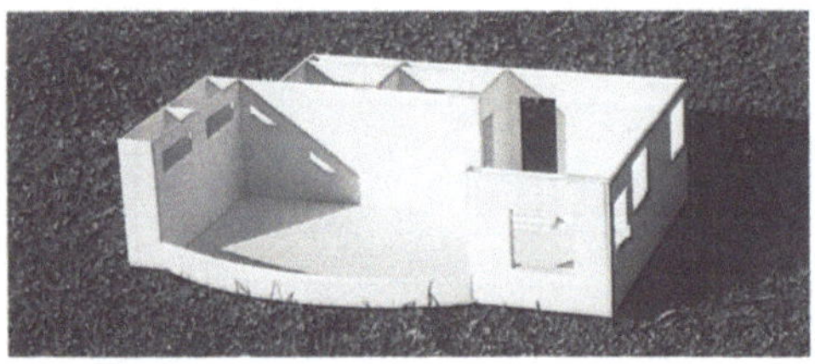

Figure 19. Scale model exterior view

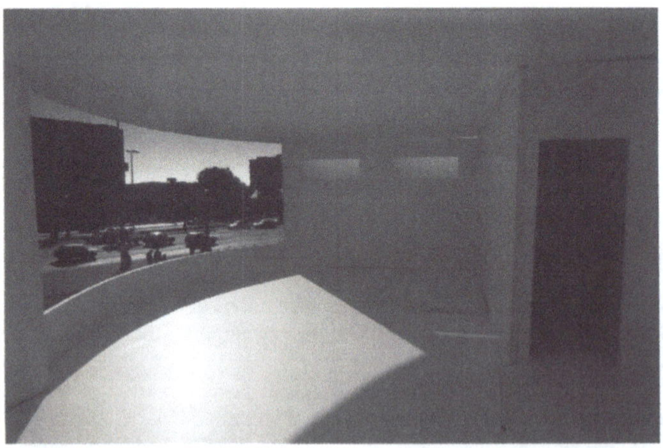

Figure 20. Scale model interior view

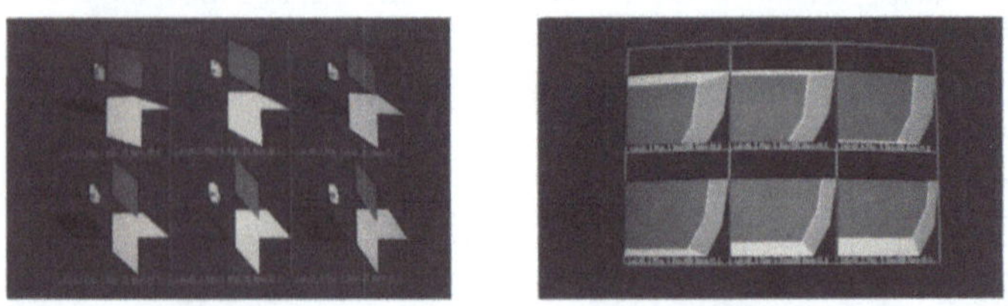

Figure 21. Computer graphics showing shadows produced in building at different times of the day

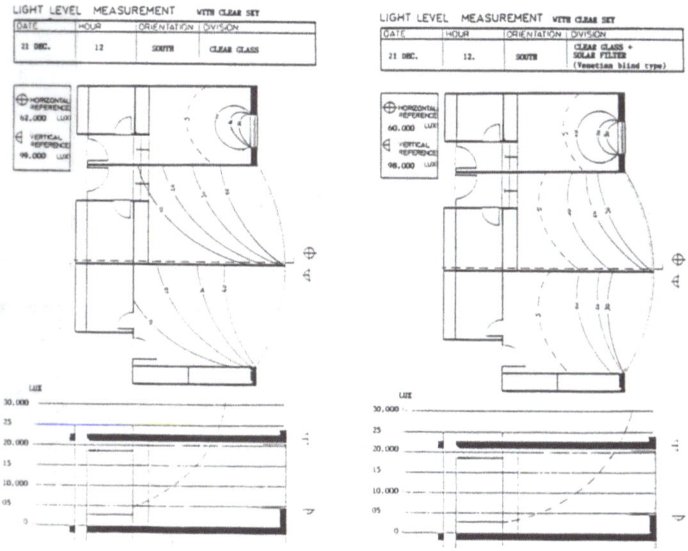

Figure 22. Results of measurements of daylight levels in scale model of three-room module

Thermal Performance

The thermal performances of two central rooms, one on the south side of the building and one on the north, on one of the intermediate floors were evaluated.

In summer, the south facade receives a considerable amount of solar radiation, making overheating a possibility. To prevent this, movable venetian blinds placed between the double glazing are used to reduce the amount of direct solar gain.

In winter, excess gains on the south side can be transferred to the north rooms by thermocirculation.

Calculation of Average Internal Temperatures

Calculations showed that in winter when the average external temperature was 8 ° C, the solar gains on the south side were 17 W/m^2 and the internal gains 2.5 W/m^2 , the average temperature in a south room was 23.7 ° C and the average temperature on the north side was 10 ° C.

In summer, with an external temperature of 24 ° C, solar gains on the south side were 15.8 W/m^2 , solar gains on the north side 3.2 W/m^2 and internal gains 2.5 W/m^2; the average temperature in a south room was calculated to be 29 ° C and the average temperature in a north room was 25.6 ° C.

Calculation of Solar Contribution to Heating Requirements

A month-by-month analysis of the solar contribution to the heating requirements of the building over the heating season is given in Table 4 and Figure 24. The total figure for the heating season is given in Table 3 and Figure 23. Solar gains make a 68% contribution to the overall heating requirements of the building. This is equivalent to 112524 kWh x 12 ptas = 1,350,288 ptas (12,000 ECU) a year. In addition, the electric lighting load will be reduced by daylighting from the light ducts.

	kWh
Total thermal requirements	164648
Contribution from solar features	112524
Auxiliary heating	52224
Solar contribution (%)	68.3

Table 3. Annual contribution of solar gains to heating requirements

MONTH	EXT TEMP (ºC)	INT TEMP (ºC)	SOLAR TEMP (ºC)	TOTAL LOOSE (kw h)	TOTAL GAIN (kw h)	REN %	UTIL GAIN (kw h)	ADIT HEAT (kw h)	SAVE %
JANUARY	8.4	21.1	16.4	34420	23855	86	20477	13943	59.5
FEBRUARY	10.1	23.2	20.1	26533	26740	68	18061	8472	68.1
MARCH	11.7	25.0	22.3	24628	31473	53	16699	7929	67.8
APRIL	14.1	26.8	24.6	16942	30269	35	10664	6277	62.9
MAY	17.5	27.7	27.7	7418	30327	24	7418	0	100.0
OCTOBER	17.8	28.1	28.1	6528	30581	21	6528	0	100.0
NOVEMBER	12.9	23.2	21.2	20388	23760	61	14429	5959	70.8
DECEMBER	10.6	22.1	18.9	27892	24552	74	18248	9644	65.4

Table 4. Monthly contribution of solar gains to heating requirements during heating season

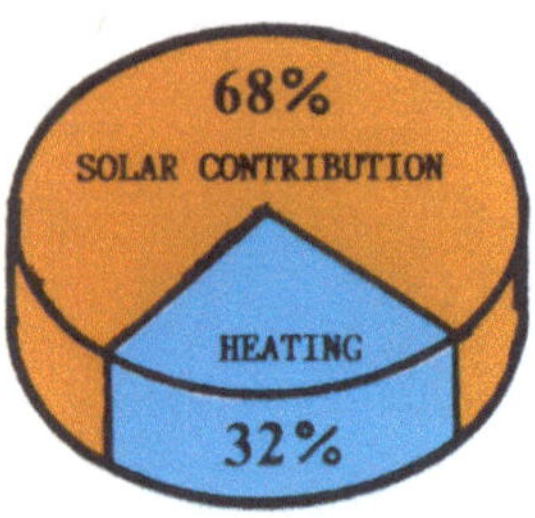

Figure 23. Annual contribution of solar gains to heating requirements

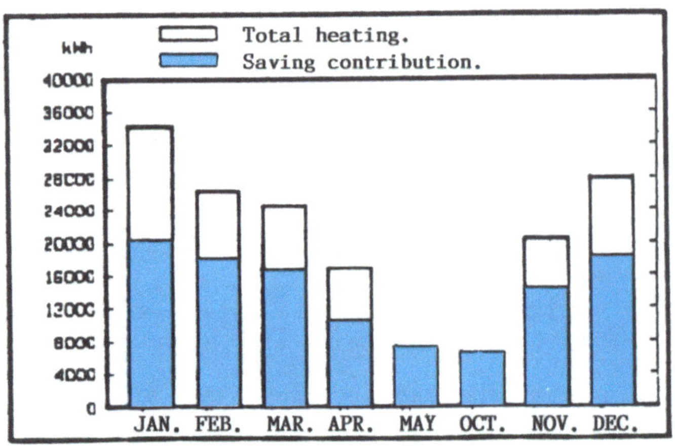

Figure 24. Monthly contribution of solar gains to heating requirements during heating season

GENERAL DESIGN GUIDELINES/POINTS OF INTEREST RESULTING FROM THIS PROJECT

In designing a hotel, the main consideration is to provide comfortable accommodation and good services. The energy requirements of such a building are intermittent and variable. It is therefore important not only to reduce energy costs but also to regulate and use energy in the most profitable and beneficial way.

In the present building, the main point of interest is the design process. This gave feedback which enabled the architects to produce a design which had a reduced energy consumption, improved natural lighting and was more suited to the site conditions and local climate.

In order to answer the questions which arose during the design process, use was made of simulations and scale models. These could easily be applied to any project. The architects' previous experience of these techniques and collaboration with other research groups in the field enabled them to obtain reliable results from such studies.

BUILDING 2000 Building 2000 brochures are published by Directorate General XII of the Commission of the European Communities to show how design studies can help architects and other building designers use passive solar principles to the best effect to produce attractive energy efficient buildings. Each brochure describes studies carried out with the support of the Commission during the design phase of one of thirty-six non-domestic buildings in the EC Member States. The studies were on such topics as daylighting, heating, cooling, ventilation, comfort, control systems and urban design. They were carried out with the help of acknowledged European experts in these fields and drew heavily on lessons learned and techniques developed through the Commission's research and development programme on solar energy applications to buildings.

Commission of the European Communities/Directorate-General for Science, Research and Development

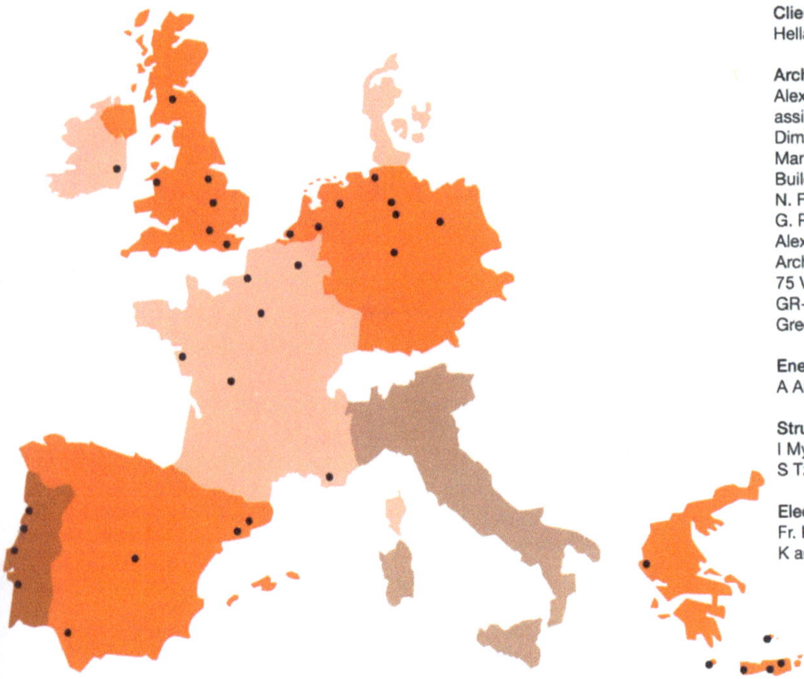

List of Design Team Participants and Advisers

Client
Hellas Holiday Hotels

Architect
Alexandros N Tombazis
assisted by
Dimitris Potiropoulos, architect
Maria Spyridaki, architect for the Building 2000 aspects of the project
N. Fletorides, L. Nella-Potiropoulou and G. Panetsos, design team architects
Alexandros N Tombazis and Associates, Architects
75 Vasilisis Sofia Avenue
GR-11521 Athens
Greece

Energy Consultant
A Argiriou

Structural Engineer
I Mylonas
S Tzivanakis

Electrical and Mechanical Engineer
Fr. Krispis
K and A Alexopoulos

This set of **Building 2000** brochures illustrates how architects and other building designers can successfully apply passive solar principles to produce energy-efficient buildings.

BUILDING 2000 Participants

Project Director
Theo C. Steemers

Coordinator
Cees den Ouden

Technical Steering Committee
Dean Hawkes
Nick Baker
Alex Lohr
Jean P. Lepoivre

Regional Liaison Agents
(D) Jörn Behnsen
(E) Vicente Sifre
(F) Michel Raoust
(GB) Alan Hildon
(GR) Matheus Santamouris
(P) Eduardo Maldonado

Further information or copies of the brochures can be obtained from prof. ir. Cees den Ouden, EGM Engineering BV, P.O. Box 1042, 3300 BA Dordrecht, The Netherlands.

BUILDING 2000

Commission of the European Communities

- Tourist hotel in shopping area near highway
- Atrium in form of pedestrian pathway between two wings of building
- Conservatory used for reception and general access area
- Heat recovery in atrium using inexpensive solar chimney
- Division of building into zones according to season and occupancy

TOURIST HOTEL

BETHUNE/FRANCE

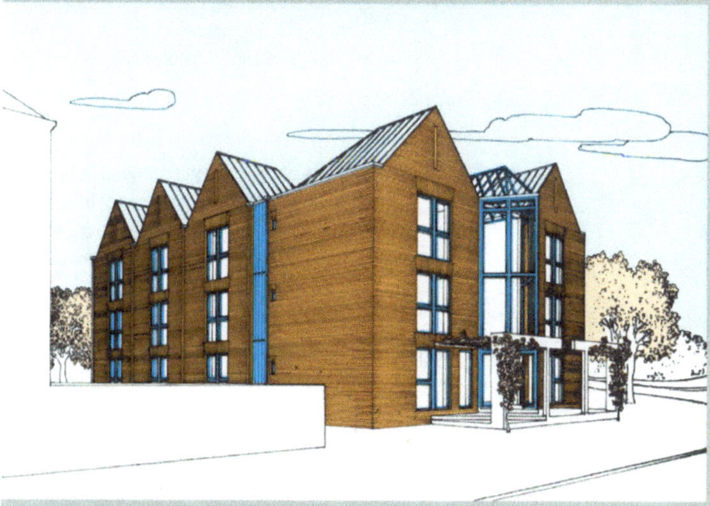

Building 2000 is a series of design studies illustrating passive solar architecture in buildings in the European Community.

ISSUE H 3

FEB 1991

2

PROJECT DESCRIPTION

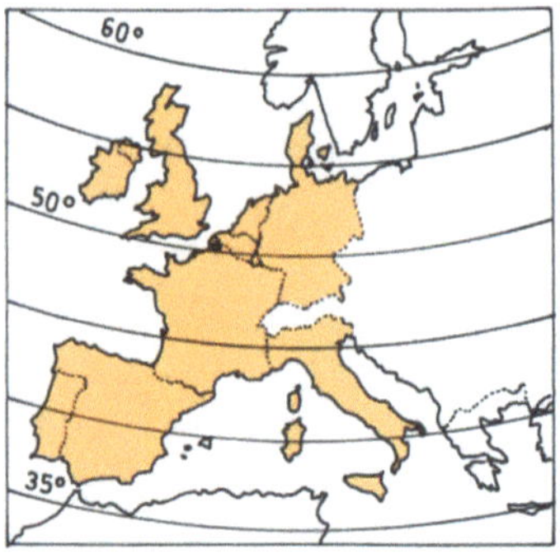

Figure 1. Location

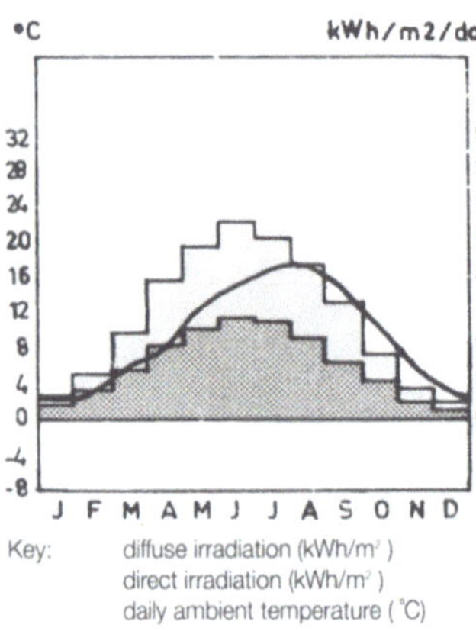

Figure 2. Average global irradiation on a horizontal plane and average ambient temperatures

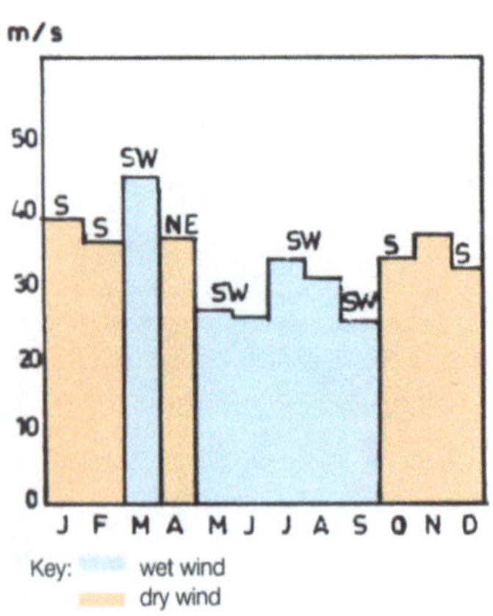

Figure 3. Prevailing winds and average instantaneous wind speeds (m/s)

Building Type

This building (which is to be opened at end May 1990) is a 45-bed hotel, one of a chain providing low-cost accommodation run by a young company called Hotellerie Blesoise.

The aim was to produce a standard design for a compact building with a low cost:quality ratio which could be run by one person. Minimum space was to be given over to movement of people. The building was to be kept comfortable by natural means, i.e. without use of mechanical equipment. Ideally, the building would be oriented north-south to optimize solar gain in winter and provide the best sunlight conditions for the general areas and 20% of the bedrooms. In practice, the orientation would be modified to suit the needs of the particular site.

Location

The site chosen for the study is in Bethune (Figure 1) in a new shopping area in an old industrial suburb. It is near a circular building, a former locomotive garage converted to a general store. The site requires the hotel to face east-south-east (see Figures 9 and 10).

Site Microclimate

The climate at the site is oceanic and is assumed to be identical to that of Lille, 33 km to the north-east (Figure 4). Average temperatures are in the range 0° C to 22° C. In winter, daily temperatures are low and there is considerable cloud cover. In summer, temperatures are higher and there is less cloud. Some climate data are given in Figures 2, 3, and 5.

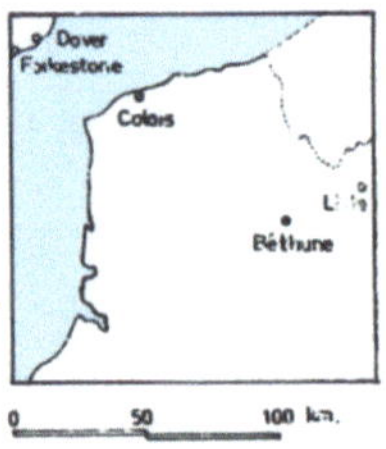

Figure 4. Bethune and Lille

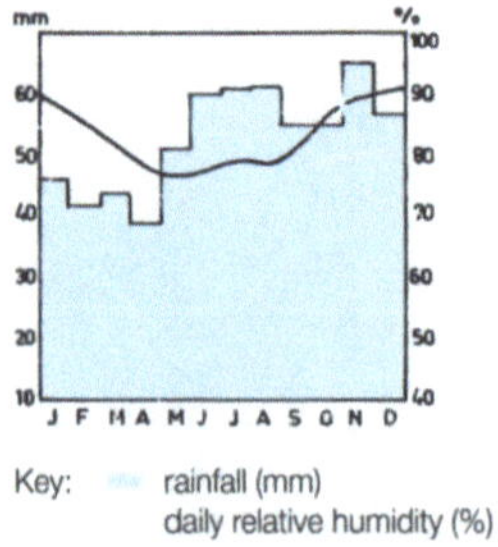

Figure 5. Average rainfall and relative humidity

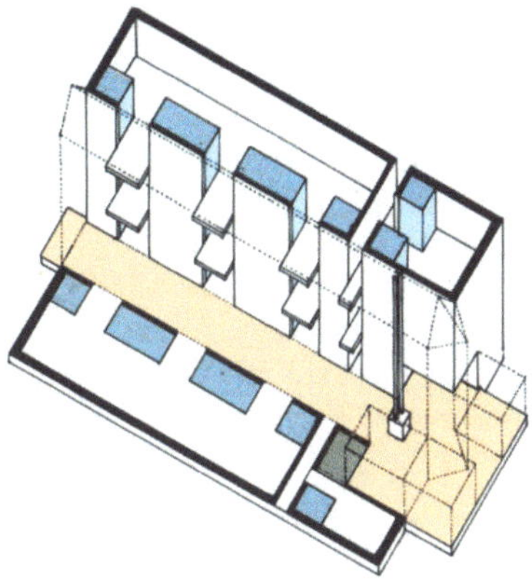

Figure 6. View into building

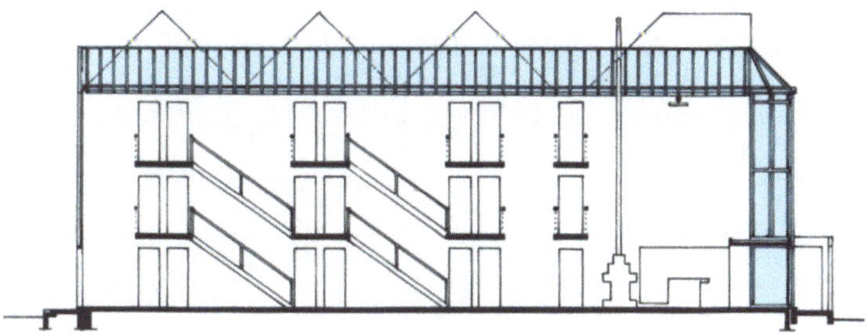

Figure 7. Longitudinal section

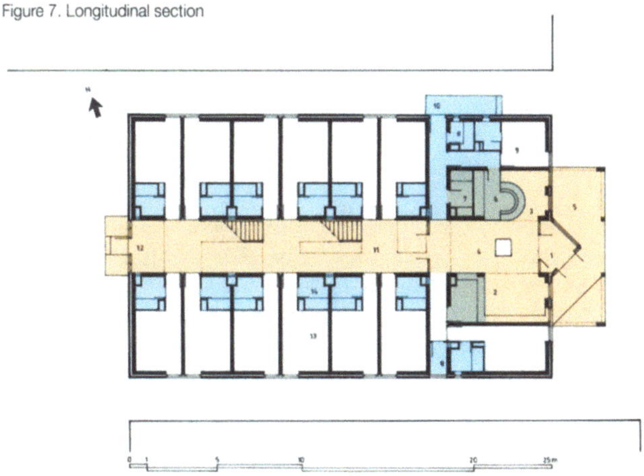

Key:
1. Public entrance
2. Reception
3. Breakfast area
4. Fireplace and lounge
5. Summer terrace
6. Kitchen
7. Laundry room
8. Linen room
9. Service bedroom
10. Service entrance
11. Access atrium
12. Emergency entrance
13. Single bedroom
14. Bathroom

Figure 8. Ground floor plan

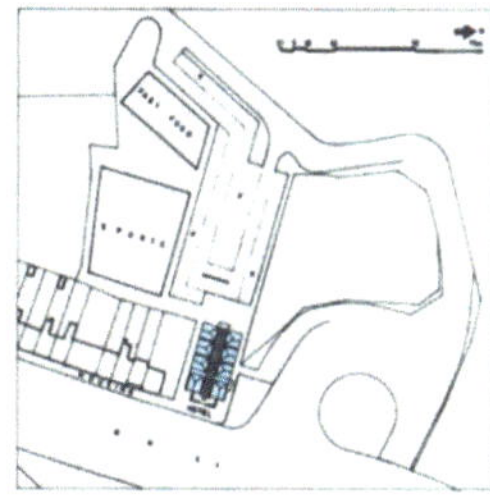

Figure 9. Site plan

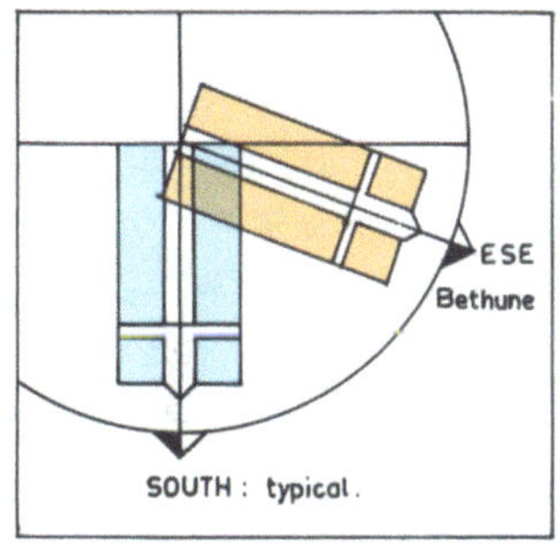

Figure 10. Orientation of ideal and Bethune buildings

Design and Construction Details

Details of the design are given in Figures 6-8. The general facilities have a lightweight and transparent structure whereas the individual rooms are built of massive and opaque materials. The external walls have an inner layer of 200 mm cement blocks and an outer layer of 100 mm brick with 90 mm insulating mineral fibre sandwiched between. The floors are 200 mm concrete. the roof is made of steel and insulated with 200 mm mineral fibre. U-values of some of the building components are shown in Table 1.

U-values in W/m² K	
roof	0.20
walls	0.37
glazing	2.8

Table 1. U-values of some building components

DESCRIPTION OF PASSIVE SOLAR FEATURES/COMPONENTS

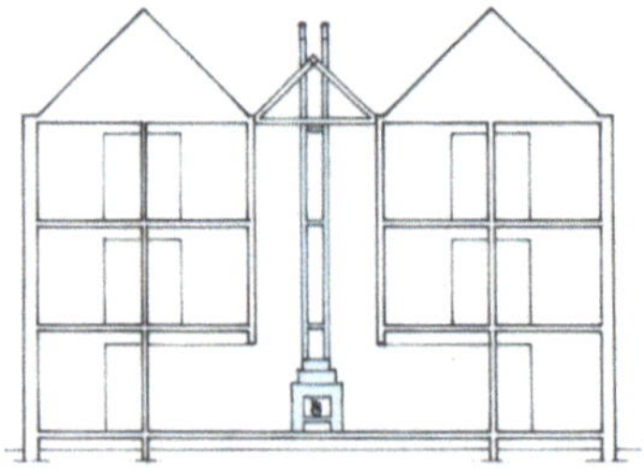

Figure 11. Heat diffusing smoke pipes

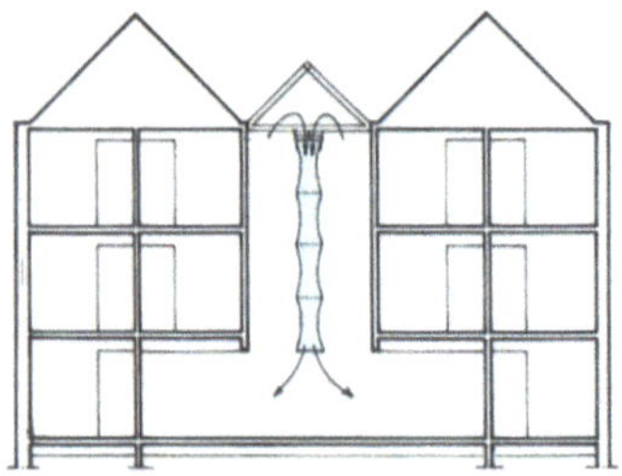

Figure 12. Solar chimney

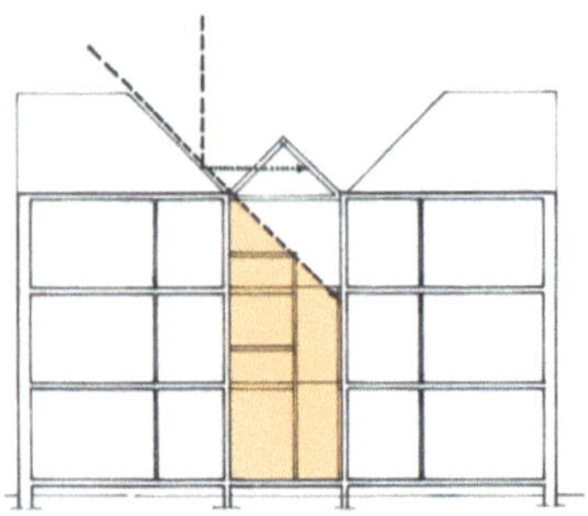

Figure 13. Sun path in atrium and reflection by roof

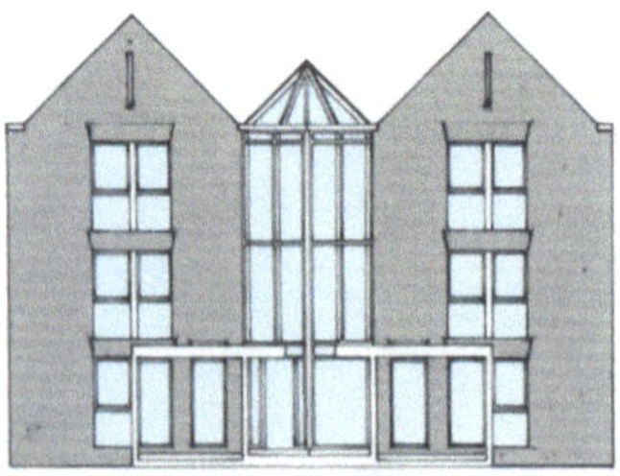

Figure 14. The winter bedrooms, conservatory and summer terrace facing south

Atrium, Conservatory and Summer Terrace

Running along the main axis of the building is a lightweight atrium structure in the form of a pedestrian pathway with a glazed roof. At the front of the building (which faces east-south-east) the atrium merges into a conservatory. In front of the conservatory there is a summer terrace. These features are shown in Figure 15.

The atrium is lit and heated by the sun through the glazing overhead and at the front of the building. It is also provided with a cast iron fireplace with heat-diffusing smoke pipes (Figure 11).

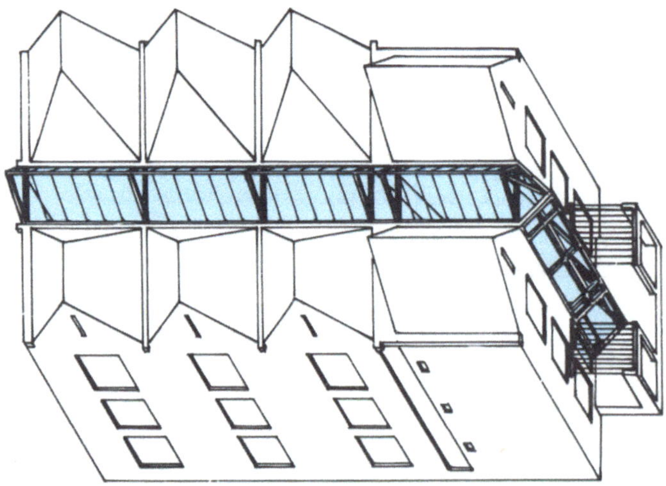

Figure 15. Axonometric view of the atrium, conservatory and summer terrace.

Air Stratification and Solar Chimney

The main disadvantages of an atrium are that the air stratifies from the bottom to the top, and there is a risk of the warm air, which has risen to the top, being cooled at the overhead glazing. To overcome this an inexpensive portable solar chimney made of fabric is placed in the atrium and warm air from the atrium is moved through it from top to bottom with the help of a fan (Figure 12).

Valley Roofing

Penetration of light into the atrium is maximized by using a valley profile roof (Figure 13). This produces diffuse radiation within the atrium and minimizes shading.

Seasonal Zoning

The building is divided into two zones, according to season and occupancy. The winter (low-occupancy) zone is confined to the general facilities and the south-facing bedrooms. The summer (high-occupancy) zone includes all the bedrooms.

ENERGY CALCULATIONS PERFORMED AND DESIGN TOOLS USED

Preliminary Analysis

The preliminary design work was carried out with the help of Givoni's building bioclimatic chart (Figure 17). This showed that the requirements of the building varied according to season and time of day. They fell between two extremes. At one extreme, there was a need for heating, solar gains and insulation; at the other, there was a requirement for solar gains and shading. This led to the use of two types of structure: solid, opaque and well-insulated forms for the parts of the building used at night (i.e. the bedrooms) and light-weight glass structures for the parts used in the day (i.e. the reception and the bedrooms and restaurant access areas).

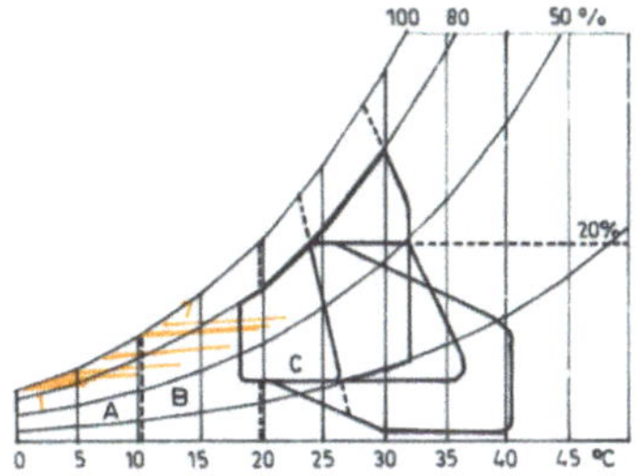

Key: A. Heating
B. Solar gains
C. Comfort zone

Figure 17. Building bioclimatic chart

Thermal Performance Studies

Orientation

The building was designed to be oriented so that its main axis runs north-south, with the entrance facing south. In practice, individual sites may require this to be modified - as it has been in Bethune. The detailed performance calculations which follow, however, relate to the ideal (north-south) orientation.

Design Tools

Detailed thermal analysis was carried out using two microcomputer-based design tools, Méthode 5000 and Oasis.

Méthode 5000 was used to evaluate the energy balance and performance of the whole building. It gives overall monthly and annual values for heat loss, internal gains, solar gains and auxiliary heating.

Oasis is a simulation model used to carry out more detailed study of the thermal behaviour of the building. It calculates thermal conditions such as internal temperatures, surface temperatures and energy balances for each room or zone on an hourly basis for selected typical days.

Design Temperatures

In carrying out the thermal performance studies, it was assumed that the bedrooms would be heated to 19° C and the entrance hall and atrium would not be heated.

Overall Thermal Performance

The annual performance of the hotel as a whole (when oriented north-south) is shown in Figure 18. Useful solar gains contribute 31% of the total energy requirements, internal gains 18% and auxiliary heating 51%. The performance is 0.007 kWh/degree day m^3. Solar gains amount to 125 kWh/m^2 aperture. The performance month-by-month is given in Figure 16.

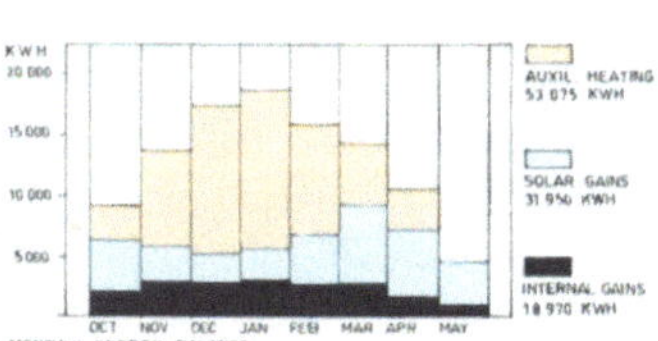

Figure 16. Monthly energy balance (kWh)

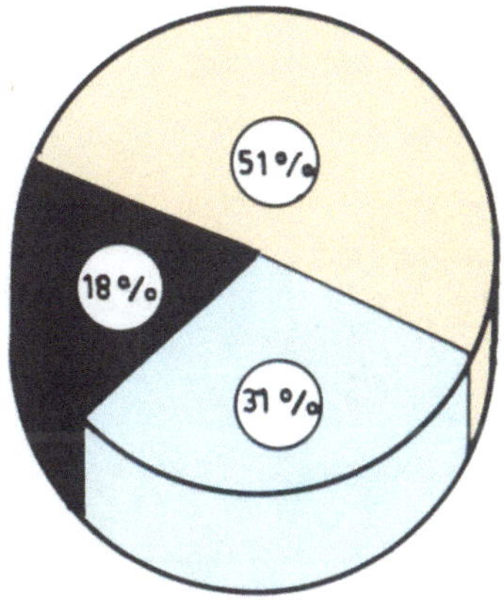

Figure 18. Annual energy balance

Detailed Thermal Analysis

In order to present as pessimistic a picture of energy requirements as possible in the detailed simulation studies, no account was taken in general of internal gains from occupants, electric lighting and equipment.

The detailed analysis showed that in winter, when the whole building is in use, temperature conditions in the atrium and entrance hall are acceptable even when there is no auxiliary heating. On an average sunless January day with an external temperature between 0.2° C and 5° C, the internal temperature is likely to be more than 12.5° C or 13.9° C (depending on internal gains). For a cold February day with a clear sky, the results are likely to be even better (see Figure 19).

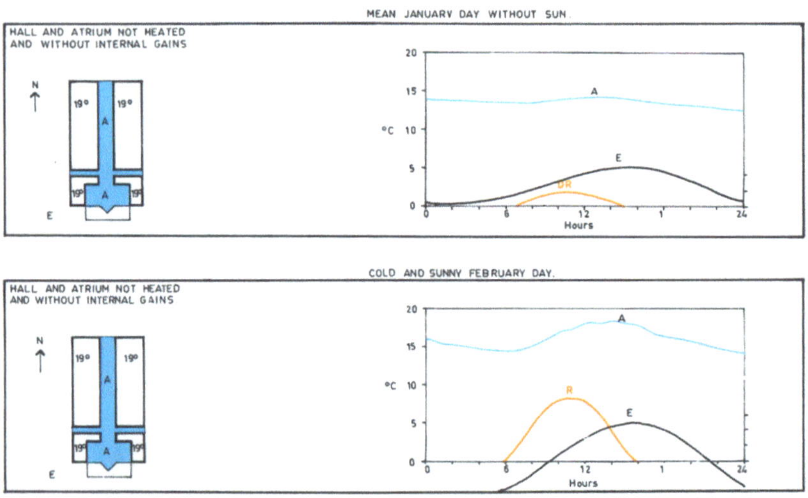

Figure 19. Atrium and entrance hall temperatures in winter

In summer, the temperature in the entrance hall, atrium and bedrooms is likely to be comfortable when internal blinds are used. For a clear sky June day (see Figure 20), the temperature in the atrium and hall will remain under 23° C when there is additional natural ventilation. In the bedrooms it should remain below 27° C without extra ventilation. The bedrooms provide thermal protection for the atrium and hall.

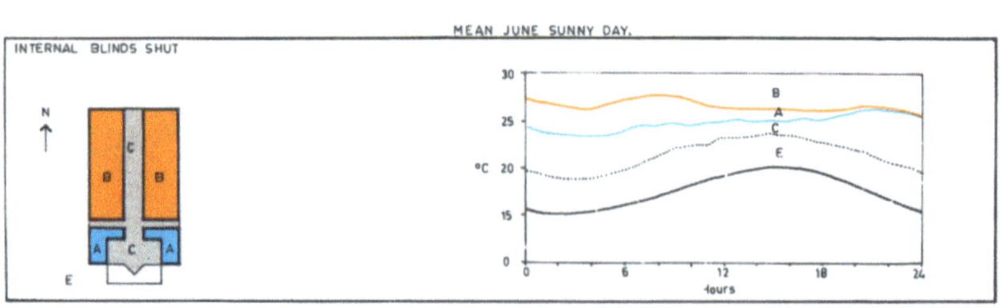

Figure 20. Atrium, entrance hall and bedroom temperatures in summer

Influence of Seasonal Zoning

As indicated above, the building is zoned so that the atrium pathway and adjacent bedrooms can be separated from the rest of the hotel in winter if necessary and not be heated. The studies showed that, when this is done, the atrium will in general be at a temperature acceptable for access purposes - or require only slight heating. For an average sunless January day, for instance, the temperature in the atrium would be between 7.4° C and 8.8° C (see Figure 21).

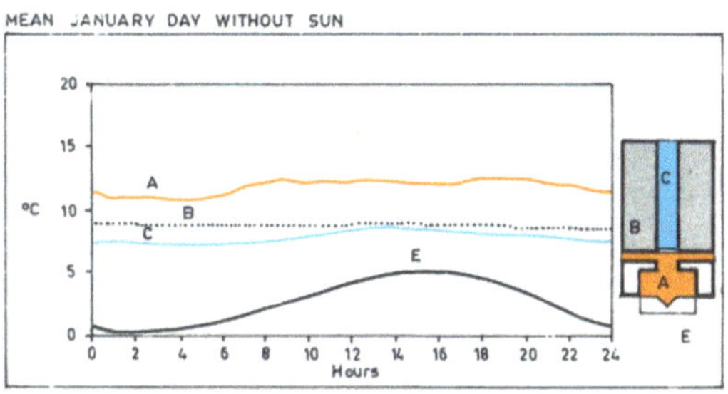

Figure 21. Atrium, entrance hall and bedroom temperatures in winter when building is zoned for partial use

Influence of Orientation

As indicated earlier, in Bethune the building is not oriented north-south but turned 75° to the east (see Figure 10). Global analysis showed that this reduced the overall energy performance by only 5%. Precise simulations, however, indicated that the south row of bedrooms would experience eight times the solar gains of the north row and could experience overheating. The atrium and entrance hall would have roughly identical solar gains to those found in the north-south oriented building.

Daylighting Analysis

The Genulux program was used to simulate the natural daylighting in the building. Analysis of the daylighting of the atrium wall (Figure 22) showed that there is shading from the passageways but that the top and side sections of the wall are very bright. The central area is darker but acceptable for circulation purposes.

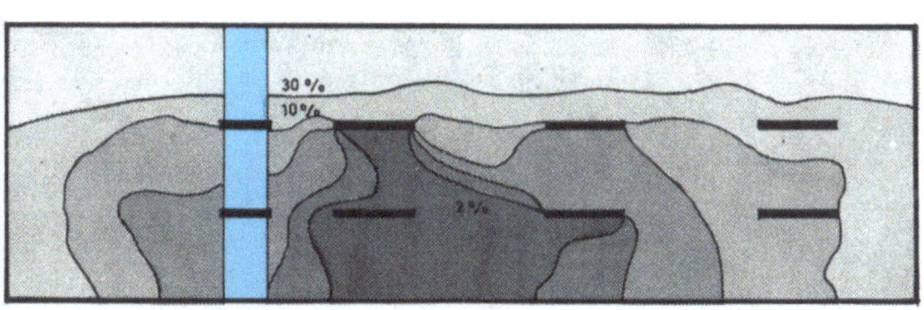

Figure 22. Illuminance distribution on the atrium wall (daylight factor - %)

GENERAL DESIGN GUIDELINES/POINTS OF INTEREST RESULTING FROM THIS PROJECT

Conclusions from Thermal Performance Analysis

Successive thermal performance analyses led to development of a design for a more compact building and a roof which was more effective as a solar energy collector.

Conclusions from Daylighting Analysis

The daylighting analysis had an influence on the final design of the atrium. It was found that:

- for good light penetration, the height had to be at least 2.5 times the width;
- to achieve a better flow and reflection of light from the top to bottom, the longitudinal passageways should be separated from the walls of the atrium;
- surfaces should be selected carefully. Floor surfaces could be matt and wall surfaces bright but nowhere should they cause glare.

BUILDING 2000 Building 2000 brochures are published by Directorate General XII of the Commission of the European Communities to show how design studies can help architects and other building designers use passive solar principles to the best effect to produce attractive energy efficient buildings. Each brochure describes studies carried out with the support of the Commission during the design phase of one of thirty-six non-domestic buildings in the EC Member States. The studies were on such topics as daylighting, heating, cooling, ventilation, comfort, control systems and urban design. They were carried out with the help of acknowledged European experts in these fields and drew heavily on lessons learned and techniques developed through the Commission's research and development programme on solar energy applications to buildings.

Commission of the European Communities/Directorate-General for Science, Research and Development

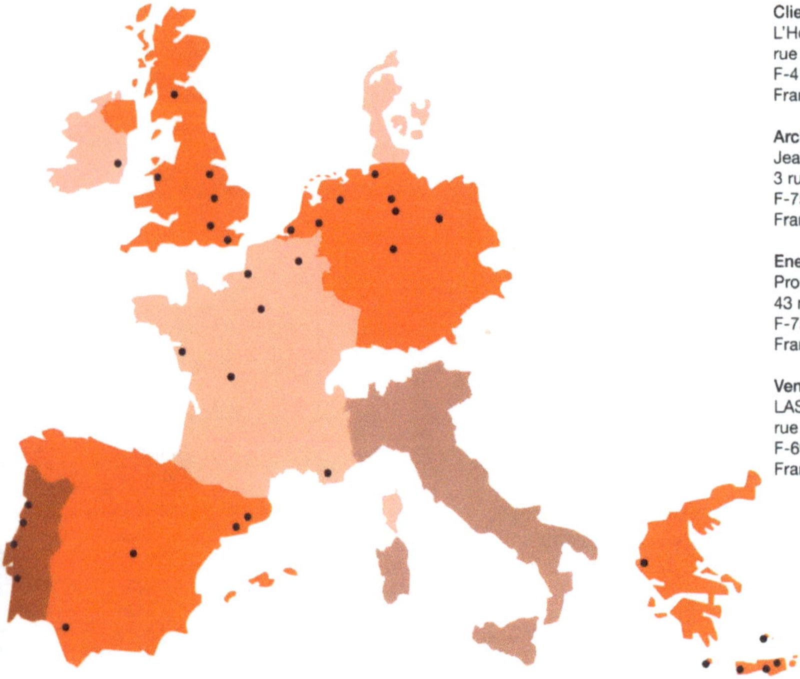

List of Design Team Participants and Advisers

Client
L'Hotellerie Blesoise sarl
rue de Montprofond
F-41260 La Chaussée St Victor
France

Architect
Jean Bouillot
3 rue de l'Agent Bailly
F-75009 Paris
France

Energy Consultant
Pronoverg
43 rue de la Victoire
F-75009 Paris
France

Ventilation and Daylighting Consultants
LASH, Ecole Nationale des TPE
rue M. Audin
F-69120 Vaulx en Velin
France

This set of **Building 2000** brochures illustrates how architects and other building designers can successfully apply passive solar principles to produce energy-efficient buildings.

BUILDING 2000
Participants

Project Director
Theo C. Steemers

Coordinator
Cees den Ouden

Technical Steering Committee
Dean Hawkes
Nick Baker
Alex Lohr
Jean P. Lepoivre

Regional Liaison Agents
(D) Jörn Behnsen
(E) Vicente Sifre
(F) Michel Raoust
(GB) Alan Hildon
(GR) Matheus Santamouris
(P) Eduardo Maldonado

Further information or copies of the brochures can be obtained from prof. ir. Cees den Ouden, EGM Engineering BV, P.O. Box 1042, 3300 BA Dordrecht, The Netherlands.

BUILDING 2000

Commission of the European Communities

- Building containing ten holiday apartments for use during the summer months and the Christmas vacation
- Cross ventilation, movable shades and a ventilation chimney system provide cooling in summer
- Direct solar gain and energy-efficient fireplaces heat the apartments over the Christmas holiday
- Thermal insulation and the thermal mass of the building helps reduce space heating demand and the size of the installed heating and cooling systems

HOLIDAY APARTMENTS

GOURNES/CRETE/GREECE

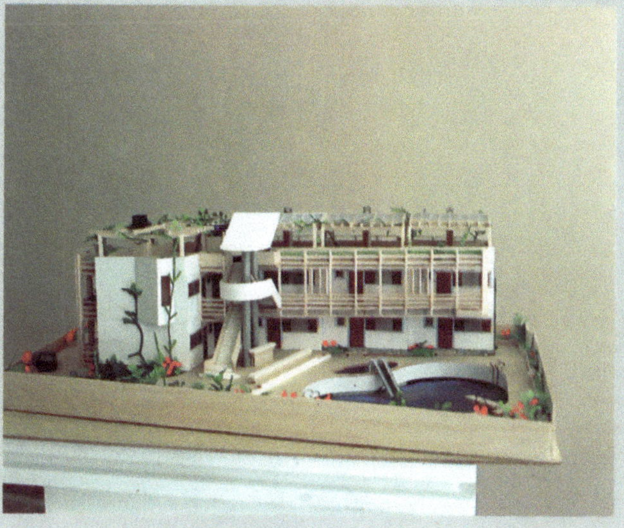

Building 2000 is a series of design studies illustrating passive solar architecture in buildings in the European Community.

ISSUE

H 4

JUN 1990

2 PROJECT DESCRIPTION

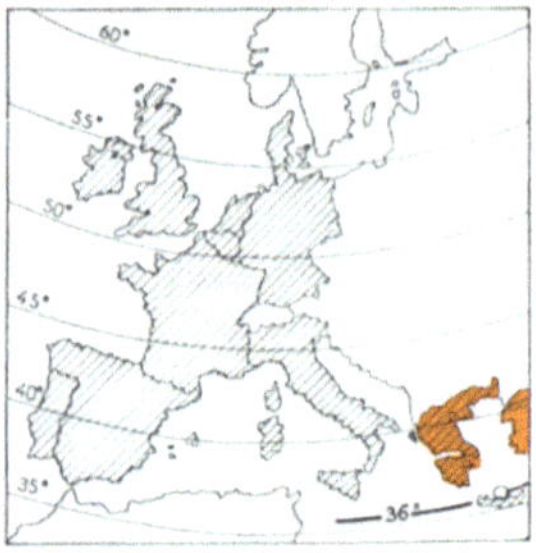

Figure 1. Location of Gournes

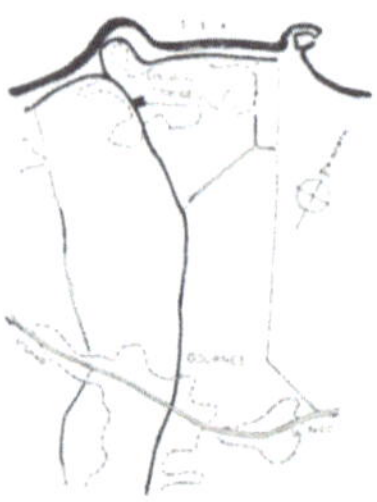

Figure 2. Location of site

Jan-Dec:	
global irradiation on horizontal plane	
in kWh/m^2	1631 kWh/m
(in MJ/m^2	5872 MJ/m^2)
sunshine hours	2794 hours
degree days (base 18 ° C)	782

Mid-Apr-Oct:	
global irradiation on horizontal plane	
in kWh/m^2	1254 kWh/m^2
(in MJ/m^2	4514 MJ/m^2)
sunshine hours	2180 hours
heating degree days (base 18 ° C)	156

Table 1. Some climate data

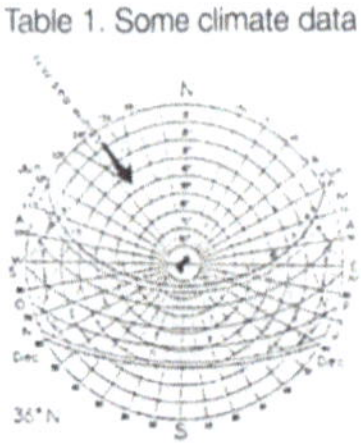

Figure 3. Solar chart

Building Type

This project concerns a building containing 10 holiday apartments for use during the summer months (mid-April to October) and the Christmas vacation.

Location

The apartments are in the holiday resort of Gournes (latitude 36 ° N), 15 km east of Iraklion in Crete (see Figures 1 and 2). The 1000 m^2 site, which is 15 m above sea level, is completely flat. It has sea views on the north, north-west and west sides and a view of the open country to the south. A model of the site is shown in Figure 4.

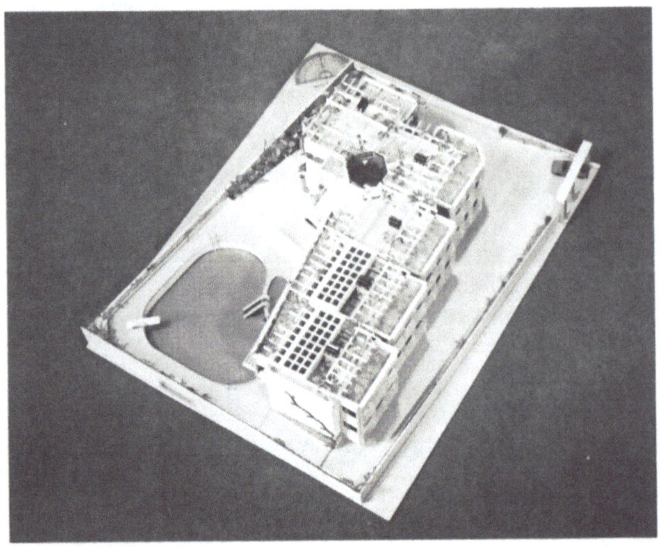

Figure 4. Model of the site

Site Microclimate

Temperature, solar irradiation and relative humidity data are shown in Figure 5, together with an indication of heating and cooling needs. Cooling is only required in July and August between midday and early afternoon when the external temperature often exceeds 27 ° C and can reach 30 ° C. At such times, the relative humidity is 58-60%. The maximum external temperature reached at the site is 36 ° C. Additional climate data are given in Table 1 and sunpath diagrams in Figures 3 and 6. The prevailing winds are cool sea breezes of medium intensity (3-4 m/s) from the north-west. They influenced the layout of the building and its siting on the lot. The wind flows on the site are illustrated in Figure 7.

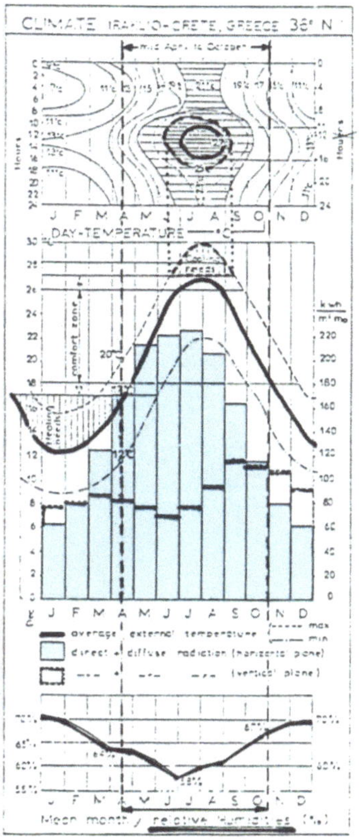

Figure 5. Temperature, solar irradiation and relative humidity data

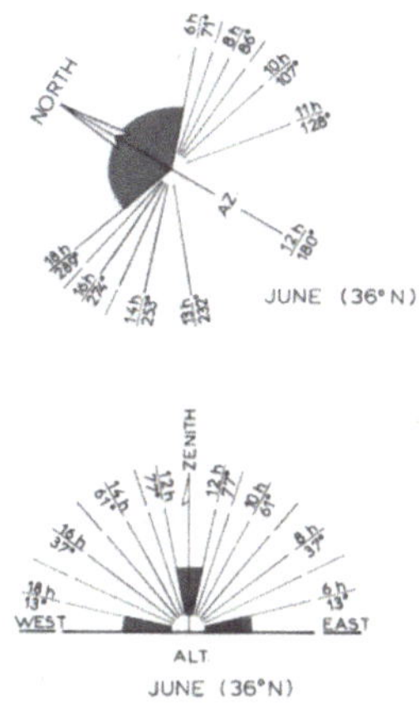

Figure 6. Sunpath diagrams

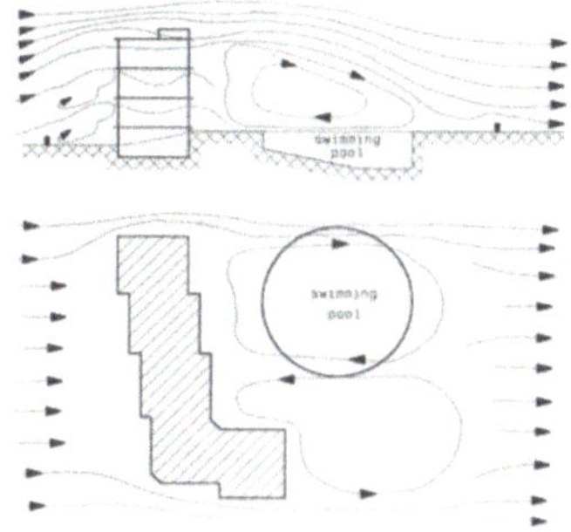

Figure 7. Wind flows on the site. (The prevailing winds are north- west sea breezes at 3-4 m/s.)

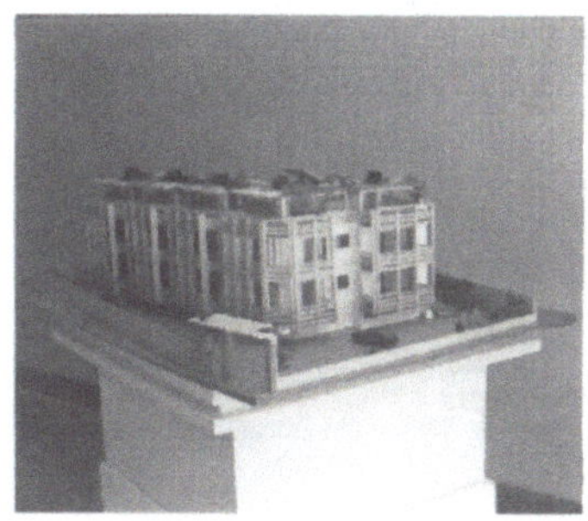

Figure 8. View of the model from the north

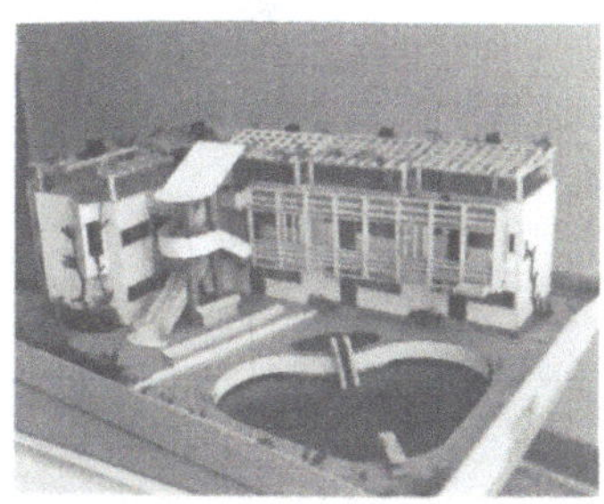

Figure 9. View of the model from the south

Design and Construction Details

Views of the model from the south and the north are given in Figures 8 and 9. The building has two storeys and a basement each of which has a floor area of 200 m^2. The ground floor and upper floor plans are given in Figures 10 and 11 and a section in Figure 12. There are five small (40 m^2) apartments on each floor, each of which has a private 9 m^2 balcony. The windows of all the apartments are on the north-west and south-east facades and the entrance doors are on the south-east side. The roof is covered with a bamboo trellis to provide shade. This makes it more comfortable for those sitting out on the roof and also reduces the cooling load of the building.

The beams, roof and floor slabs of the building are constructed of reinforced concrete. The U-values of some of the building elements are given in Table 2. The overall heat transfer coefficient of the building is lower than that required by local regulations. The walls consist of two layers of brick with thermal insulation between. This construction, together with the positioning of the walls, gives the building a high thermal inertia.

Concrete walls	0.51
External brick walls	0.50
Roof	0.35
Floor (basement roof)	0.48
Doors/windows	3.6/5.8

Table 2. U-values of some building elements in $W/m^2 K$

DESCRIPTION OF PASSIVE SOLAR FEATURES/COMPONENTS

Introduction

A number of passive solar features have been incorporated into the building to keep it cool in summer and warm in winter. Because of the quick turn around of tenant, the systems in each apartment have been designed to be autonomous, of simple construction and straightforward to operate.

The cooling systems include cross ventilation, use of shades of various kinds and ventilation with air cooled by passage over water flowing through the basement. Their mode of operation is illustrated in Figure 12.

When the building is occupied in winter, heating is provided by a special energy-efficient fireplace.

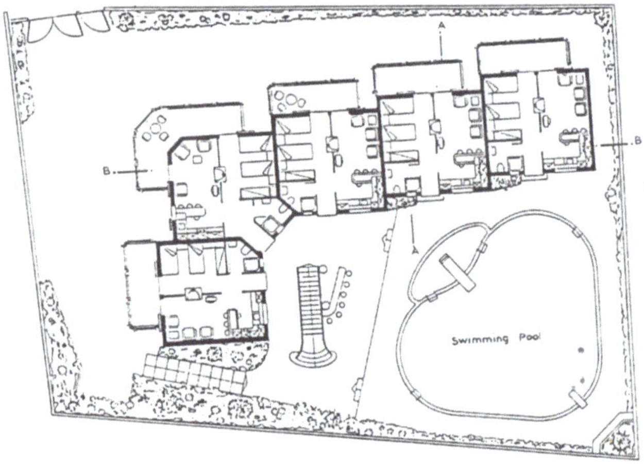

Figure 10. Ground floor plan

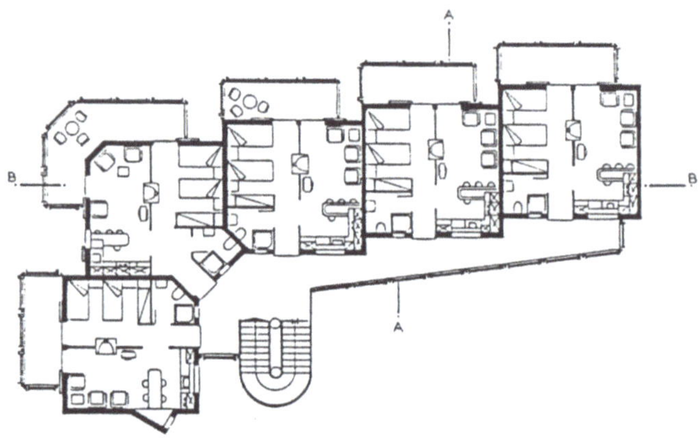

Figure 11. Second floor plan

Cooling Systems

Cross Ventilation

All the apartments have a double exposure to the prevailing cool sea breezes. The openings on the windward and leeward sides were designed to give comfortable air speeds in the living room.

Movable Devices on Balcony to Provide Shade and Improve Ventilation

Movable shading devices which can be rotated vertically and horizontally are placed on the parapet of each balcony (see Figure 13). These can be used to perform three operations, as required. They can obstruct direct solar radiation during the hottest part of the day. When necessary, they can minimize the intensity of the natural light entering the building. By manual adjustment of the fins, the air flow rate can be changed and air streams deflected to one direction or another so that the natural ventilation is improved.

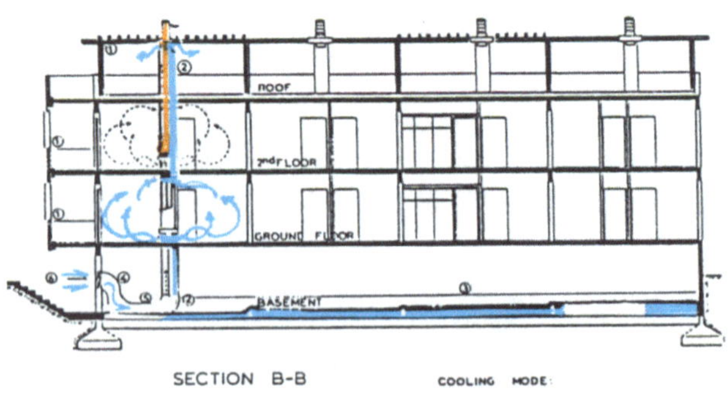

Figure 12. Section B-B

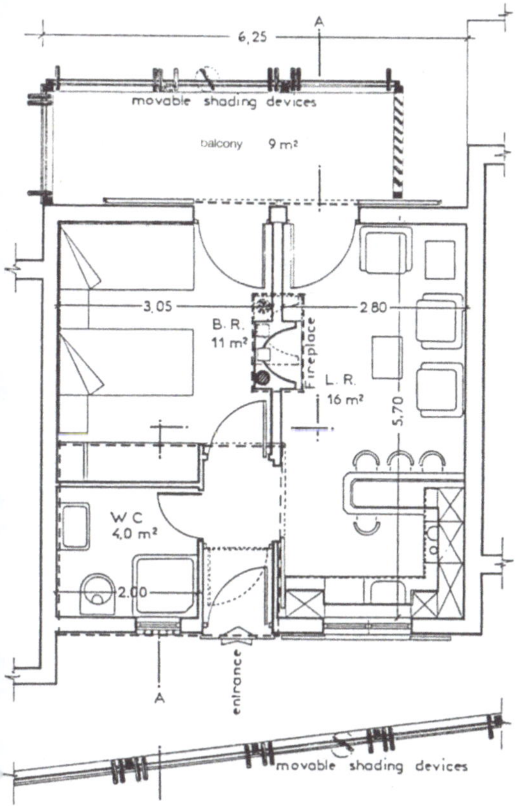

The south-east side of the basement is partly buried while the north-east side is 1.5 m above ground level. Water from a nearby well runs freely over the floor of the basement at a constant temperature of 14 ° C. Fresh air enters the basement from five openings on the north-west side. Each of these openings houses an axial fan, sized to deliver the air at an appropriate flow rate to two apartments, one on the first floor and one on the second floor of the building.

The axial fans distribute the air over the wetted surface of the basement and the cooled air is delivered to the two apartments through a diverging pipe system. A plastic cover is spread immediately above the wetted surface to minimize the movement of air throughout the basement. As the dry air flows past the wetted surface, there is a transfer of both sensible and latent heat. The air mass temperature drops and its relative humidity rises. The cooled fresh air is delivered to each apartment separately from the basement. The exhaust (warmer) air leaves each apartment via openings near the top of the fireplace. From there it is piped to a point some 3 m above the covered roof of the building. The end of the chimney consists of a 1 m^2 black metallic sheet which, when it reaches high temperatures, enables the temperature and density differences needed for operation to be created. Alternatively, an air circulator can be used to achieve proper operation of the system.

If the apartments are occupied in winter, the outlet openings are kept closed with dampers to prevent heat loss.

Shading of Roof

The roof is covered with a sloped double-layered bamboo trellis. This provides tenants with a shaded 200 m^2 area with beautiful sea views and protects the roof from the intense solar radiation experienced in Crete during the summer months.

Ventilation Chimney

The above-mentioned shading and ventilation systems, coupled with the thermal insulation and high thermal inertia, result in a building with a low cooling load. This can be met by a hybrid cooling system involving what is termed a ventilation chimney. This is described opposite and illustrated in Figures 12 and 14.

Heating System

Energy-Efficient Fireplace

To provide heating during the Christmas holidays, each apartment contains an energy-efficient fireplace (see Figure 16). This is built in the partition wall between the two main rooms and consists of metallic hollow slabs (60 x 120 mm in cross section and 4 mm thick) in contact with high thermal capacity bricks.

The metallic heat exchangers are arranged in the shape of a parabola (see Figure 15). Room air passes naturally through them, gains heat, and is conveyed to the bricks, delivering to them part of the heat gained from the heat exchangers. Part of the air returns to the room from openings 1 m above the floor (see Figure 14). In addition, the combustion gases passing through the chimney deliver some of their heat to the brick wall surrounding the chimney.

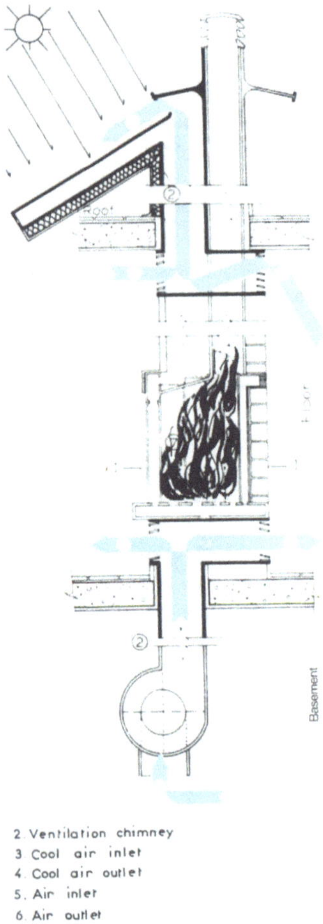

Figure 14. The ventilation chimney cooling system and the energy- efficient fireplace

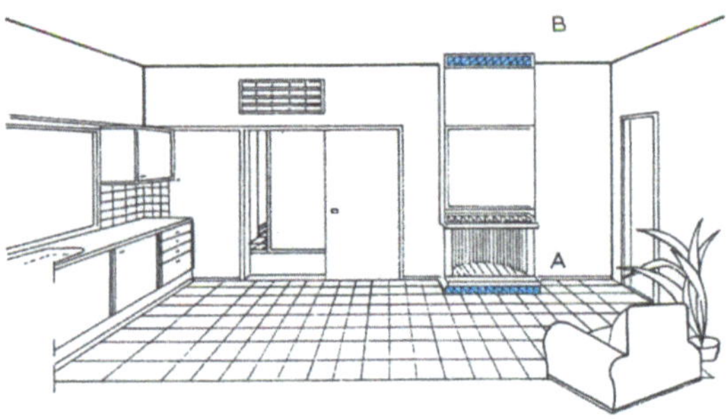

Key:- a. cool air inlet
b. warm air outlet

Figure 16. The interior of the living room showing the energy-efficient fireplace

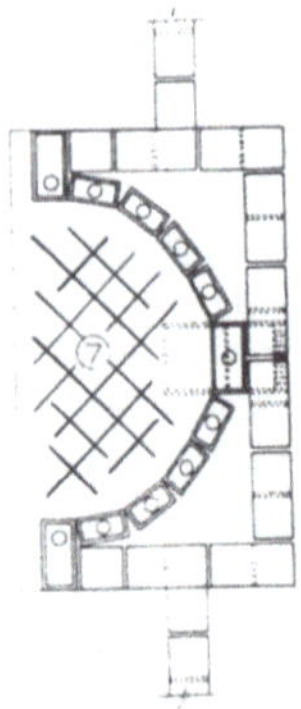

Figure 15. Section I-I of the energy-efficient fireplace

ENERGY CALCULATIONS PERFORMED AND DESIGN TOOLS USED

Introduction

The energy performance of the building was simulated on a computer using a simplified thermal network method. The ten apartments were divided into four zones according to their exposure to the weather (see Figure 18). For each zone, hour-by-hour simulation runs were carried out using mean monthly hourly values of solar radiation, ambient temperature and internal gains resulting from the occupants' activities. The internal temperature distribution was calculated with and without the effect of the passive techniques incorporated in the building and the cooling and heating loads computed whenever the temperature was above 26 ° C or below 20 ° C respectively. This enabled the efficiency of the passive measures to be evaluated under mean monthly conditions. In addition, the ventilation chimney system was sized under extreme summer conditions (36 ° C and 55% relative humidity) and full occupancy to ensure thermal comfort even under these rare conditions.

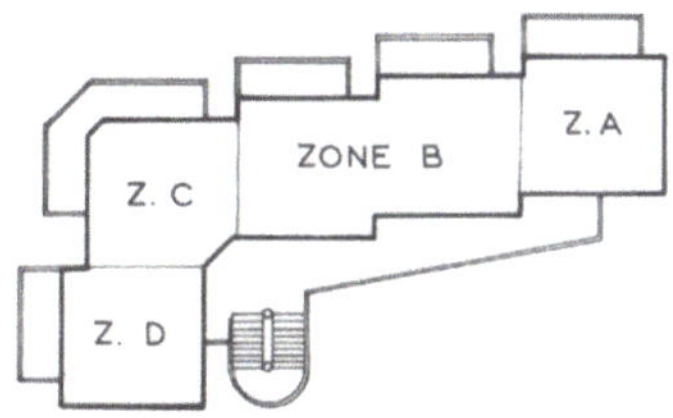

Figure 18. The four zones

In addition, the daylighting performance was assessed using the MICROLITE computer program.

Performance of Passive Solar Systems in Cooling Season

Shading Devices

With the shading devices for the roof, balcony, windows and doors in operation (see Figure 17), the total mean annual cooling load of the whole building was calculated to be 60 GJ. Without shading, the annual cooling load was 99 GJ (see Figure 19). The peak cooling load to achieve an internal temperature of 28 ° C under extreme summer conditions was in the range 2.5 kW to 3 kW per apartment, depending on the zone.

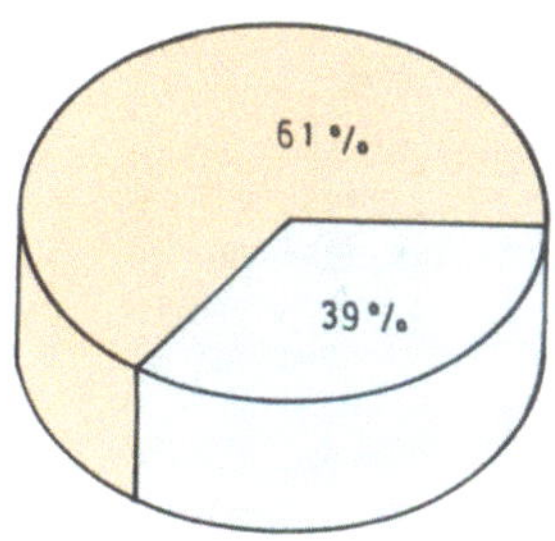

Figure 19. The shading devices reduce the cooling load by 39%

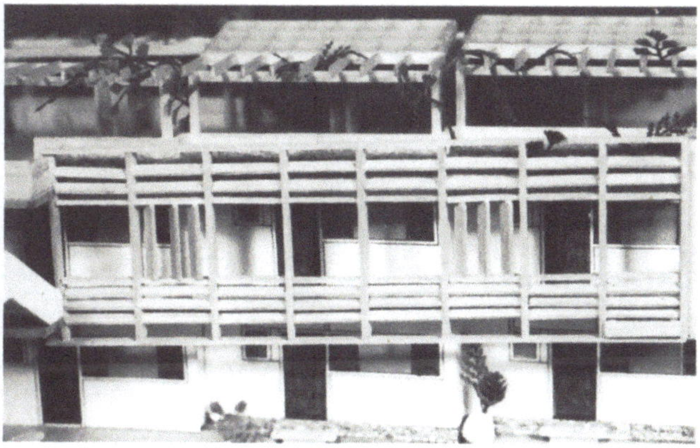

Figure 17. Movable shading devices

Cross Ventilation

The effect of cross ventilation was investigated under mean monthly conditions. Air flow rates were calculated using ASHRAE standard procedures. The average air speeds achieved in the living space were computed with a semi-empirical equation produced from Givoni's experimental data. Figures 20-23 show the variation in internal temperature with and without natural ventilation for a typical apartment in zone B. In late morning and afternoon when wind speeds are high the internal temperature nearly equals the ambient temperature. This is because the large ventilation openings give rise to high air flow rates. To reduce the maximum air speeds produced in the living space, the size of the windward and leeward openings were adjusted to make them equal. In the bedrooms, for instance, out-lets were incorporated in the loft on the leeward side of the building.

Average air speeds in the living space range from 0.14 m/s to 0.97 m/s. The higher value is encountered in the warmest months (July and August), when they are most desirable. In these months the internal temperature of the ventilated rooms reaches 29 ° C with a relative humidity of 55%. The temperature index chart (Figure 24) shows that, under these conditions, the temperature can be reduced to just below 26°C with an air speed of around 1 m/s. It can also be seen that, for dry bulb temperatures from 26 ° C to 29 ° C (the range of internal temperatures found in the ventilated space) and relative humidities varying from 55% to 60%, the desired temperature of 26 ° C can be produced with air speeds of 0.14 m/s to 1 m/s. This shows that, under mean monthly conditions, the cross ventilation system can create thermally comfortable conditions for occupants.

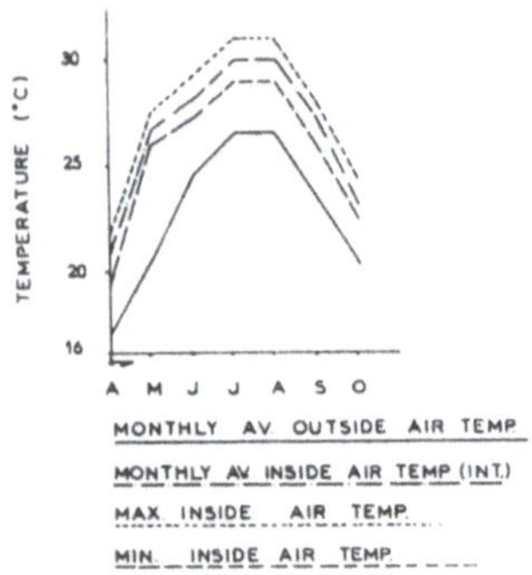

Figure 20. Variation in internal temperatures in zone B in absence of natural ventilation

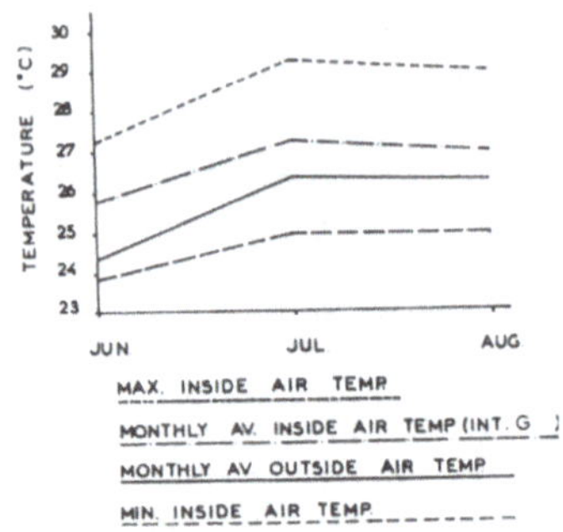

Figure 21. Effect of natural ventilation on zone B internal temperatures

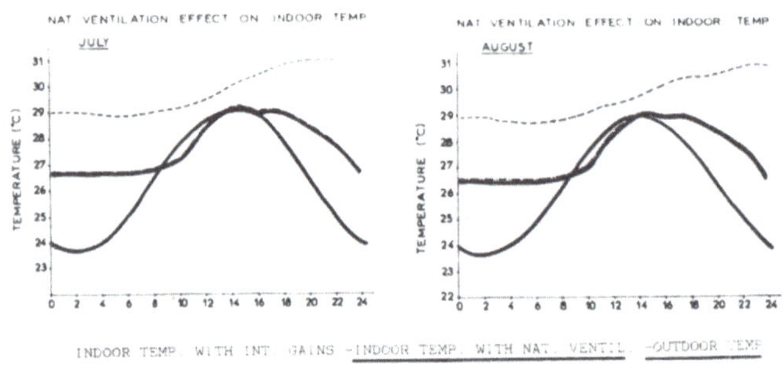

Figure 22. Effect of natural ventilation on zone B internal temperatures in July

Figure 23. Effect of natural ventilation on zone B internal temperatures in August

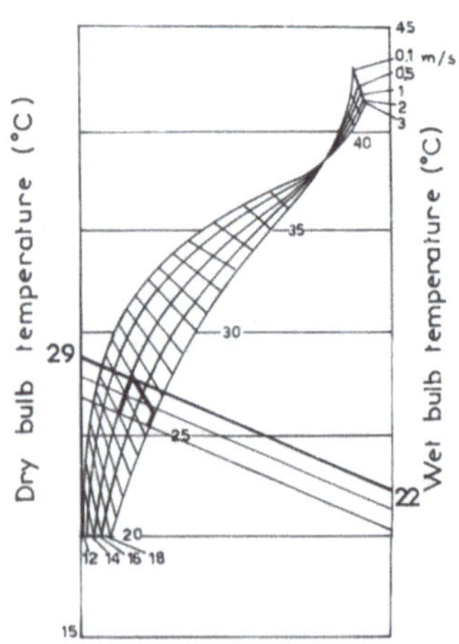

Figure 24. Temperature index chart

Ventilation Chimney System

The ventilation chimney system was designed to work when the window shutters are closed. The incoming ventilation air is passed over water from a nearby well running at a constant temperature of 14 ° C over the basement surface. The preliminary analysis showed that the unforced natural movement of the air in the basement was not enough to provide any real cooling of the apartments. After a careful study of the parameters involved, the scheme depicted in Figure 25 was developed.

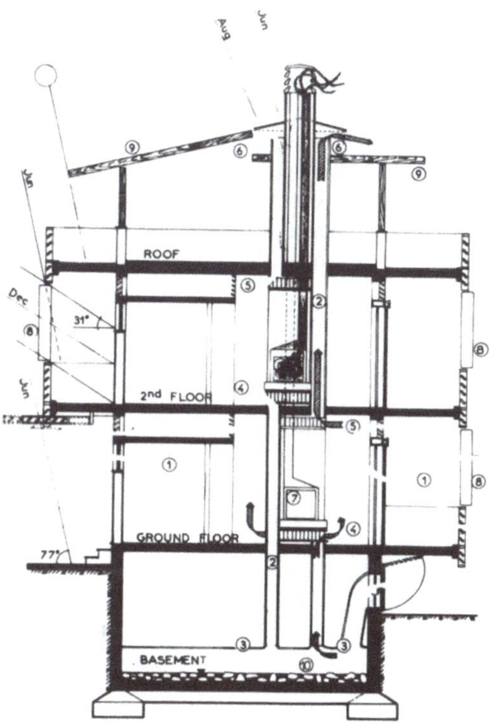

Key:

1. Natural ventilation
2. Ventilation chimney
3. Cool air inlet
4. Cool air outlet
5. Air inlet
6. Air outlet
7. Fireplace
8. Movable shading devices
9. Roof Shades
10. Cool water

Figure 25. The ventilation chimney system

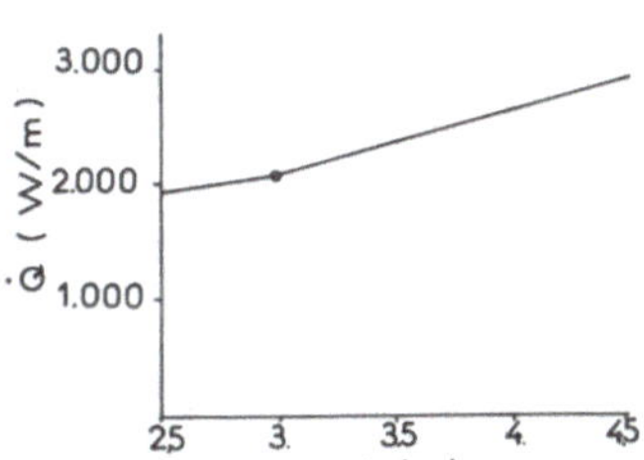

Figure 26. 3 m of surface will create required cooling load with fan velocity of 11.3 m/s

It was found that the sized axial fan is suitable for cooling each apartment when the dry air passes over a wetted covered surface 3 m long. Five openings house the fans which are connected to a flexible piping system to deliver air at the needed flow rates to a couple of apartments. The energy performance analysis showed that the maximum cooling load for each apartment, under extreme conditions, is 3 kW when the maximum temperature of the outside air is 36 ° C and the relative humidity 55%. The calculation is based on determination of the flow of air, on the local Nusselt Number, and the average convection coefficient. Figure 26 shows the length of surface to create the required cooling load with a fan velocity of 11.7 m/s. Data analysis showed that, under the above extreme external conditions, the temperature of the air entering the apartment is 25 ° C and its relative humidity 76% - values judged to be comfortable for the occupants. Further simulation analysis for a range of different external temperatures (33-36 ° C) and relative humidities (55-75%) showed that, if the initial relative humidity was above 60%, the apartment could become uncomfortable because, although the air entering the room would be at an acceptable temperature (i.e. in the range 23- 25 ° C), its relative humidity would be intolerably high (i.e. in the range 85-93%). This led to elimination of the idea of incorporating an evaporative cooling system at the loft of each apartment: the temperature and relative humidity of the air entering the room would make living conditions uncomfortable.

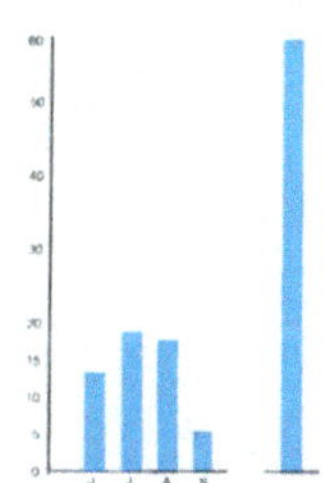

Figure 27. Mean monthly and annual cooling load

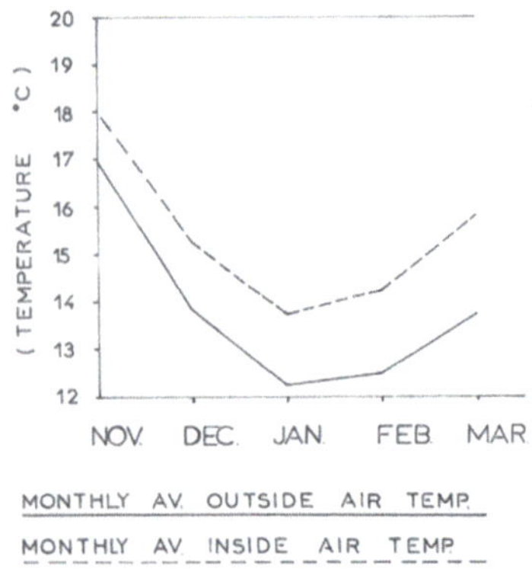

Figure 28. Variations in monthly average internal temperatures due to solar gains

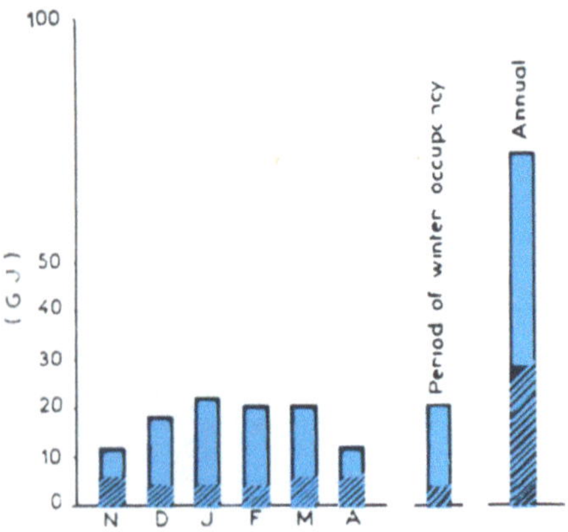

Figure 29. Mean monthly and annual heating load and solar contribution

Performance in Heating Season

The mean annual heating load of the complex was calculated to be 74.5 GJ, of which the solar contribution was found to be up to 28.5 GJ. For the only winter period when it is expected that the building will be occupied (i.e. the one month around Christmas), the load is 16 GJ with a solar contribution of 26% (see Figure 29). These calculations assumed a thermostat setting of 20 ° C. For extreme winter conditions (i.e. an ambient temperature of 3 ° C and no solar gains) the capacity of the heating system installed in each apartment would need to be in the range 2.2 kW to 2.6 kW, depending on the apartment's exposure to the weather. The heating capacity of the dynamic fireplace installed between the two rooms of each flat is 4 kW, which is more than meets this requirement.

A detailed analysis was carried out on the internal configuration of the fireplace, the damper location and the height and location of the lintel. The best chimney size was established from the system resistance coefficient and the pressure loss. The convection heat rate from the fireplace to the room when a glass firescreen is used was also estimated.

Daylighting Performance Analysis

The daylighting was assessed using the MICROLITE computer program. All four zones were examined for an overcast sky in December and a clear sky in July. The analyses showed that there are relatively few spots in the bedrooms where daylighting is poor, i.e. the daylight factor is less than 1%. Since these positions are where the head of the bed is located this could be helpful, rather than otherwise. All the other living spaces have daylight factors in the range 2-6%. Both window shutters and shading devices can be adjusted to meet occupants' individual requirements.

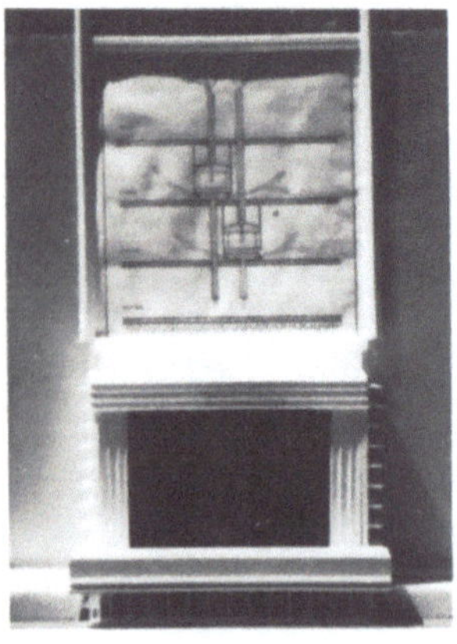

Figure 30. The energy-efficient fireplace

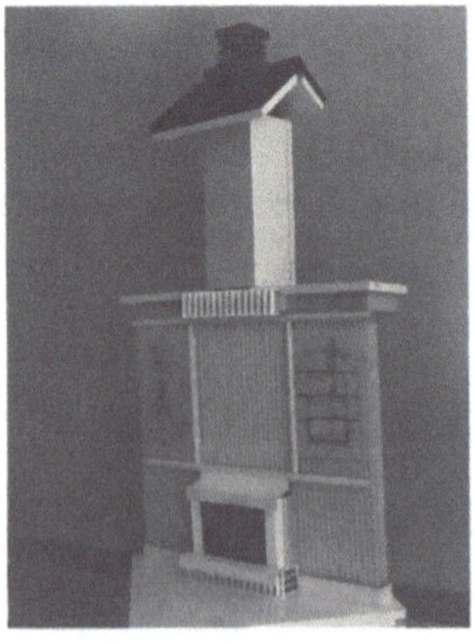

Figure 31. The energy-efficient fireplace

GENERAL DESIGN GUIDELINES/POINTS OF INTEREST RESULTING FROM THIS PROJECT

The investigation of the effect of cross ventilation on thermal comfort showed that ventilation of the two zone D apartments (whose openings are on the leeward side of the complex) is not as effective as that of the other zones. If the building were turned on the site so that it was oriented slightly more towards the north (i.e. at an angle of about 10 ° W of N) ventilation in all the zones would be improved, even though it is not necessary for the other zones.

Ventilation can also be improved by adjusting the vertical fins of the shades at the north-west parapet of the building so that they are turned due west. This allows the air streams to be deflected towards the balcony doors so that greater air flows can enter the rooms.

Positioning the fins due west also improves shading. The shading devices proved to be very effective for shading, ventilating and reducing daylight levels when desired - for instance, during a siesta.

The studies on the ventilation chimney showed that the important parameters are the sizing and technical characteristics of the fan, the length of the wetted surface and the temperature of the water running through the basement.

The fireplaces proved to be simple to construct and operate, appropriate for the heating requirements of the flats and energy-efficient.

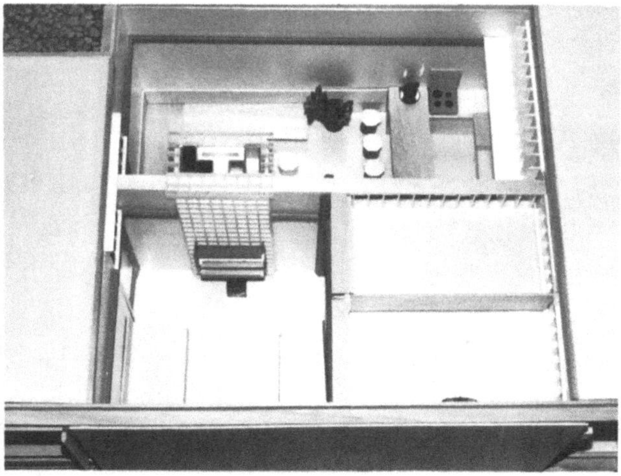

Figure 32. Interior view of model of apartment

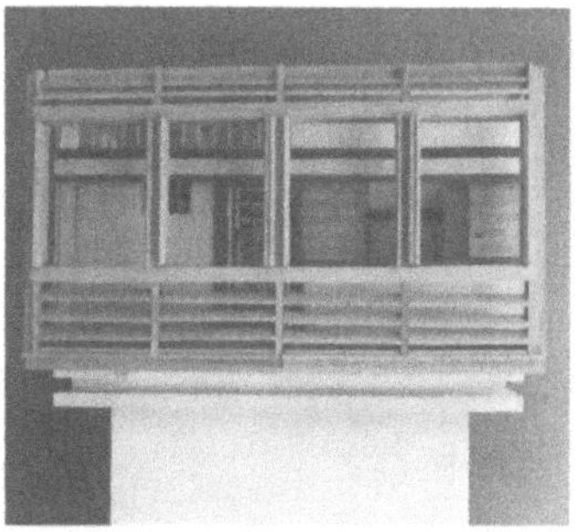

Figure 33. The movable shading devices

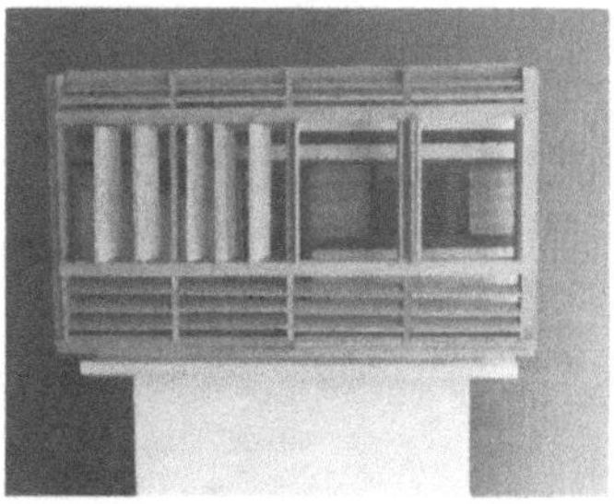

Figure 34. The movable shading devices

BUILDING 2000 Building 2000 brochures are published by Directorate General XII of the Commission of the European Communities to show how design studies can help architects and other building designers use passive solar principles to the best effect to produce attractive energy-efficient buildings. Each brochure describes studies carried out with the support of the Commission during the design phase of one of thirty-six non-domestic buildings in the EC Member States. The studies were on such topics as daylighting, heating, cooling, ventilation, comfort, control systems and urban design. They were carried out with the help of acknowledged European experts in these fields and drew heavily on lessons learned and techniques developed through the Commision's research and development programme on solar energy applications to buildings.

Commission of the European Communities/Directorate-General for Science, Research and Development

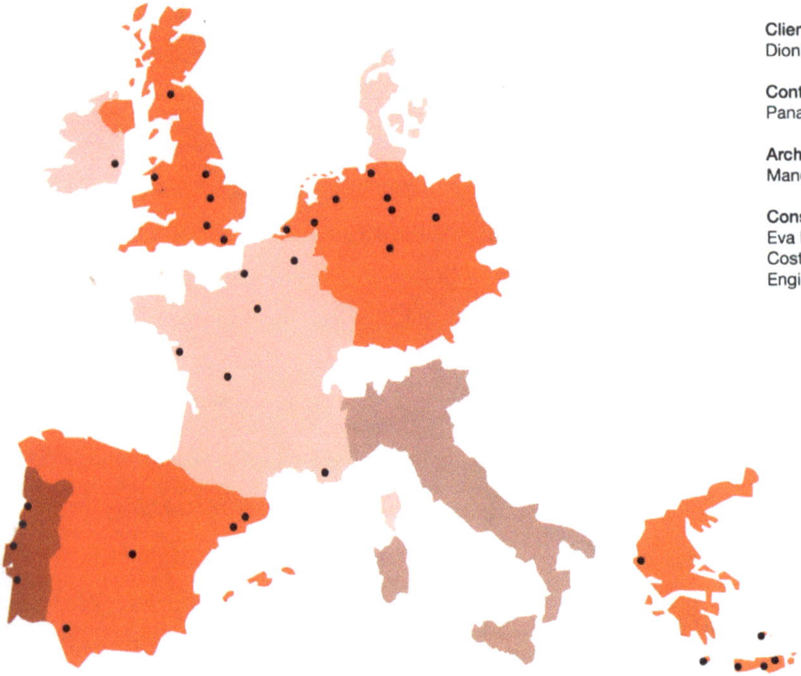

List of Design team Participants and Advisers

Client
Dionisios & Dionisia Kladis

Contractor
Panayiotis Kladis

Architect
Manolis N Kopidakis, Civil Engineer

Consultants
Eva K Athanassakos, Physicist
Costas G Theofylactos, Mechanical Engineer

This set of **Building 2000** brochures illustrates how architects and other building designers can successfully apply passive solar principles to produce energy-efficient buildings.

BUILDING 2000 Participants

Project Director
Theo C. Steemers

Coordinator
Cees den Ouden

Technical Steering Committee
Dean Hawkes
Nick Baker
Alex Lohr
Jean P. Lepoivre

Regional Liaison Agents
(D) Jörn Behnsen
(E) Vicente Sifre
(F) Michel Raoust
(GB) Alan Hildon
(GR) Matheus Santamouris
(P) Eduardo Maldonado

Further information or copies of the brochures can be obtained from prof. ir. Cees den Ouden, EGM Engineering BV, P.O. Box 1042, 3300 BA Dordrecht, The Netherlands.

BUILDING 2000

Commission of the European Communities

Fully-furnished apartments suitable for short- and long-term stays throughout the year

Summer-time comfort achieved by natural ventilation, shading, earth cooling, evaporative cooling, mechanical ventilation using air cooled in underground tubes, and radiant cooling

Winter-time comfort is improved by high thermal mass, south-facing orientation, direct solar gains and solar chimneys

Solar systems contribute 71% of annual heating demand

Good daylighting in all buildings

TOURIST COMPLEX

CHANIA/CRETE/GREECE

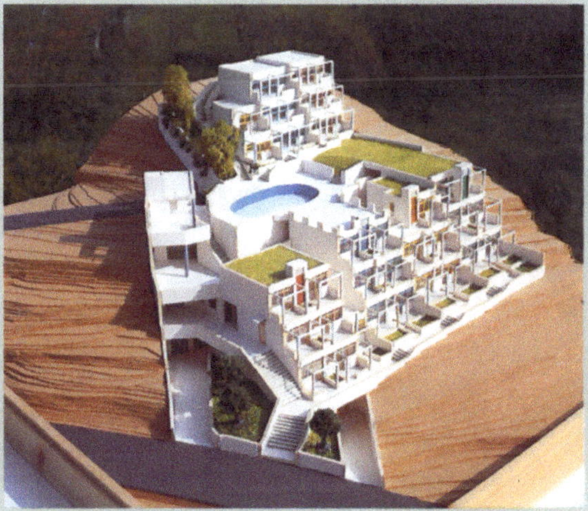

Building 2000 is a series of design studies illustrating passive solar architecture in buildings in the European Community.

ISSUE H 5

FEB 1991

PROJECT DESCRIPTION

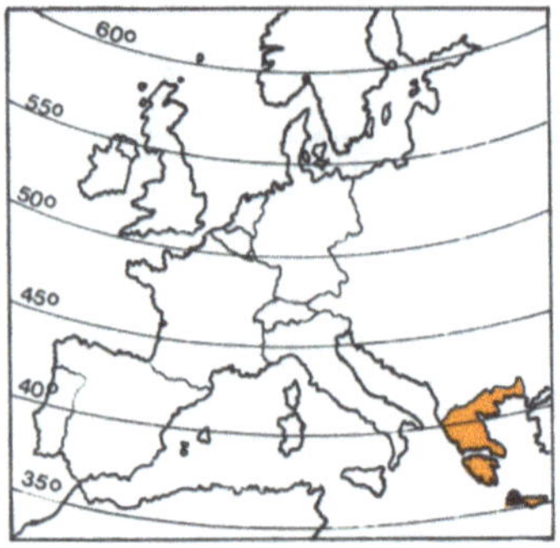

Figure 1. Location of Crete

Building Type

This project concerns a tourist complex open all the year round. It consists of three buildings containing 16 single-room studios, 7 two-room apartments, 4 split-level apartments, communal facilities and services. The objective has been to provide a range of 2-4 person accommodation for those wishing to stay in the area for a few days or much longer, at any time of the year. Because the complex is near a main town, it is expected to attract all kinds of visitors, not just holiday-makers. Special attention, therefore, has been paid to making the apartments comfortable throughout the year. The communal facilities have been designed to encourage social contact between visitors.

Location

The complex is in a rural area 4 km from the town of Chania on the north coast of Crete (latitude 35° 30' N) - see Figure 1. The 1738 m² site has a 22% slope to the south and is on the steep side of a small hill not much above sea level. From the site there is an unobstructed view to the south. The site is not overshadowed. The site plan is shown in Figure 3.

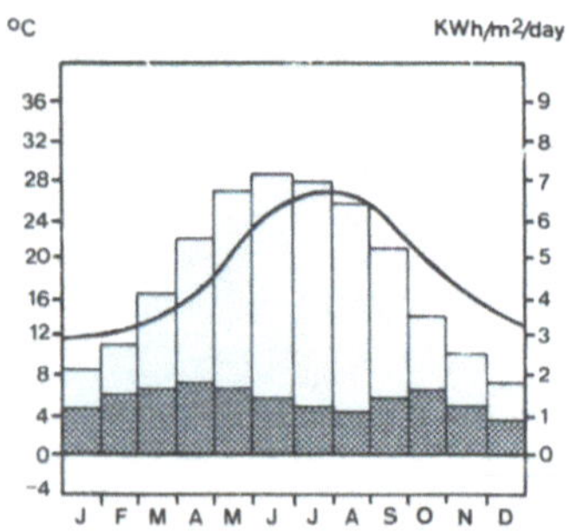

Key: diffuse irradiation
direct irradiation
daily external temperature

Figure 2. Mean solar irradiation and temperature data

Site Microclimate

The climate is typically Mediterranean. Average solar irradiation and temperature data are given in Figure 2. Additional climate data are given in Table 1.

Average ambient temperature	
Jan	11.9 ° C
Jul	26.9 ° C
Degree days	
(base 18 ° C)	681
Sunshine hours	
per year	2809
Global irradiation per year	
in kWh/m²	1630
(in MJ/m²	5868)

Table 1. Some climate data

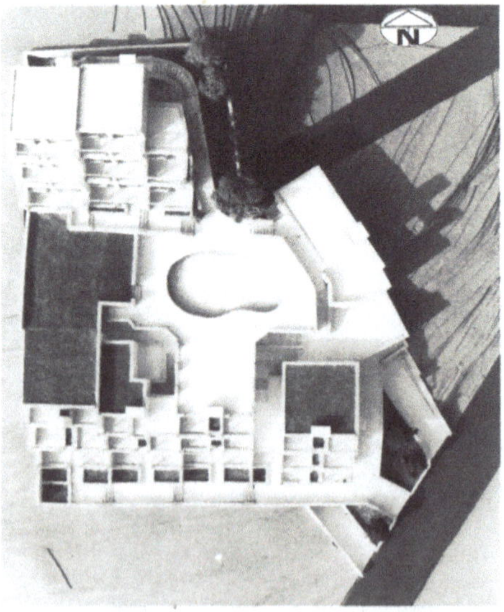

Figure 3. Site plan

Design and Construction Details

The complex consists of three buildings with a total floor area of 1005 m^2 and a volume of 2920 m^3, built along the slope of the hill (see Figure 4). By taking advantage of the site slope, it has been possible to put sections of the buildings underground, which makes them both thermally comfortable and less noticeable. There are three entrances - one from the parking lot and two for pedestrians. The buildings face south and are built on the slope in such a way that none obstructs the view of the others. The reception, bar, breakfast room and other communal facilities are located in the centre of the eastern side of the site. The swimming pool is in the central open air plaza. Movement of pedestrians horizontally across the complex is via a series of open and covered galleries; vertical movement is achieved by staircases and (in the case of the communal facilities building) a lift. South-facing apertures in all the apartments and communal areas enable occupants to see the view and provide enough direct and diffuse natural light to reduce the need for artificial lighting. Because there are a limited number of open air spaces in the complex, part of the roof of one of the buildings is used as an extension of the central courtyard. Privacy is achieved by placing much of the construction underground and arranging the various levels one above the other in stepped terraces.

The buildings have a massive construction with frames of in situ reinforced concrete. The external and internal walls are of hollow 190 mm bricks. The external walls and roofs are insulated externally; the floors have perimeter insulation. All the insulating material is extruded polystyrene - except for the movable roof panels, which are made of polyurethane/aluminium sandwich. Hollow bricks (60 mm thick) protect the insulation on the walls; the roofs are covered with soil or tiles, whichever is most suitable. The U-value of the roof is 0.39 W/m^2 K and that of the external walls 0.48 W/m^2 K. The doors and window frames are made of wood.

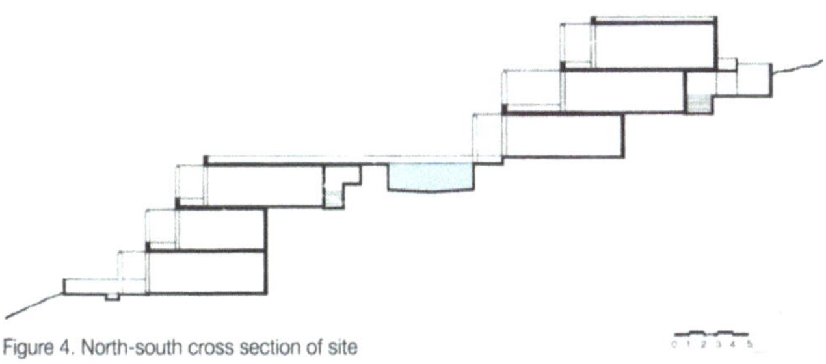

Figure 4. North-south cross section of site

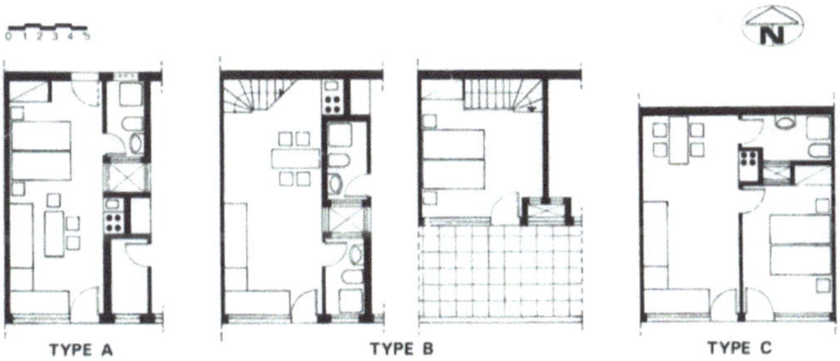

Figure 5. Typical apartment plans

DESCRIPTION OF PASSIVE SOLAR FEATURES/COMPONENTS

Introduction

Passive solar systems have been used to create comfortable living conditions inside and out and to minimize consumption of fossil fuels. The systems were selected for their ease of operation and maintenance and low capital cost. Special attention was paid to achieving effective cooling in the peak tourist season. Light colours are used on external surfaces. The slope and orientation of the site led to the idea of placing sections of the buildings underground and achieving good daylighting. The microclimate round the buildings is improved by evaporative cooling and shading. The buildings' ventilation air is cooled by passage through an underground pipe. Solar chimneys have been incorporated into most of the apartments to facilitate circulation of cool and warm air, as required. Movable roof panels help keep the buildings cool in summer. Domestic hot water is supplied by a solar water heating system.

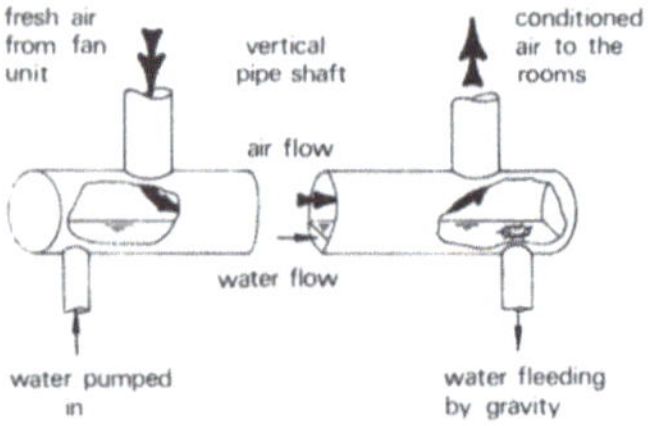

Figure 6. The earth-tube cooling system

Passive Solar Systems

Direct Solar Gain and Movable Shades

All apartments face south. Light-coloured venetian blinds are fitted to the inside of all south-facing doors and windows. Curtains on all the glazed surfaces provide extra thermal insulation from the inside. Retractable horizontal awnings are installed on the outside for use when there is a need for shading.

The Earth-Tube Cooling System

Incoming fresh air for the buildings is cooled by passage through a horizontal concrete pipe set into the ground 6-8 m below the surface (see Figure 6). Water runs freely through the tube up to a level of a third of the radius and, on leaving, is collected by an overflow pipe. The air is cooled by convection and evaporation and passes to the apartments free of the odours associated with many ground cooling systems.

Solar Chimneys

The solar chimneys incorporated into most of the apartments have a rectangular cross-section and are formed by painting the south-facing wall of the apartment a dark colour and covering it with glass. The chimney is connected to the apartment by ducts and manually-operated adjustable dampers. In summer, the damper at the top of the chimney is fully open to enable hot air to leave the apartment. In winter, this damper is shut off and two other dampers opened. One, at the bottom of the external wall, enables external fresh air to enter the chimney to be preheated. The warmed air enters the apartment by the other damper, which is at the top of the internal wall.

Radiant Cooling Using Roof Panels

Rows of movable insulated panels are placed on some of the roofs. In summer, these protect the roof surface during hours of sunshine; at night they are moved to an upright position so that heat can be lost from the warm roof to the cooler sky. The panels are made of polyurethane sandwiched between aluminium sheet and can be rotated on a horizontal axis.

Auxiliary Systems

Heating

Auxiliary space and domestic hot water heating is provided by a gas-oil fired boiler and radiator system. The space heating is controlled by a room thermostat set at 21 °C. Each room thermostat controls the radiator's electromechanical valve.

Mechanical Ventilation

A mechanical ventilation system is installed capable of providing the rooms with up to 3 air changes per hour for summer cooling.

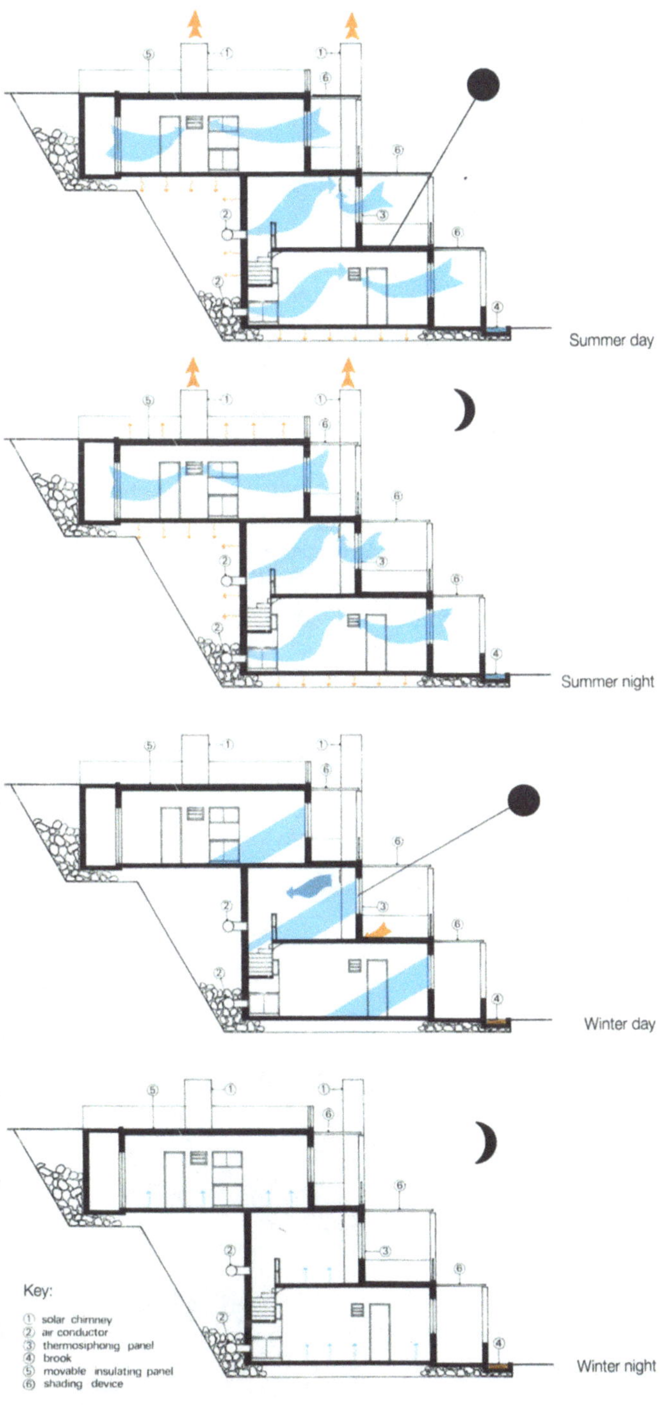

Figure 7. Operating modes on summer day, summer night, winter day and winter night

Operating Modes

Summer

In the daytime, incoming fresh air is cooled by passage through the underground ducts, which are cool because of the water running through them. Circulation within the buildings is achieved by the solar chimneys using fans and adjustable dampers. Direct solar radiation and a large part of the diffuse radiation is deflected away from the building by shading devices and planting in the surrounding area. The swimming pool and streams of water running in front of the apartments cause evaporative cooling. At night, cross ventilation is achieved by proper positioning of the north and south openings. The underground ducts supply the buildings with cool fresh air. The thermal energy which has accumulated in some of the roofs is released to the night sky by removal of part of the roof insulation; this allows coolth to be collected for the next day.

Winter

In the daytime, large south-facing windows allow solar radiation to enter each apartment for absorption by the building's mass. The solar chimneys operate like thermosyphonic collectors, replenishing some apartments with warm air. At night, the heat collected in the walls, floor, etc., warms the apartments. Dark curtains reduce loss of heat to the outside.

ENERGY CALCULATIONS PERFORMED AND DESIGN TOOLS USED

Introduction

Because of their extensive south-facing glazing the apartments would, unless protective measures were taken, be unbearably hot in summer and uncomfortable even in October and April. Particular care, therefore, was paid to the incorporation of adequate shading, cooling and ventilating measures into the complex. Because the site faces south and is protected from the sea by the wake of the hill, the buildings will not experience cool sea breezes. Quite often, therefore, natural ventilation systems will not be enough. At such times, a mechanical air blowing system takes over. All the systems were designed with the help of energy calculations.

A simulation model, TCM-Heat, was used to determine the thermal performance of the buildings in winter. This is a simplified model suitable for direct gain systems. It creates its own output file from the weather and building data supplied. The building file contains a number of zones plus a full description of all the building elements making up the zone. The program is limited to handling a maximum of nine zones.

The daylighting was assessed using the MICROLITE program version 1.0.

Calculations Relating to Design of Cooling Features

Window Openings

For good natural ventilation, it was found that 15% of the total window area should be used for incoming fresh air; such openings should be located at the bottom. 20% of the total window openings should be used for removal of warm room air from the buildings; these openings should be located high on the wall opposite, to create cross ventilation.

Exterior Shades

The calculations showed that retractable horizontal overhangs extending 1.8 m over the doors and windows would have a shading factor of 19%. Extra shading would be provided to some of the south glazing by the protruding west wall. This would protect the building from the annoying afternoon sun.

Internal Shades

The light-coloured internal venetian blinds fitted on all south-facing glazing would give a shading factor of 50%. They would also improve the daylighting factor by diffusing sunlight onto the ceiling and walls.

Evaporative Cooling

Because of the nearby sea, relative humidities at the site are average to high. Therefore, evaporative cooling alone is not as effective a cooling device as would be liked. However, during extremely hot days when the dry bulb temperature is above 35 °C the relative humidity tends to drop 8-12 points below the monthly average. On such occasions, evaporative cooling will come into its own.

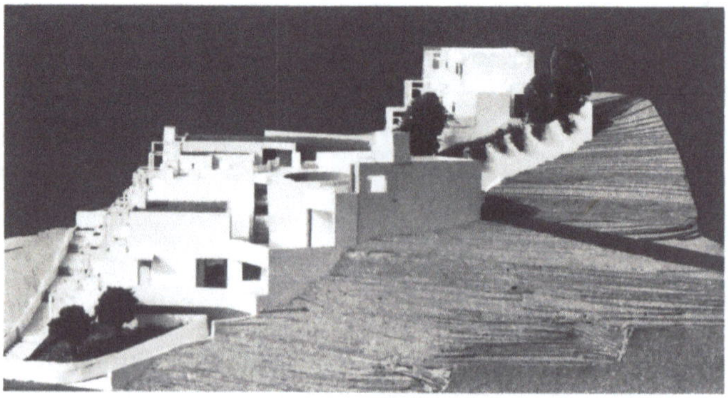

Figure 8. East elevation of model

Earth-Tube Convection-Evaporation Cooling System

When evaporative cooling alone is not effective, a mechanical system is necessary. In the system used here, air is cooled by evaporation and convection when forced over the surface of running water in tubes set nearly horizontally (actually it has a 0.3% negative slope) into the earth some 6-8 m below the surface. A centrifugal fan is used to pump fresh air down through a vertical pipe to the earth tube. After passing along the length of the tube the conditioned air enters another vertical pipe shaft which brings it to the air distribution system. At the same time a water pump drives water from a little stream near the building at the top end of the earth tube. The water flows by gravity down the tube and is collected by an overflow pipe at the other end and returned to the stream, cooler than when it started. Calculations showed that the water level should be around a third of the tube's radius.

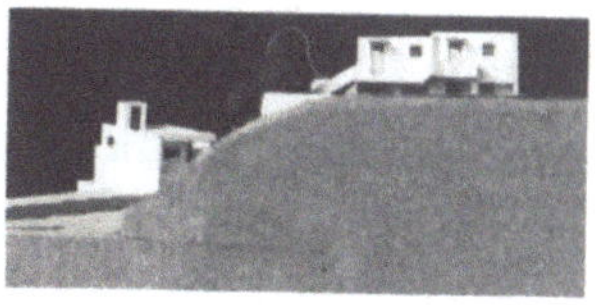

Figure 9. North elevation of model

To establish the diameter and length of the earth tube which would give the desirable conditions for the cool air entering the apartments, step-by-step calculations were carried out. First, the diameter of the tube and vertical pipe shaft was determined which would give an air speed of 10 m/s - and hence 3 air changes per hour in the apartments. Second, the length of tube was calculated which would give exit air a relative humidity of 75% due to evaporation. Any heat losses due to convection were ignored. Finally, the properties of the air entering the apartments were determined using this tube length and the relationship between heat and mass transfer which would give turbulent flow of fluids in pipes.

Table 2 gives estimated values of the temperatures and relative humidities of the conditioned air produced by the earth-tube system under different ambient conditions. The effect of the high relative humidities is modified by increasing the air movement inside the rooms mechanically to give 3 air changes per hour.

Ambient conditions		Delivered air	
temperature °C	relative humidity %	temperature °C	relative humidity %
36	48	27	80
34	50	26	81
32	55	25	83
30	57	24	85
28	60	23	86

Table 2. Estimated temperature and relative humidity of air produced by earth-tube system under different ambient conditions

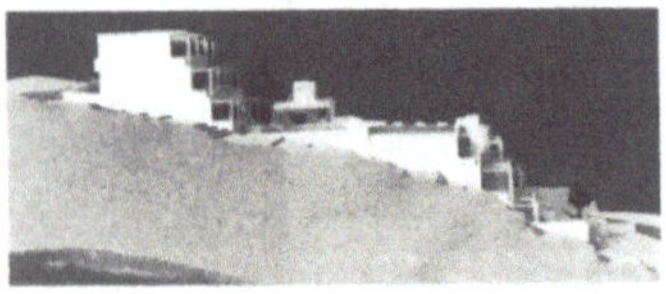

Figure 10. West elevation of model

Radiant Cooling Using Roof Panels

Calculations showed that, during a typical summer night, radiant cooling will occur at rates ranging from 20 W/m^2 to 40 W/m^2, depending on the vapour content of the atmosphere and the cloud cover. Thus the 400 m^2 or so of exposed roof surface would lose (400 x 9 x 30) - 1000 = 108 kWh/night during a typical June night, assuming an average rate of net total losses of 30 W/m^2. If the average thermal capacity of the roof and internal walling is 140 Wh/m^2 ° C, then the resulting temperature drop after overnight cooling will be 108 ÷ (0.14 x 400) = 1.9 ° C.

Solar Chimneys

The solar chimneys servicing all the apartments have a rectangular cross section with faces made of concrete and the long axis along the south facade. The south-facing wall is painted black and covered by glass. To prevent back-flow of air in summer, care had to be taken to design a duct of the correct width. The width was determined by first calculating the thickness of the boundary layer of the vertical flow of air formed on the inside wall of the duct. The inside width of the duct should be twice the thickness of this boundary layer. The amount of time during which the chimney operates in summer can be estimated from the relationship of the heat capacity of the chimney's mass and the heat carried away by the air flow inside the chimney. Air movement will continue while the air density gradient keeps on feeding the chimney's inlet port with room air. Air dampers at the top and the inlet port control the movement of air and prevent the chimney operating when it is not needed. It was estimated that each chimney can move air upwards with a speed of 0.1 m/s, thus removing a total of 108 m^3/h. The air removed from each flat is therefore 54 m^3/h.

Daylighting Performance

All the units have rooms with south-facing windows and glazed doors. The room areas are either 4.7 m^2 or 5.0 m^2. In addition, some units have a bathroom oriented north or south and/or an extra glass door facing due north or east.

Simulations were carried out using the program MICROLITE version 1.0 in order to estimate the daylight factors 1 m above the floor inside three representative rooms at 12 noon on 21 December. The results showed that all the rooms had daylight factors throughout which were greater than or equal to values considered good for a hotel room.

Thermal Performance in Heating Season Under Average Monthly Conditions

Monthly Temperatures, Heating Loads and Solar Contribution

The mean monthly temperatures, heating loads and contributions from solar gain were determined using the simulation model, assuming that ventilation occurred at a rate of 2 air changes per hour, internal gains amounted to 100 W/unit, the ambient temperature was the average for each month and the room thermostats were set at 21 ° C. The results are given in Figures 13 and 14.

Annual Heating Demand

A breakdown of the annual heating demand under the same conditions is given in Figure 15.

Thermal Characteristics Under Extreme Weather Conditions

Evaluations were also carried out to see how the complex performed under the most extreme weather conditions found in Chania. Two cases were taken. The first was for a cold cloudy day with no direct radiation, ambient temperatures 3 ° C below the monthly average and ventilation rates of 2 air changes per hour. The results are given in Figure 11. The second case was for a warm and sunny day with a clear sky on the 21st of each month, temperatures 2 ° C above the monthly average and 2 air changes per hour. The results are given in Figure 12. On warm and sunny October days overheating will occur unless shading and/or ventilation is provided in the daytime.

Energy Use for Artificial Lighting

The energy consumed for artificial lighting was estimated from the number of hours between late afternoon and early morning for each month, aggregated over the whole year. The energy used in the living quarters in the low and middle seasons was assumed to be a proportion of that required in the high season. The calculations predict that the total amount of energy used for lighting is 15700 kWh/year.

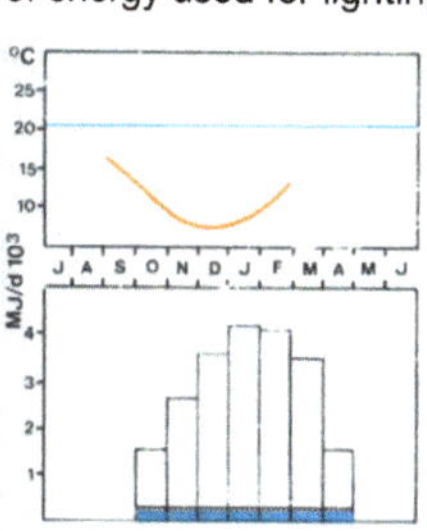

Figure 11. Average thermal characteristics of building on cold, very cloudy days

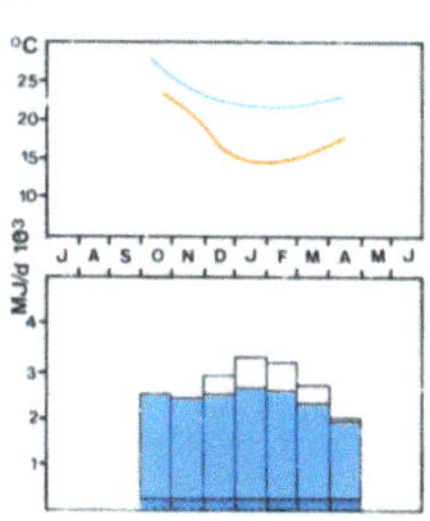

Figure 12. Average thermal characteristics of building on warm and sunny days

Key to Figures 11 and 12:

internal temperature
external temperature
solar gains
internal (casual) gains
auxiliary heating demand

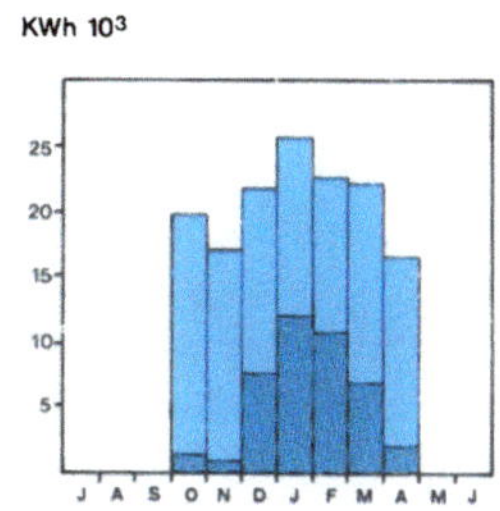

Key:
solar gains
auxiliary heating

Figure 13. Net monthly space heating energy use

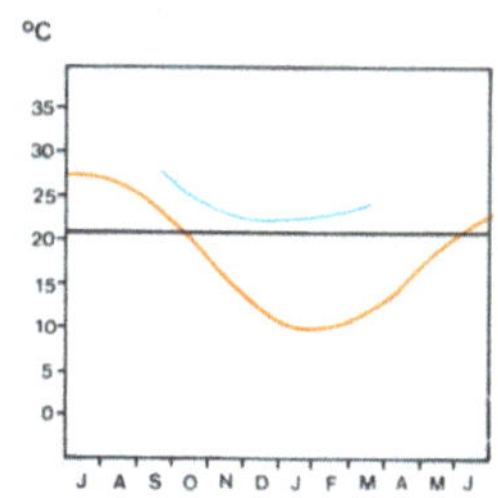

Key:
internal design temperature
internal temperature
external temperature

Figure 14. Mean monthly internal temperatures

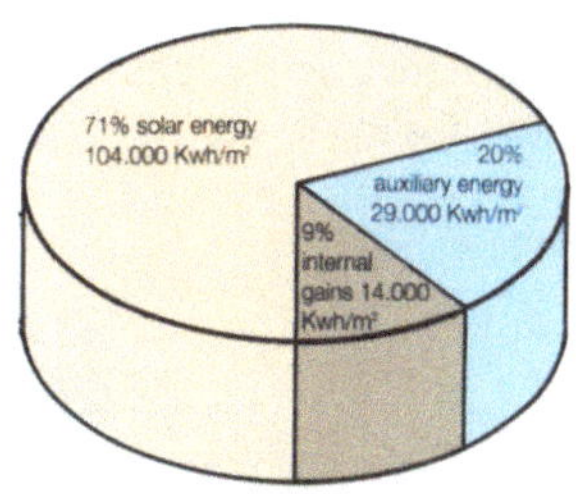

Figure 15. Breakdown of annual heating demand

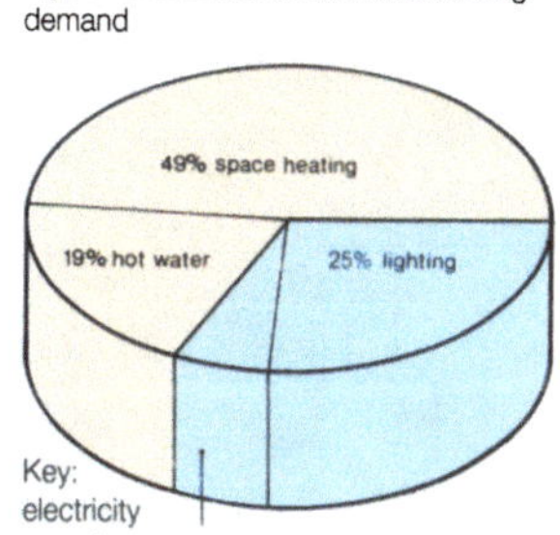

Key:
electricity
gas-oil

Figure 16. Breakdown of auxiliary fuel use

Cost-Effectiveness

It is difficult to compare the total energy consumption of this building, with its passive solar and other energy saving features, with that of a building of conventional design. There are three reasons for this. First, the occupancy of the building in winter cannot be predicted accurately. Second, Building Regulations do not specify the heating standards to be achieved in different types of buildings. Third, it is difficult to compare the results achieved by a passive cooling system with those from a conventional air conditioning unit. It can, however, be said that the energy savings from passive cooling in this complex were calculated to be roughly equivalent to those achieved by passive heating. The building construction cost should be no higher than that of a traditional building because of the absence of an air conditioning installation and the need for a smaller capacity central heating system.

Thermal Comfort Analysis

In order to find out the likely views of occupants on thermal comfort, the Predicted Mean Vote (PMV) and Predicted Percentage Dissatisfied (PPD) were determined using the Bruël & Kjaer demonstration program of the comfort equation. This expresses PMV and PPD in terms of room air temperature, mean radiant temperature, relative air speeds, water vapour pressure (or relative humidity), the thermal insulation provided by clothing, the occupants' metabolic rate (determined by his activity level) and external work (if there is any). The results under monthly average conditions during the heating season are shown in Table 3. They suggest that quite satisfactory thermal comfort conditions would be achieved under these circumstances.

	Oct	Nov	Dec	Jan	Feb	Mar	Apr
Air temp. (° C)	27.3	23.6	21.5	21.3	21.4	21.6	22.8
Mean radiant temp. (° C)	24.8	21.3	18.9	18.2	18.4	19.0	20.6
Relative air speed							
Water vapour pressure (Pa)	1610	1277	1117	971	984	1104	1224
Clothing (clo)	0.5	1.0	1.0	1.0	1.0	1.0	1.0
Metabolic rate (met)	1.2	1.2	1.2	1.2	1.2	1.2	1.2
External work (met)	0	0	0	0	0	0	0
PMV	0.2	0.09	-0.32	-0.44	-0.41	-0.30	-0.08
PPD (%)	6	5	7	9	9	7	5

Table 3. Average thermal comfort conditions inside the buildings during the heating season

The same program was also run for an extremely hot July day and a very cold and cloudy January day. It gave a PMV of +0.39 and a PPD of 8% for a hot July day and a PMV of -0.84 and a PPD of 20% for a cold and cloudy January day. Thus, even under extreme conditions the PMV is between -1.0 and +1.0.

GENERAL DESIGN GUIDELINES/POINTS OF INTEREST RESULTING FROM THIS PROJECT

General Findings from the Project

This study has shown that, for passive solar buildings in a Mediterranean environment on sites like this one:
- thermal inertia plays an important role in smoothing out room temperature variations;
- retractable horizontal shading overhangs prove their worth in early autumn and late spring days;
- south-facing vertical windows provide an easy and aesthetically acceptable way of collecting solar radiation;
- in hot and humid summers, day and night ventilation is needed to keep the building comfortable;
- in the summer natural air cross ventilation using windows is satisfactory only in places which experience strong north winds. In other areas, a mechanical ventilation system is necessary.
- a mechanical ventilation system using earth tubes is capable of providing all the cooling needed by the building;
- radiation cooling is almost negligible in areas where relative humidities are greater than 55-60%;
- where possible, buildings should be set in the earth, partly underground;
- good thermal insulation is essential for reducing heating loads in winter and only slightly affects the summertime natural cooling of a heavy mass construction.

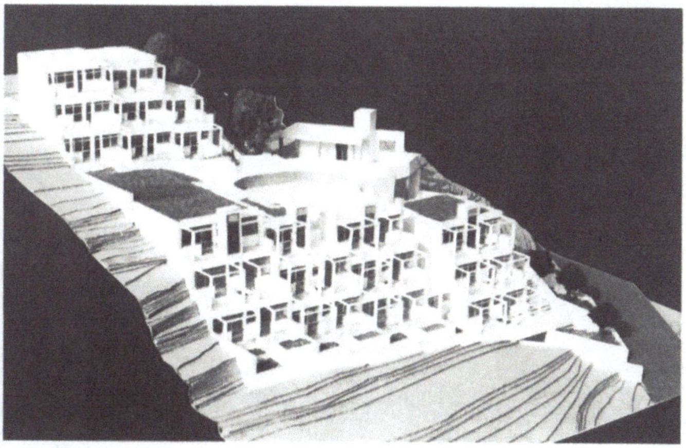

Figure 17. South elevation of model

General Comments on Avoiding Pitfalls in Passive Solar Design

Passive solar building construction is an ancient technique updated with modern ideas on design and materials. For its effective use, problems like overheating, large day-night temperature variations and the need for personal involvement in operating some of the passive systems have to be considered and, where possible, avoided.

Overheating can be minimized by use of a heavy mass construction and provision of satisfactory external shading devices. Light coloured walls and roofs should be used where possible. A feeling of discomfort can be reduced or removed by greater-than-usual air ventilation speeds.

Large variations in room temperature can be avoided by use of a heavy mass construction.

Occupants' involvement in operation of the passive solar features (curtains, overhang, window opening, etc.) should be kept to a minimum. Any action which has to be taken by occupants should be very simple and similar to actions required in conventional buildings.

BUILDING 2000 Building 2000 brochures are published by Directorate General XII of the Commission of the European Communities to show how design studies can help architects and other building designers use passive solar principles to the best effect to produce attractive energy-efficient buildings. Each brochure describes studies carried out with the support of the Commission during the design phase of one of thirty-six non-domestic buildings in the EC Member States. The studies were on such topics as daylighting, heating, cooling, ventilation, comfort, control systems and urban design. They were carried out with the help of acknowledged European experts in these fields and drew heavily on lessons learned and techniques developed through the Commision's research and development programme on solar energy applications to buildings.

Commission of the European Communities/Directorate-General for Science, Research and Development

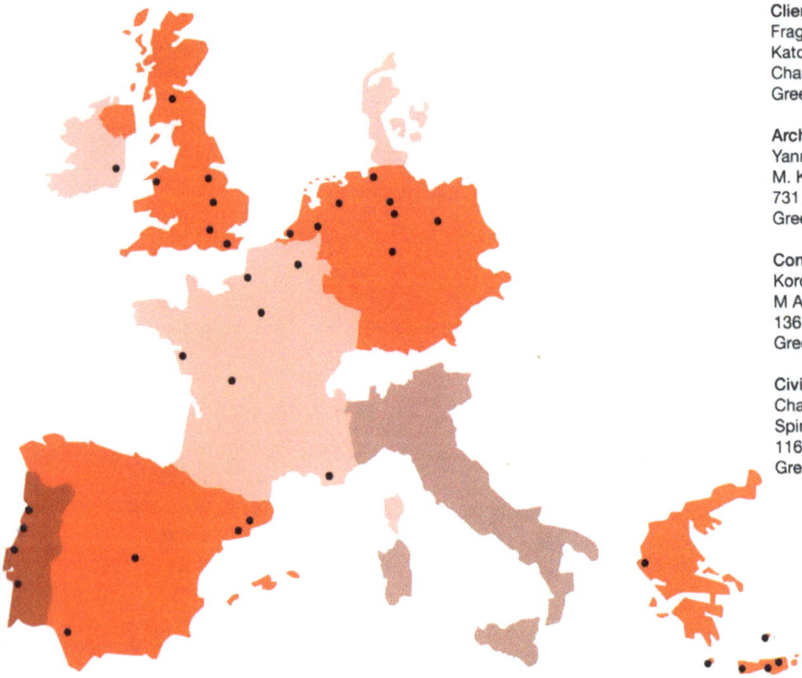

List of Design Team Participants and Advisers

Client
Fraghiadakis Christos and Georgios
Kato Daratso
Chania
Greece

Architect
Yannis Kalligeris
M. Koundouros 22
731 33 Chania
Greece

Consultants
Koronakis Periklis
M Alexaxndrou 77
136 71 Thrakomakedones
Greece

Civil Engineer
Chazidakis Aris
Spirou Merkouri 10
116 34 Athina
Greece

This set of Building 2000 brochures illustrates how architects and other building designers can successfully apply passive solar principles to produce energy-efficient buildings.

BUILDING 2000 Participants

Project Director
Theo C. Steemers

Coordinator
Cees den Ouden

Technical Steering Committee
Dean Hawkes
Nick Baker
Alex Lohr
Jean P. Lepoivre

Regional Liaison Agents
(D) Jörn Behnsen
(E) Vicente Sifre
(F) Michel Raoust
(GB) Alan Hildon
(GR) Matheus Santamouris
(P) Eduardo Maldonado

Further information or copies of the brochures can be obtained from prof. ir. Cees den Ouden, EGM Engineering BV, P.O. Box 1042, 3300 BA Dordrecht, The Netherlands.

BUILDING 2000

Commission of the European Communities

Tourist complex containing 17 furnished apartments for use during April to October

Shading, cross ventilation, ventilation chimneys, radiant barriers on roofs, high thermal mass construction plus careful planting and site layout produce comfort conditions in summer

Good insulation and thermal mass enables heating requirements during occupancy period to be met by direct solar gains through windows

TOURIST COMPLEX

DRIOS/PAROS/GREECE

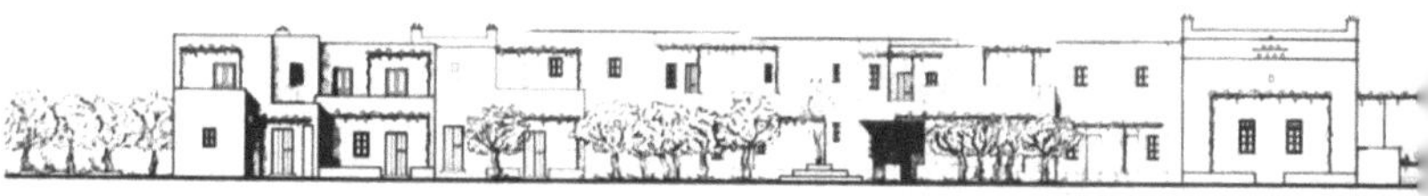

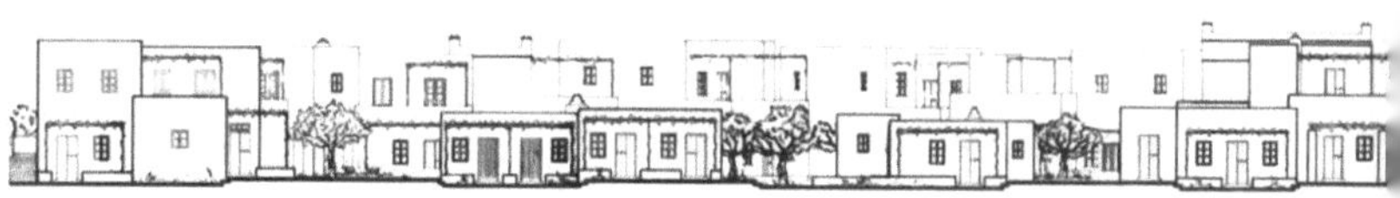

Building 2000 is a series of design studies illustrating passive solar architecture in buildings in the European Community.

ISSUE

H 6

FEB 1991

PROJECT DESCRIPTION

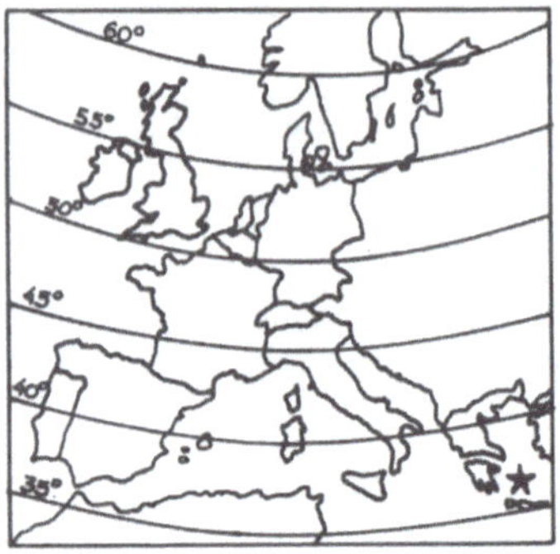

Figure 1. Location of Paros

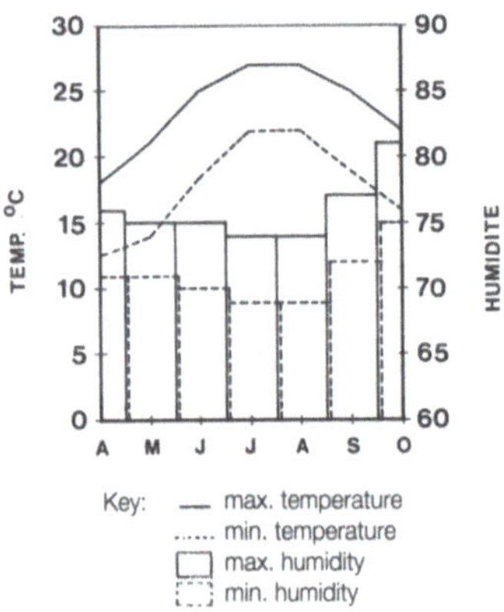

Figure 2. Temperature and humidity data

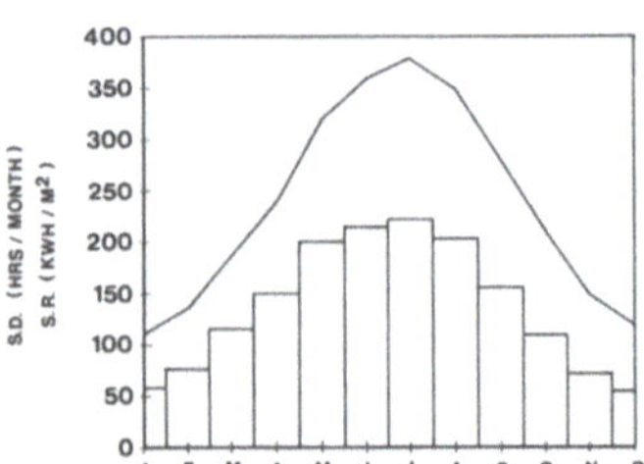

Figure 3. Solar irradiation (kWh/m^2) and sunshine duration (hours/month) data

Figure 4. Wind speeds (m/sec)

Building Type

This project concerns a complex for use from April to October by tourists of all ages. It consists of eight buildings containing 17 furnished apartments with a total of 54 beds plus a service building which houses the reception and all the communal facilities.

Location

The complex is located in a peaceful spot 250 m from the sea-shore near Drios (latitude 37 ° N), a village in south-east Paros, one of the Cyclades group of islands in the Aegean sea (see Figure 1). The 4070 m^2 site is completely flat. There is an uninterrupted view of the sea to the south- east.

Site Microclimate

The climate is characterized by high levels of solar radiation throughout the year, average winter temperatures which rarely fall below 10 ° C and north and south winds all year round. Monthly temperature, humidity, solar irradiation, sunshine hour and wind speed data are given in Figures 2-4.

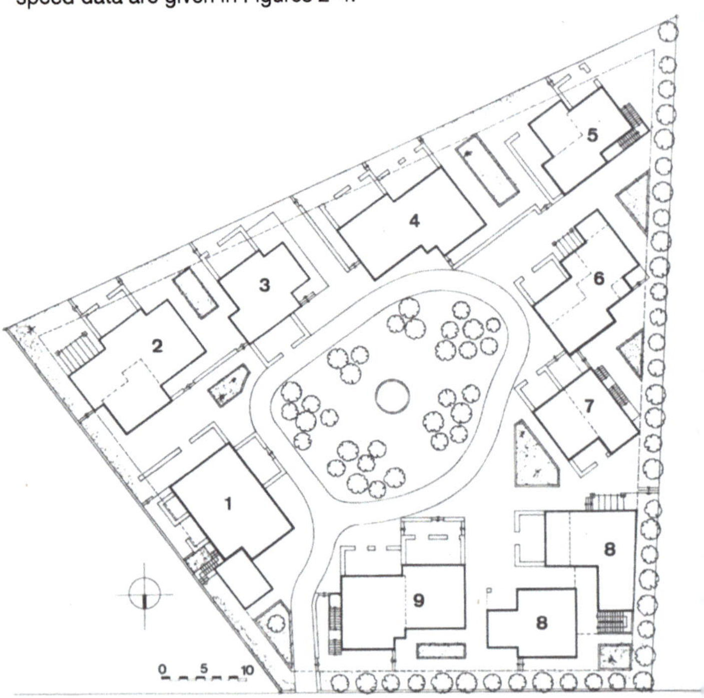

Key:
1. Reception building
2-9. Apartment buildings

Figure 5. Site layout

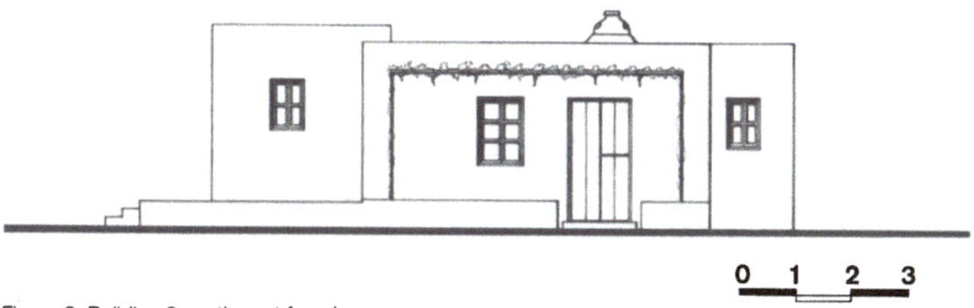

Figure 6. Building 3 south-east facade

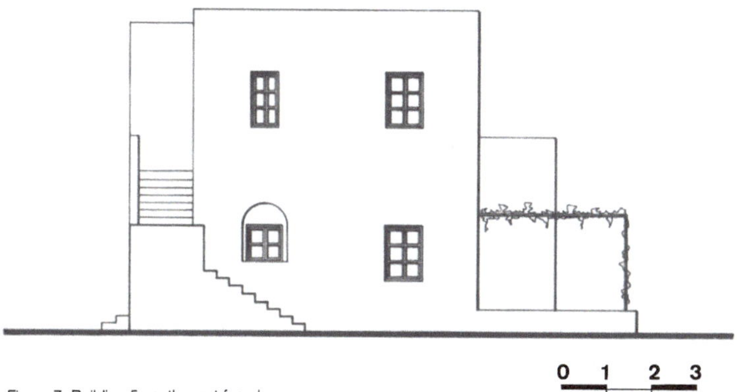

Figure 7. Building 5 south-west facade

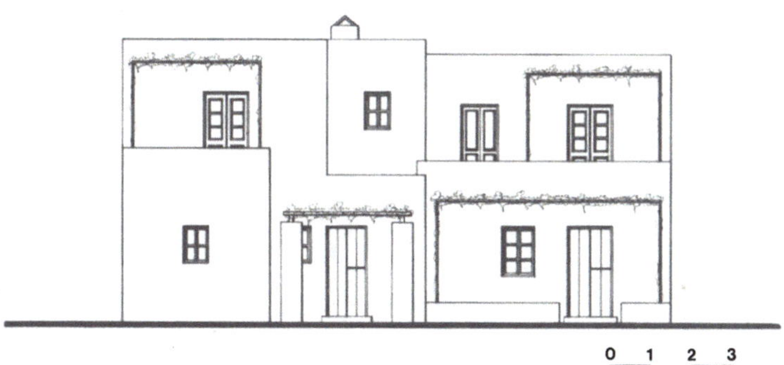

Figure 8. Building 6 south-east facade

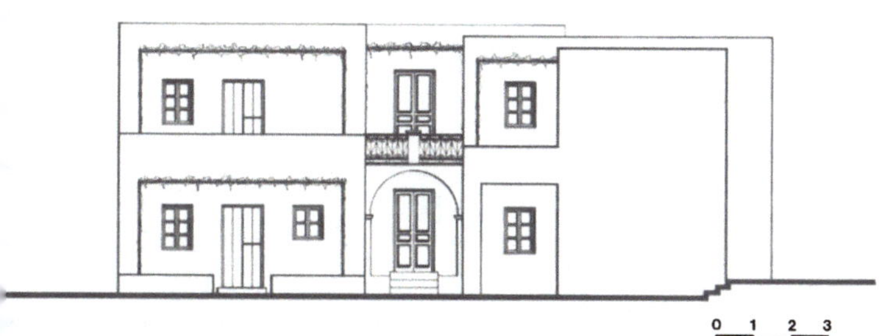

Figure 9. Building 8 east facade

Design and Construction Details

Design Details

The complex consists of 9 detached single- and two-storey buildings. The site layout is given in Figure 5. Building 1 serves as a reception building and houses all the communal facilities for recreation, etc. Buildings 2-9 contain between them the 17 furnished apartments.
The total floor space of the buildings is 1480 m^2 and the total volume 5500 m^3. There are 2590 m^2 outdoor space containing pergolas, trees and gardens.

The buildings are arranged in a circular fashion facing south-east so that they all have an unobstructed view of the sea. The central area has been kept free of buildings and contains trees and water. Care has been taken that no building inhibits the flow of natural ventilation air through any of the others.

There are two types of apartments. Type A (of which there are 14) contains a bedroom, bathroom, WC, hallway, kitchen, dining area and a sitting room which can be used as an extra bedroom. The 3 Type B apartments have an extra bedroom with a private bathroom.

The main axis of most of the buildings is virtually parallel to the sea-shore so that the main facade faces south-east. The entrances are on the north-west. Each building has at least one pergola covering an outdoor sitting area.

Local Design Requirements

The site layout and design of the individual buildings were chosen to be in keeping with local requirements.

In the Cyclades islands, the style of all new buildings has to fit in with the vernacular architecture. In general, buildings must not be more than two storeys high and their overall height must not exceed 7.5 m. In Greece, the architectural drawings of all buildings related to tourism must be approved by the Greek Tourist Organization (EOT) before being submitted to the local planning authority. In addition, plans for buildings on the islands must be examined by a local architectural council to ensure that they comply with design requirements related to the local traditional style.

To be in keeping with the local architecture of Paros, the buildings in the complex incorporate the following design features:
- the external walls are made of local stone;
- the individual buildings are small and at different levels;
- they are designed using simple cubes of different heights;
- open-air staircases relate to and give emphasis to the different volumes;
- chimney-pots, some of which have the form of a traditional jar, are short and blend in with the different volumes;
- pigeon-holes are included where the volume and height of a wall makes this appropriate:
- pergolas with short stone walls serve as extensions to the buildings and create outdoor sitting areas.

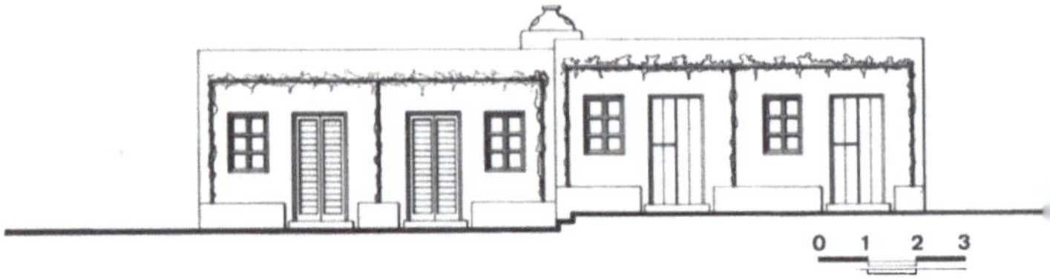

Figure 10. Building 4 south-east facade

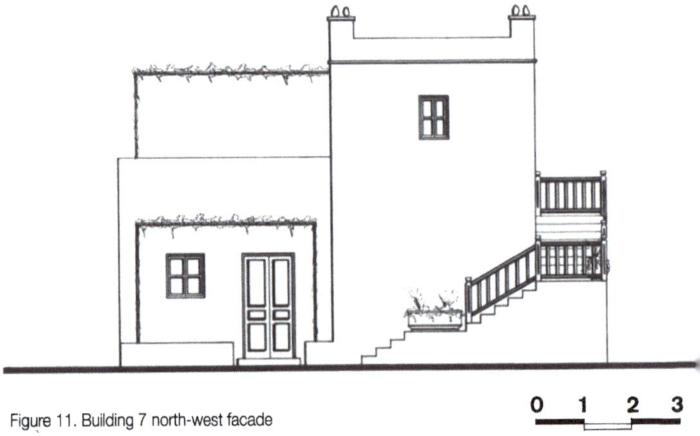

Figure 11. Building 7 north-west facade

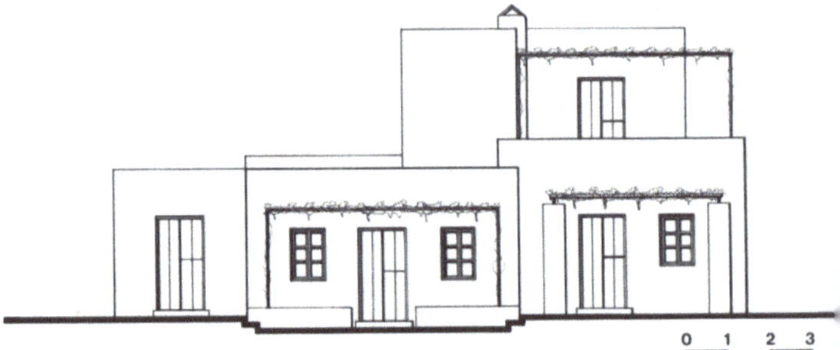

Figure 12. Building 2 south-east facade

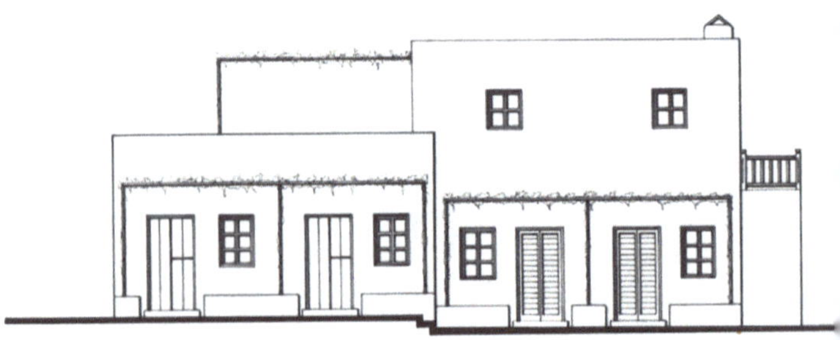
Figure 13. Building 1 south facade

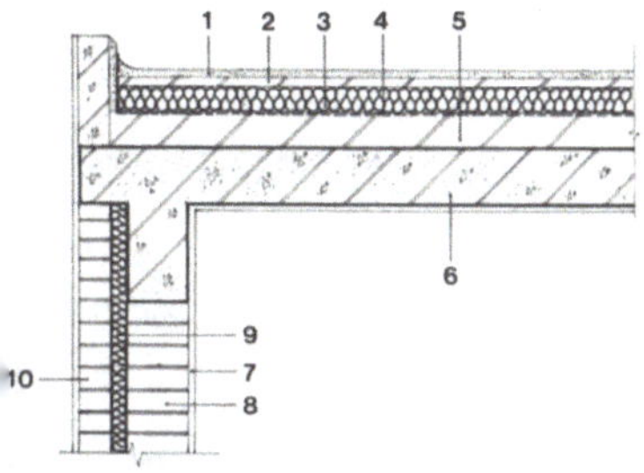

Key: 1. cement
2. concrete
3. expanded polystyrene
4. waterproof layer
5. slope
6. reinforced concrete
7. plaster
8. brick
9. glass wool
10. brick

Figure 14. Building 1 construction details

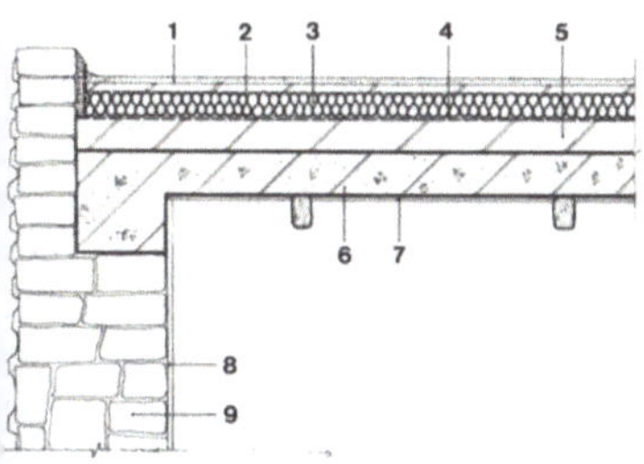

Key: 1. cement
2. concrete
3. expanded polystyrene
4. waterproof layer
5. slope
6. reinforced concrete
7. bamboo
8. plaster
9. stone

Figure 15. Buildings 2-9 construction details

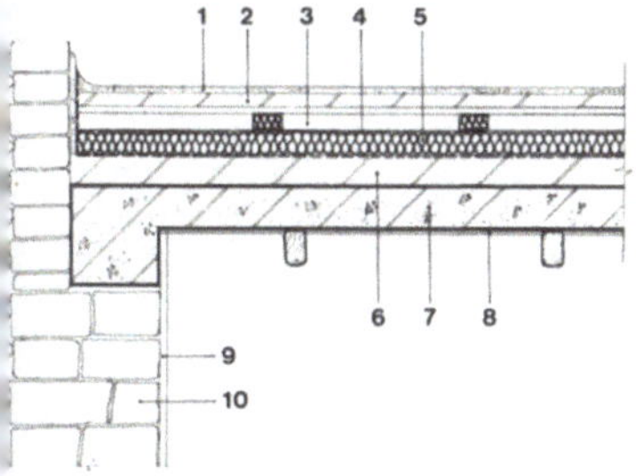

Key: 1. cement
2. concrete
3. air gap
4. aluminium foil
5. glass wool
6. slope
7. reinforced concrete
8. bamboo
9. plaster
10. stone

Figure 16. Radiant barrier construction details

Construction Details

The construction details of Building 1 are shown in Figure 14. All beams and columns are of reinforced concrete. The external walls consist of an external layer of 100 mm brick, 50 mm glass wool and an internal layer of 200 mm brick. The roof is of reinforced concrete, above which is a slope with an 80 mm layer of expanded polystyrene. The windows are double glazed.

The construction details of the apartment buildings (Buildings 2-9) are shown in Figure 15. The structural walls are 500 mm thick and built by local workmen using stone from nearby quarries. The roofs are similar to that of Building 1. The partition walls between the apartments are 200 mm thick and made of hollow brick. The walls between rooms within each apartment consist of 100 mm hollow brick.

In all the buildings, the doors, window frames and external shutters, etc., are made of wood.

DESCRIPTION OF PASSIVE SOLAR FEATURES/COMPONENTS

Introduction

Because the complex is open from April to October and the buildings are well-insulated and have a high thermal mass, the space heating requirements can be satisfied entirely by direct solar gains through the windows. The critical issue is cooling. The main objective, therefore, has been to incorporate various passive cooling strategies into the design so that the buildings are comfortable most of the time without the use of mechanical cooling systems. All windows are equipped with exterior shutters. Deciduous planting on the pergolas also contributes to solar control. Natural cross ventilation is created in all buildings by placing openings on opposite facades. For times when this is not sufficient, a ventilation chimney has been placed in the centre of each unit. The well-insulated roofs have radiant barriers and the vertical walls have a high thermal mass which smooths out internal temperature variations. The natural vegetation and water in the centre space of the complex provides evaporative cooling.

Passive Solar Cooling Features

Pergolas

In the Aegian islands during April to October, people usually spend most of their time outdoors. Therefore a vine-covered pergola has been placed in front of the living room balcony door of each apartment to create a fully-shaded extension to the living space. Because the vine is deciduous, it does not inhibit penetration of solar radiation into the building in winter when direct solar gains are required for heating.

Cross Ventilation

Because the apartments are divided into small units and steady breezes are always present, natural cross ventilation is an effective cooling strategy. This has been created in all the apartments by placing openings on opposite facades (see Figure 17). Where the layout of the apartment allows it, these openings are on the north and south (i.e. the windward) sides. Where possible, the principle has been applied to both the living rooms and the bedrooms. The distribution and sizing of the windows have been optimized to increase airflow within the building but reduce excess solar gains.

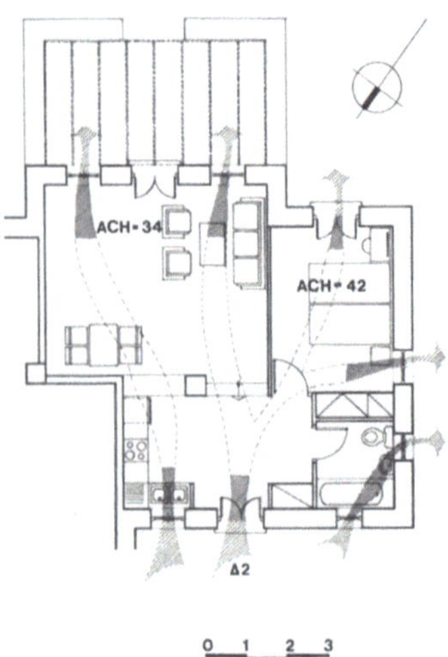

Key: ACH = air changes per hour

Figure 17. Apartment layout showing cross ventilation

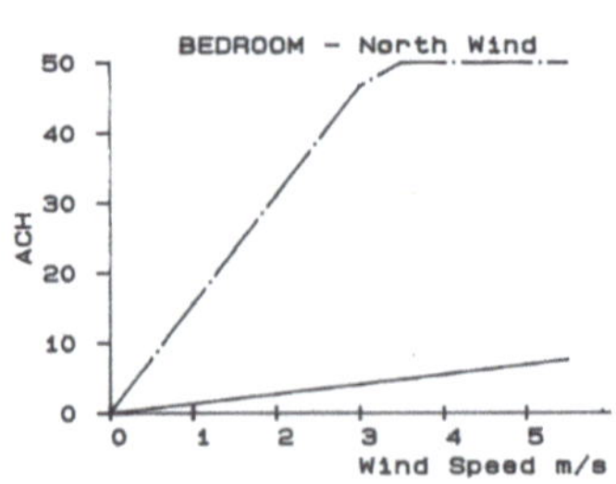

Figure 18. Cross ventilation in a living room with a north wind

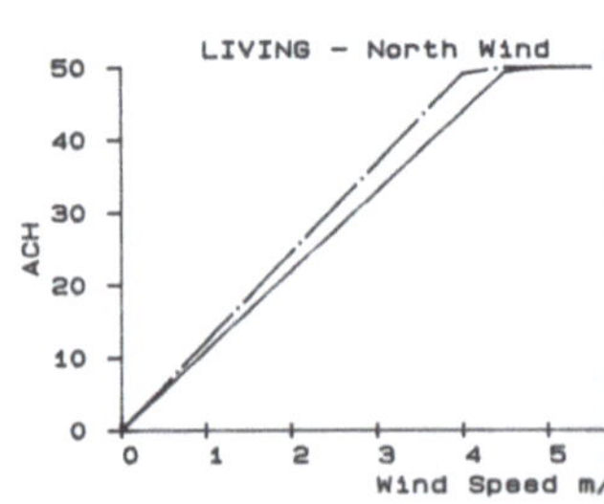

Figure 19. Cross ventilation in a bedroom with a north wind

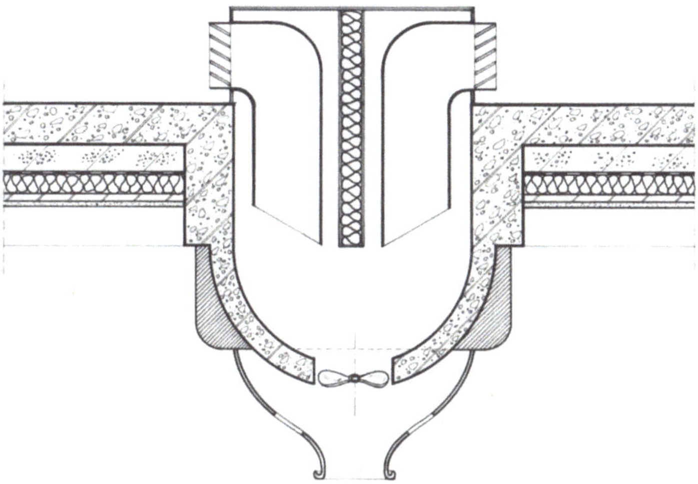

Figure 20. Section of ventilation chimney

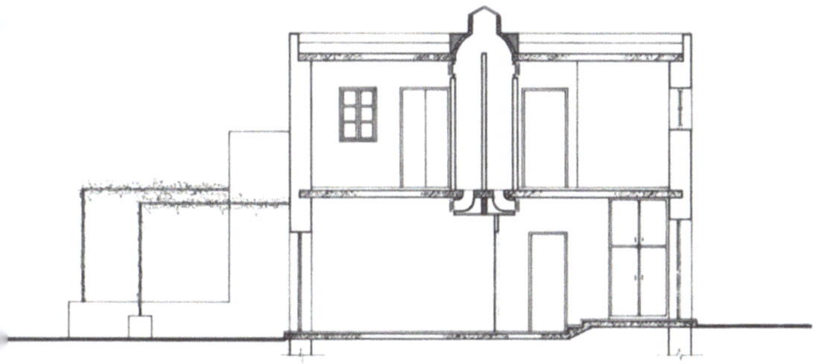

Figure 21. Section of two-storey building showing ventilation chimney

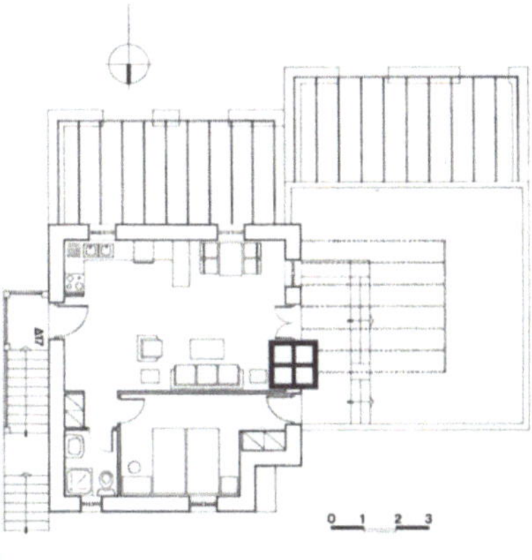

Figure 22. Use of chimney to create ventilation in apartments when natural cross ventilation through window openings is insufficient

Ventilation Chimney

The arrangement of interior spaces in some of the apartments is not conducive to good cross ventilation. To overcome this and to create sufficient air flows at times when there is not much wind, a ventilation chimney has been placed in the centre of each unit (see Figures 20-22). This encourages thermocirculation of air through the apartment using the stack effect. The ducts have been designed to minimize noise transmission. The devices can be fan-assisted so that, if there is insufficient wind, air change rates of 30 ACH are achieved.

In the single-storey buildings, the chimneys extend from just below the ceiling (where the warm air is taken in) to the top of the building. As indicated earlier, the top of the chimney has been designed to look like the stacks found traditionally in this area.

In the two-storey buildings, the lower opening of the chimney is located just below the ground floor ceiling (Figure 21).

In one of the buildings an attempt has been made to further encourage the thermocirculation of air by glazing the south-west side of the chimney between the ground and upper floors.

Radiant Barrier

The roof of one of the buildings is insulated externally with expanded polystyrene faced with aluminium foil. This, together with an adjacent air gap, decreases the transmission of long wave radiation to the building. This technique is seldom used in Europe but in the US it has been found to reduce cooling loads by 10% for no additional capital cost. The system has been installed in one of the buildings in this complex for demonstration purposes.

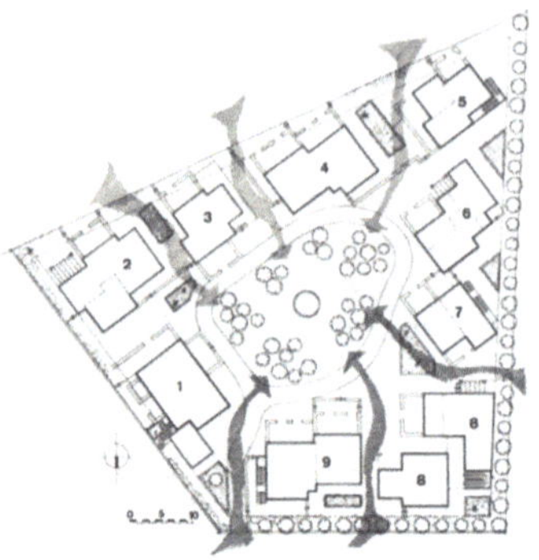

Figure 23. Plan showing circulation of air through the site

Thermal Mass

All the buildings are constructed with materials such as stone, concrete and brick, which have a high thermal mass. This prevents the build-up of high temperatures in the buildings.

Oasis Effect

Special attention has been paid to landscaping in order to create a favourable microclimate on the site. The circular arrangement of the buildings gives rise to a central space which has been planted with olive and lemon trees. There is a small pond with a water fountain in the middle. The trees and water bring about a reduction in air temperature from water evaporation. This precools ventilation air for the buildings. In addition, the arrangement of trees, patios and terraces encourages the circulation of air throughout the site (see Figures 23 and 24). The planting on the pergolas protects the buildings from direct solar gain in summer (see Figure 25).

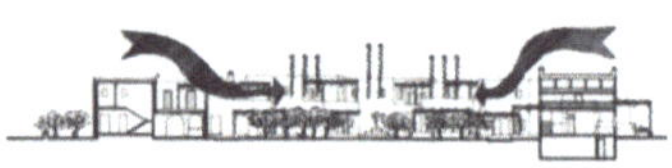

Figure 24. Section showing circulation of air through the site

Evaporative Cooling

At the initial design stages it was proposed that the ventilation air be further precooled by passage over water circulated through a porous clay pipe. This was eventually rejected on the grounds that, with the high humidity levels experienced at the site (they can reach 70%), such a system could produce poor quality air and cause structural and medical problems.

Passive Solar Heating Systems

Direct Gain

In most of the buildings, the main facade faces south-east and contains glazed apertures. The central building, the main facade of which faces north-east, has a sloped clerestory along its south-east side. These openings enable direct solar gains to provide all the heating required in the complex in the April to October period.

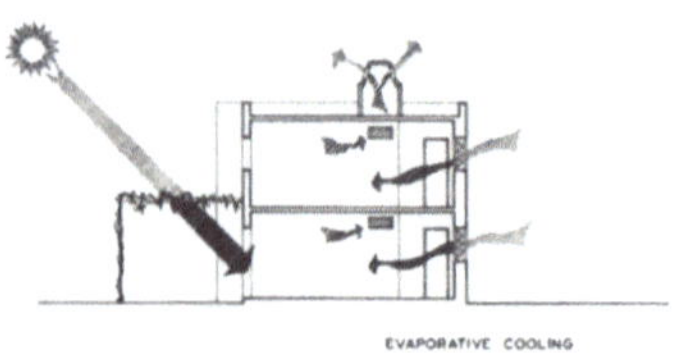

Figure 25. Mode of operation of cooling systems in summer

Thermal Mass

Because of their high thermal mass, the buildings absorb the excess solar energy collected during the day and release it slowly back to the living spaces at night.

Daylighting

Provision of good lighting in the daytime is not of prime concern in the apartments and adequate levels of daylight are easily achieved through the windows. The reception building, however, is used extensively all day and particular attention has been paid to designing good daylighting systems for this in order to create a pleasant and visually comfortable environment and avoid the overheating which would occur with artificial lighting. Diffusers under the clerestory create high levels of diffuse illumination, reducing contrast and glare. An external movable shading device reduces excess solar gains and controls lighting levels. The glazing can be opened to create cross ventilation and release excess heat in summer when necessary (see Figure 15).

Mechanical Ventilation Systems

In a few rooms, the use of openable windows will not provide adequate ventilation because of their poor orientation with respect to the prevailing winds. In such cases, air velocities are increased by use of ceiling fans 1 m in diameter with a maximum power consumption of 75 W (see Figure 26).

It was not possible to achieve cross ventilation in all the bathrooms. To prevent the spreading of water vapour from such rooms to the rest of the apartments a duct running to the roof has been installed together with a small exhaust fan (see Figures 27 and 28).

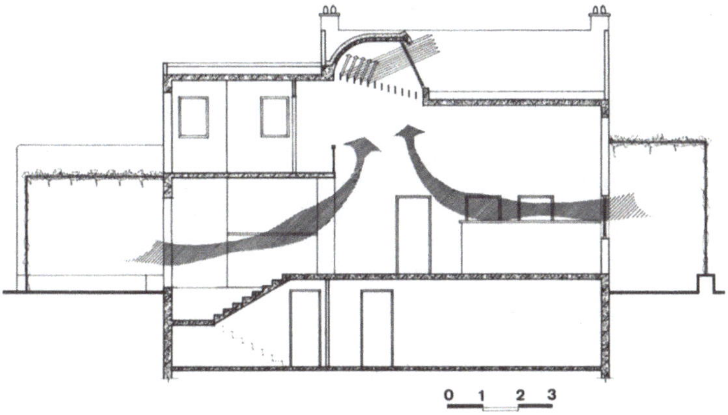

Figure 26. Use of ceiling fan to increase air flows in rooms without adequate natural ventilation

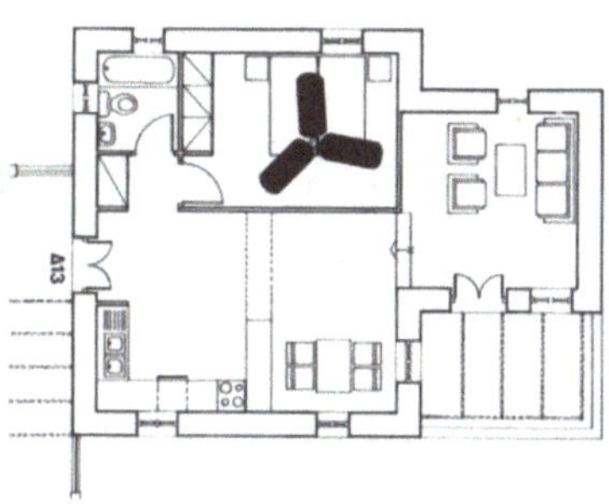

Figure 27. Mechanical ventilation in the bathroom - fan

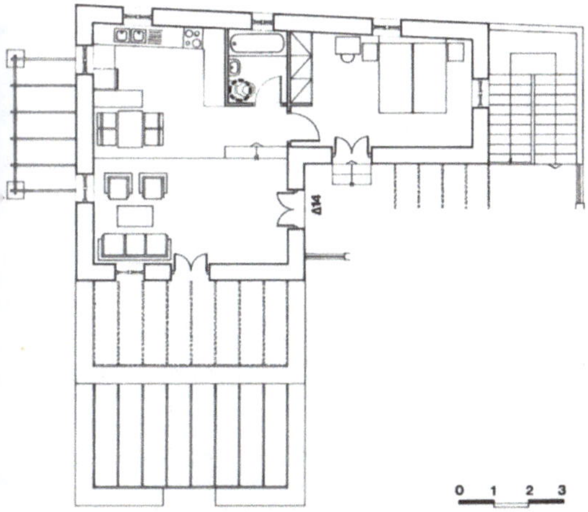

Figure 28. Mechanical ventilation in the bathroom - position of duct

ENERGY CALCULATIONS PERFORMED AND DESIGN TOOLS USED

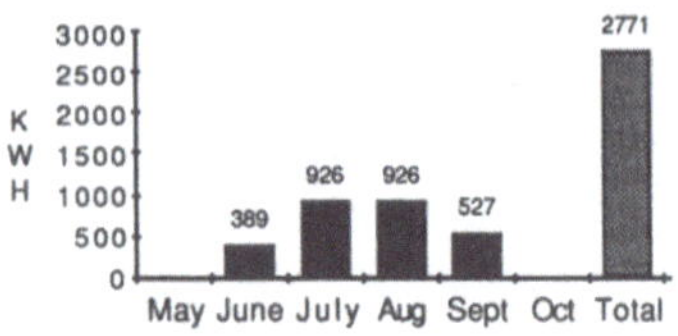

Figure 29. Cooling loads for reception building with no passive features (i.e. no shading or night ventilation)

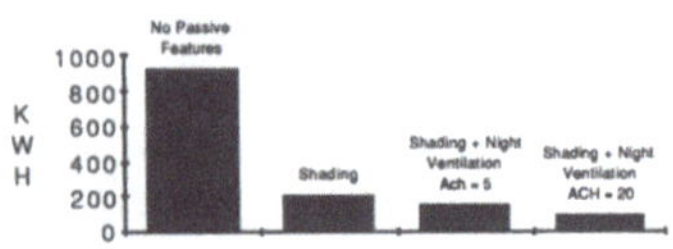

Figure 30. Cooling loads for reception building in July

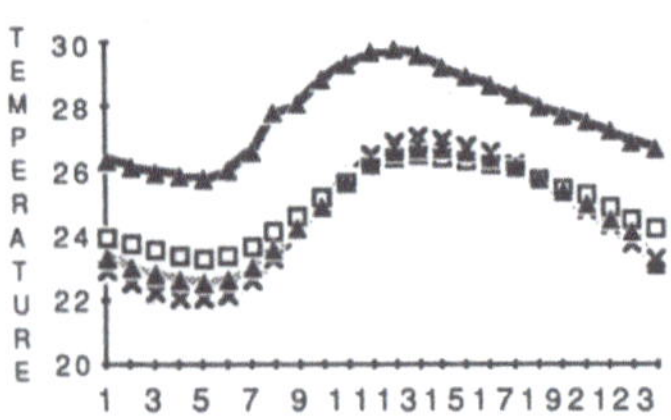

Key: outside temperature
temperature in hall with no passive features
temperature in hall with shading and night ventilation at 5 ACH
temperature in hall with shading and night ventilation at 20 ACH

Figure 31. Hourly temperatures in reception hall in July

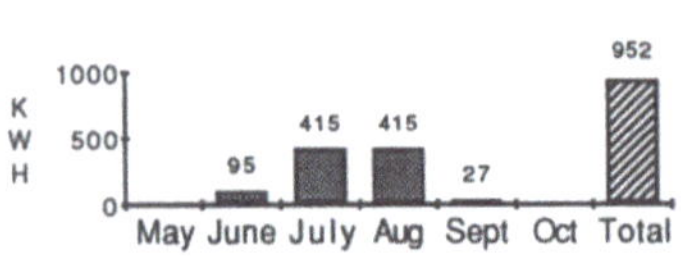

Figure 32. Cooling loads for apartment with no passive features (i.e. no shading or night ventilation)

Design Tools

The design of the passive solar systems was carried out with the help of a number of design tools.

A module of the ESP simulation model known as ESP-AIR was used to evaluate the air flow patterns. The output presented the air flow patterns within the buildings for different wind speeds and directions. The air velocity within the occupied space was then used to calculate the comfort index according to the Fanger or Gagge theory.

The dynamic thermal model CASAMO-CLIM was used to evaluate the comfort conditions in the buildings and calculate the cooling loads when a mechanical system was used from internal gains, solar gains, building envelope, air change rate, building inertia, etc. Hourly simulations for a typical apartment and the reception building were made for a design day every month.

Results for Reception Building

The reception building has a large volume and high occupancy. The annual cooling load if the building has no shading or night ventilation amounted to 2770 kWh (see Figure 29) with a peak power consumption of 2.8 kW. In July or August (which have similar climates and are critical from the cooling point-of-view) use of shading devices will reduce the cooling load by 77% (see Figure 30). The greatest contribution to this is made by the clerestory shading. Night ventilation at 5 ACH would save another 5%. Night ventilation at 20 ACH would save 12%.

With night-time ventilation and shading but no mechanical cooling, the temperature in the reception hall varies from 23 ° C to 27 ° C (see Figure 31). The comfort diagram shows that, provided the air velocity is maintained at a sufficiently high level, the hall will be comfortable. The required air velocity can be achieved by cross ventilation and it may be enhanced locally by use of a ceiling fan.

Results for Typical Apartment

The annual cooling load for a typical apartment with no shading or night ventilation was found to be 950 kWh (see Figure 32). The peak cooling power requirement in July or August was at 2 pm and amounted to 0.72 kW.

As with the reception building, July and August are regarded to be the most critical months for cooling in the apartments. Therefore the effectiveness of various cooling strategies was evaluated for these months. It was found that shading reduced cooling loads by 25% (see Figure 34). Night ventilation can save another 32%. Peak cooling loads can be reduced by 20% - mainly by solar protection. Even with ratios of glazing area to floor area as low as 9%, shading is a very effective way of reducing cooling loads. With shading and night ventilation the high cooling loads occur in the daytime when the apartments are empty. It can be deduced, therefore, that the absence of a mechanical cooling system does not matter. With the temperature range found in the apartment, good ventilation (either natural or mechanically-assisted) will give comfortable conditions without a mechanical cooling system.

GENERAL DESIGN GUIDELINES/POINTS OF INTEREST RESULTING FROM THIS PROJECT

In this project the design guidelines were as follows:

- to design buildings which are in harmony with the natural and built environment of the island;
- to create by means of shading and the oasis effect microclimates in the centre of the complex and around each building which promote outdoor activities;
- to integrate several passive cooling systems into a scheme which has to be approved by a local architectural council;
- to integrate the passive solar devices in such a way that they are not noticeable when the building is viewed from the outside;
- to integrate elements from traditional Parian architecture into a building which complies with the standards and requirements of a modern holiday complex.

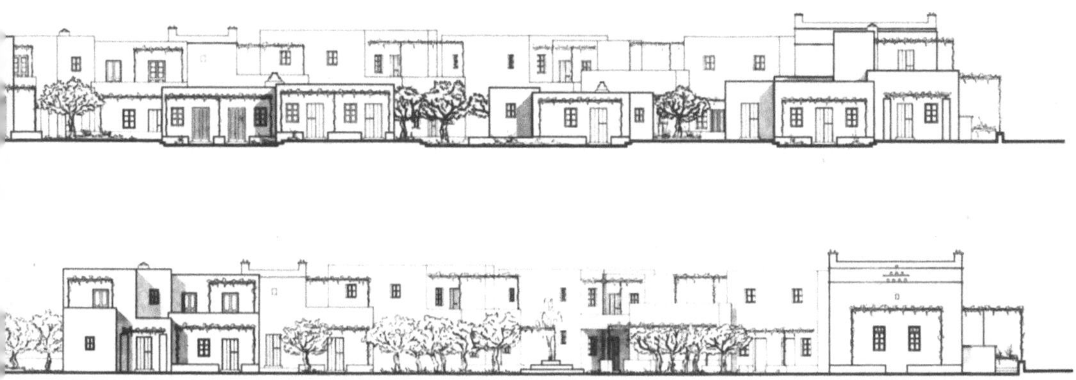

Figure 33. Panoramic views of the complex

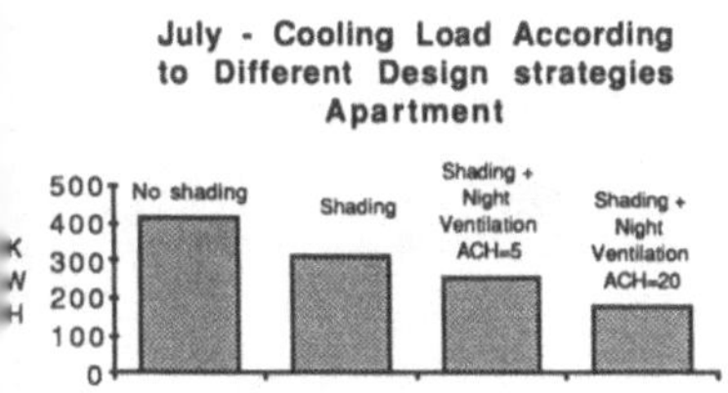

Figure 34. Cooling loads for apartment in July

BUILDING 2000 Building 2000 brochures are published by Directorate General XII of the Commission of the European Communities to show how design studies can help architects and other building designers use passive solar principles to the best effect to produce attractive energy-efficient buildings. Each brochure describes studies carried out with the support of the Commission during the design phase of one of thirty-six non-domestic buildings in the EC Member States. The studies were on such topics as daylighting, heating, cooling, ventilation, comfort, control systems and urban design. They were carried out with the help of acknowledged European experts in these fields and drew heavily on lessons learned and techniques developed through the Commision's research and development programme on solar energy applications to buildings.

Commission of the European Communities/Directorate-General for Science, Research and Development

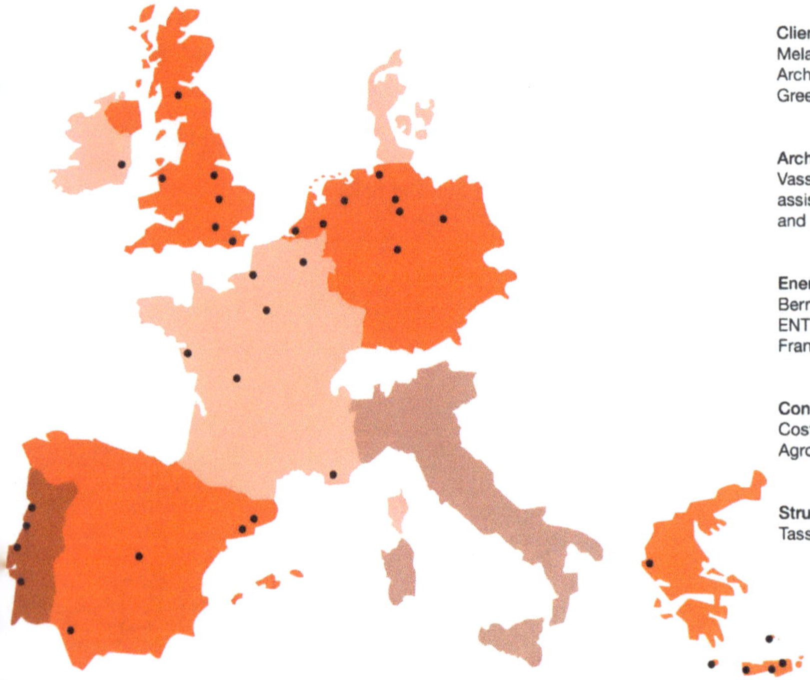

List of Design Team Participants and Advisers

Client
Melaniti M.
Archilochos,
Greece

Architect
Vassilis Bouriotis
assisted by Sofia Vassiliadi (architect)
and Andrianos Sakas (designer)

Energy Calculations
Bernard Fleury
ENTPE - LASH
France

Consultants
Costis Stambolis
Agroclimatic Ltd

Structural Engineer
Tassos Kavelaras

This set of **Building 2000** brochures illustrates how architects and other building designers can successfully apply passive solar principles to produce energy-efficient buildings.

BUILDING 2000
Participants

Project Director
Theo C. Steemers

Coordinator
Cees den Ouden

Technical Steering Committee
Dean Hawkes
Nick Baker
Alex Lohr
Jean P. Lepoivre

Regional Liaison Agents
(D) Jörn Behnsen
(E) Vicente Sifre
(F) Michel Raoust
(GB) Alan Hildon
(GR) Matheus Santamouris
(P) Eduardo Maldonado

Further information or copies of the brochures can be obtained from prof. ir. Cees den Ouden, EGM Engineering BV, P.O. Box 1042, 3300 BA Dordrecht, The Netherlands.

Introduction in Design Support Activities and summaries of Design Support Reports.

C. den Ouden

During the period between the start of Building 2000 (officially 1-1-1988, but effectively since May 1988) and June 1990, a wide range of Design Support Activities has been carried out for almost all Building 2000 projects (only a few projects were delayed and, therefore, design support was not useful).

Three different types of design (advice) support activities have been offered to the Building 2000 design teams:

a. The consultation of an individual European Passive Solar Expert or a National Design Support Expert on Passive Solar and related topics.

Note:
Since several of the participating projects were already in the detailed design phase, this seemed to be the only effective way to improve the design on certain component detailing and/or minor changes in the passive solar and daylighting aspects already integrated in the design concept.

b. The carrying out of a small case-study by a recognised Design Support Unit having experience with certain passive solar/ daylighting design tools. In this case-study a sensitivity analysis for certain design parameters to be subject to change could be made so as to build up to the selection of the final size of these parameters.

Note:
This seemed to be of value for those projects in which the preliminary design phase had not yet been concluded.

c. The conducting of a more detailed modelling study. In several Building 2000 projects detailed simulation models might have to be used to provide answers on specific design aspects, such as air movements in atria, approaches to passive cooling etc.

Note:
When these requirements were identified to the coordinator, he contacted the right modelling expert to conduct a more detailed study on a subject preferably of use to a group of similar Building 2000 projects.

Moreover, several Design Support Workshops were organised during the Building 2000 programme during which Building 2000 design teams could directly exchange design ideas with the various design advice experts represented at these workshops.

In the second part of the Building 2000 final report a summary of a selection of many reports is presented, as produced by Design Support experts. Most of the design support activities resulted in changes of the various designs, as have been reported by the design teams in the brochures presented in the first part of the book.

The willingness to improve the building concepts by collaborating with R & D experts was encouraging for the Commission's action in this field. It is to be expected that design aids and simulation tools for passive solar options, daylighting concepts, comfort criteria etc. under development in the Commission's R & D-programme will be utilised more frequently in the future.

This will result in a better exchange of information between the actual design practitioners and the European R & D community, and most importang that this technology transfer, as initiated by this Building 2000 programme, will result in buildings with a higher quality with respect to energy and environmental issues.

REPORT ON CONSULTANCY FOR THE UNIVERSITY OF SEVILLE / EXPO BUILDING, B2000 PROJECT 01

This report describes the main technical issues which were the subject of three consultations between the Architect Jaime Lopez de Asiain and Nick Baker of Cambridge Architectural Research (CAR). A summary of the main results of a laboratory flow modelling study carried out by P Linden and G Lane-Serf of Cambridge Environmental Research Consultants.

There have been three main consultancy sessions -

i) at the Barcelona workshop in December 88, a brief introduction to the project,

ii) at a three day visit to Seville in January 89, introduction of a rather modified brief and a detailed consideration of certain elements, and

iii) a brief meeting at Dublin in March 89 to review progress and discuss proposals for further study.

1.1 Meeting at Barcelona

At this stage, the brief was only concerned with the use as the Expo Comissariat Building. The proposal then was for the central patio/atrium to be non-glazed, with an openable water-shedding opaque roof. (fig 1)

Discussions were mainly concerned with temperature conditions within the atrium and how the atrium might give thermal benefit in summer, ie cooling, to the surrounding spaces. The question of whether the atrium could be cooled by radiant cooling and convective cooling at night sufficient to ventilate the surrounding offices with cool air was addressed, but not answered. The conflict here is that as soon as air in large quantities is drawn from the atrium, to ventilate the surrounding offices, hotter air will replace it. the amount of "coolth" storage would be critical. It was realised that analysis would be necessary to answer these questions.

Other considerations were the degree to which vegetation in the atrium would cool the air. We were fairly confident that vegetation and shading would result in the patio/atrium itself being significantly cooler than ambient.

Depending on the permanent use of the building, the need for passive heating in winter was discussed. It seemed that there was now more interest in glazing the roof and having a closable structure for winter.

Shortly after this meeting, the brief for the final project was changed somewhat when the final use for the building became clear. It was to become the University of Seville Humanities Faculty.

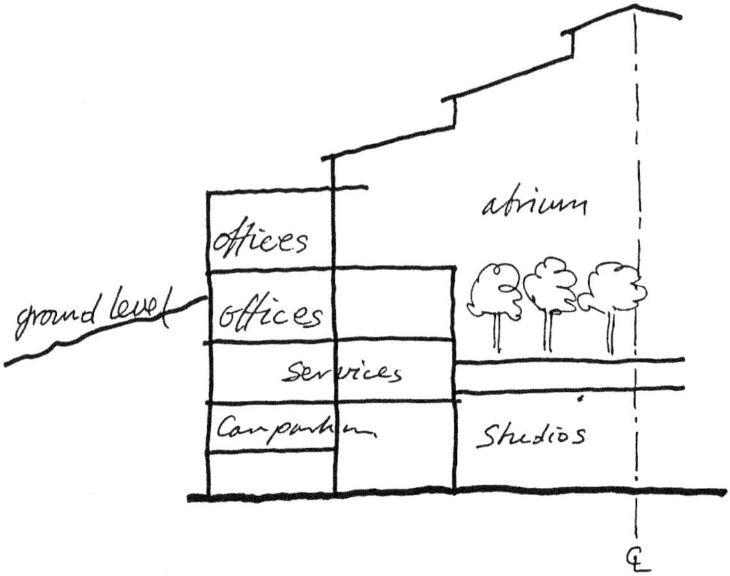

Fig 1 Original proposal for Expo Commissariat building

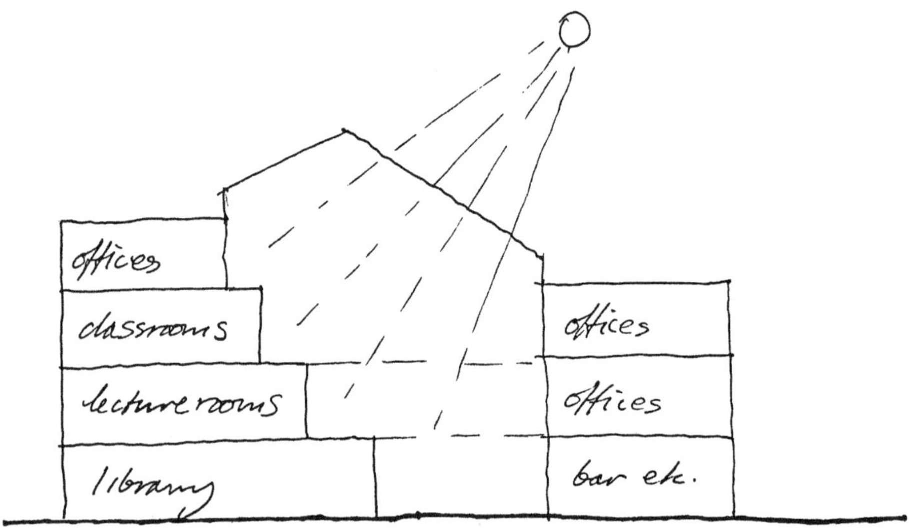

Fig 2 Proposed section at Seville meeting Jan 89

2.1 Meeting at Seville

There had been a significant change in emphasis in the use of the building from the Expo Comissariat Building which would have been used at the hottest time of the year to the University building which would be used for most of the year but excluding the two hottest months, July and August.

This had prompted Escula Architectura Bioclimatica - Architect Jaime Lopez, to change the section to an asymmetric one, more optimised towards winter passive heating. (fig 2)

However during the discussion, on examining the climatic data of the site and estimating the actual size of the annual heating load, we agreed that the summer performance was the most critical, and that the provision of daylight and the avoidance of mechanical cooling, were probably the most energy-conserving measures.

It was realised that any measures aimed to allow winter passive heating, must not in any way increased the risk of overheating.

2.2 Ventilation strategy

With daytime maximum temperatures in June of around 31°C, it seems that in spaces with low internal gains, and minimum daytime ventilation, room temperatures might be able to be held significantly below ambient by the absorption gains into the massive structure. This would not be possible, however in high gain spaces such as the library, lecture rooms and classrooms. Here, high ventilation rates would be necessary in order to remove the gains, and the only viable cooling strategy would probably be direct physiological cooling by air movements between 0.5 and 1.0 m/s.

Various passive solutions were proposed for the spaces, most of which used the atrium to provide a ventilation extract, fig 3. This would normally be driven by the rather weak thermal buoyancy or stack-effect, but would be enhanced by a breeze, from the south side, provided the exit vent design had the appropriate aerodynamic property.

2.3 The Library

This being on the north side of the building had good access to daylight without the need for shading. High ceilings, (4.5m) resulted in good light penetration suggesting that the main seating area should be naturally lit on the perimeter, whilst the book stacks should be in the central section. This zone would be artificially lit with a high efficacy system which was to be operated only when and where required.

At this stage there was a further reading/seating area adjacent to the atrium. Cross-ventilation was available via openings in the north wall, across the top of the book stacks, and out into the atrium at first floor level.

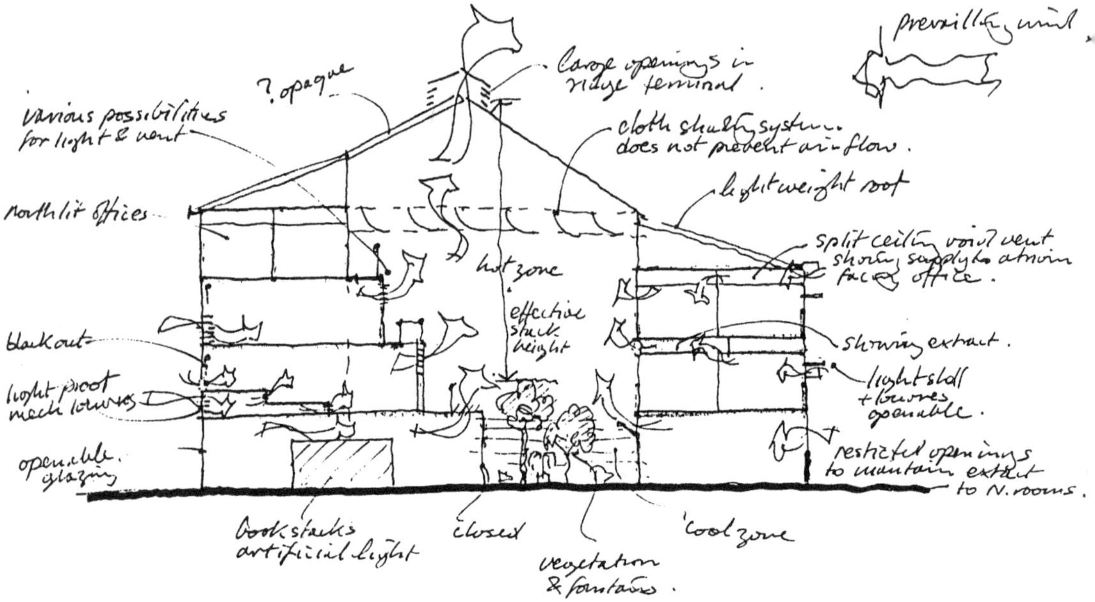

Fig 3 Modifications and proposed modes of operation Jan 89

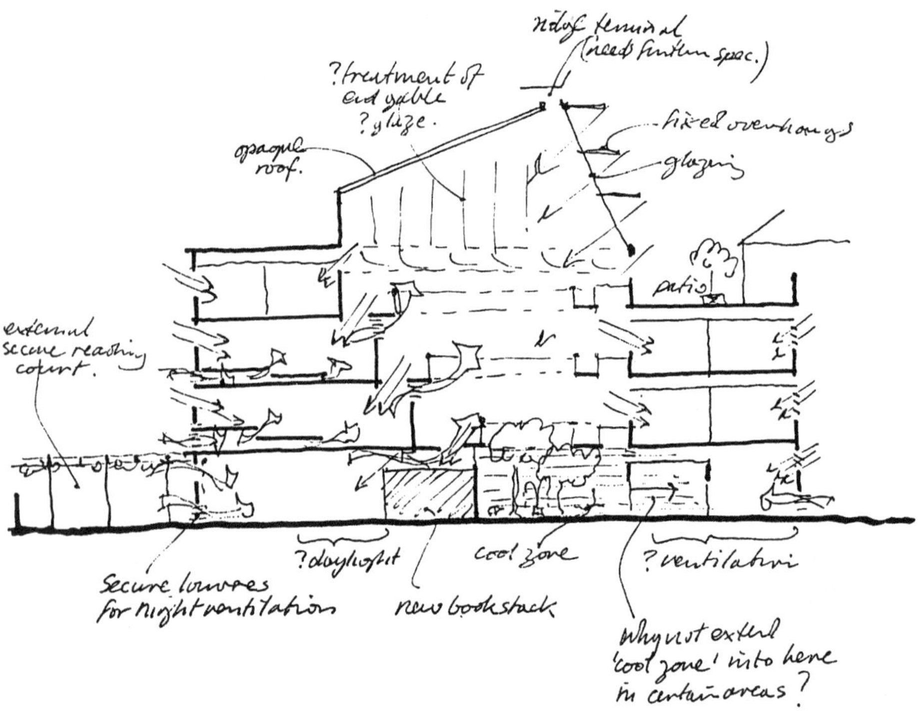

Fig 5 Final technical proposal as at Dublin meeting March 89

2.4 The Lecture rooms and Classrooms

The lecture rooms presented the biggest challenge to the passive design, partly due to the dense occupation and hence high loads, and partly due to the fact that they would have to be blacked out for some of the time but still receive natural ventilation.

Our proposed solution was to bring air in from the north wall under a raised, stepped floor, and extract into the atrium via light-proof louvres at the lecturer's end of the room. Separate glazed areas could be blacked out, but could also be openable for extra ventilation when blackout was not required. The delivery of air from the floor void would have the advantage of bringing the fresh air close to the occupants, also making some benefit from perceptible air movement.

Calculations (fig 4) showed that an openable free area of about 3.8m^2 at both inlets and outlets, would with the prevailing conditions in the atrium, give sufficient buoyancy ventilation to keep the temperature increment in the room down to 3.5^oC above above ambient. With an ambient temperature of 31^0C but a wet bulb temperature temperature of only 23^oC, this gives a corrected effective temperature of 29^oC at 0.5m/s which is just 2^oC above the normal comfort zone. We felt that as this represented a worst case it was just acceptable.

This ventilation analysis is not straightforward, and the testing of stack-induced flow is the subject of a physical modelling study proposal.

The ventilation analysis of the smaller classrooms would be similar in principle, but no underfloor voids would be in use. In winter in both types of space the principle would be much the same, but ventilation rates would be much lower, the minimum necessary for fresh air.

2.5 Computer controls

Since the ventilation dampers would be mechanical and have to be remotely controlled, we felt that there was an opportunity to introduce "intelligent control" from a computer. This would allow ventilation to be optimised - i.e. to minimise or maximise ventilation according to the relative thermal conditions of ambient air, room air, and the thermal mass associated with the room. This would automatically permit useful night ventilation.

The computer would also control other environmental components in the building, e.g. the atrium shading devices. The computer would probably hold a thermal model of the building so that it could be predictive. This is useful for an intermittently occupied heavyweight building.

2.6 The Offices and Study rooms

South-side perimeter offices will need effective shading and openable windows for cross ventilation. Shading systems combining fixed light shelves with movable louvres as employed at the school at Guillena by

Design conditions, June worst case, July and August building not occupied.

Ave T_{max} = 31°C RH = 46% wet bulb T_w = 23°C
diurnal variation = 16°C approx

Corrected Effective Temperature CET -

26.0 °C at 0.5 m/s
26.5 °C at 0.1 m/s

For lecture rooms allow temp rise ∂T = 3°C

Gains to lecture rooms - 150 people @ 100 watts 15 kW
900 lux daylight over 123m2 @ 110 lumens/W 1 kW
total 16 kW

Vent loss $W_v = .33\ V\ \partial T$

$16000 = .33\ .\ V\ .3$ where V is vent rate m^3h

V = 4.44 m^3/s room vol = 492m^3 vent rate = 32.5 ac/h

Stack pressure $\partial P = 0.043\ H\ \partial T$

for ∂T = 5°C H = 16 m **∂P = 3.44 Pa**

flow rate $V = A_1.A_2 / \sqrt{(A_1^2 + A_2^2)}\ .\ 0.827\ .\ \partial P^{.5}$

If A1 = A2 $V = A / \sqrt{2}\ .\ 0.827\ .\ \partial P^{.5}$

If V = 4.4 m^3/s and ∂P = 3.44 Pa then **A = 4.05m^2**

Fig 4 Calculations of lecture room ventilation

Alberich and Lopez de Asiain could be used though strictly the light shelf would not be necessary over such a shallow plan. South side, atrium-facing offices will need no shading, but openable windows.

In order to attain cross ventilation in double banked offices we agreed that a false ceiling void, split in plan, could act as exhaust for the outside office and supply to the atrium -facing office.

Other possibilities considered, included wind towers on the south facing wall, providing extract for the south facing offices and supply (possibly with evaporative cooling) to the atrium-facing offices.

It was suggested that a sloping lightweight roof be used over the concrete slab over offices to improve aerodynamic flow (to generate more reliable negative pressure at ridge of atrium to under prevailing wind conditions) and to offer improved solar protection.

The north side office, (top floor only). No particular problem from direct radiation (shading will be in atrium roof). Cross-ventilation pattern a little uncertain, and could reverse under prevailing south west breeze if the atrium is positively pressurised by large openings on the south side.

2.7 The Atrium Patio

This has an important function in driving the ventilation of the rooms (and providing daylight). However it must be realised that the atrium will only act as an extract if the air temperature in the atrium is greater than ambient. This is already very high under extreme conditions.

However the upper part of the atrium is not generally occupied, except for the bridges and walkways. Physical hydraulic modelling studies (see proposal) of other buildings, and monitoring, have already suggested that a distinct horizontal interface between cool air and warm air can exist. Moreover the vertical position of this can be controlled by the distribution and size of the ventilation openings in the top of the atrium and in the lower zone.

It is then our intention to allow the upper unoccupied part of the atrium to be about 5°C to 7°C above ambient and to be ventilated at a relatively high rate, providing exhaust for the library, lecture rooms and offices.

The atrium floor will be extensively planted, and will contain fountains. Both of these will have a cooling effect. Solar gains will be minimised by atrium roof shading, and the ventilation rate of this lower atrium zone can be relatively small, since there will not be a high density of gains.

We propose to stabilise this volume of cool air by minimising openings at low level on the south external wall (facing the prevailing wind), and minimising openings on the north side of the atrium at low level. The vent openings to the library will be at first floor level and will be carefully designed to direct air upwards. Studies on other buildings already indicate that this cool volume will only be disturbed from openings above when the incoming air is at a lower temperature than the air already at the base of the atrium. This would be the condition at night when it would then provide useful structural cooling.

On the cooling effect of vegetation and fountains, so far, no detailed analysis has been carried out, but it seems that it should be possible to reduce air temperature at least half way to the wet bulb temperature, and there would be further cooling from massive surfaces cooled by night ventilation. If asked to guess, we would say that the temperature could be kept at least 5°C below the maximum daytime temperature.

The proposed physical model study will assist in assessing the plausibility of this strategy.

Other matters concerning the atrium, discussed at this meeting, were the need for shading and ventilation in the atrium roof. There would be an advantage in extending the volume of the roof above the occupied spaces, giving stack effect without the penalty of bringing hot air in contact with the higher level atrium -facing offices.

It was agreed that the ventilation openings should be in the ridge, to give maximum stack height and the best aerodynamic extract effect. Also, if sufficient ventilation was provided, internal shading of a variable shading factor, with a minimum transmission of about 15% would be appropriate.

It was suggested that the north sloping atrium roof could be translucent, with a transmission of about 10%. It could be supported by an external structure, which at the ridge would support the vent element, and on the south sloping glazing would support the fixed overhangs.

Certain environmental problems could result from having this large, long atrium one single volume. Noise would be a problem, and fire safety might be an issue. Also it was felt that the length of space might cause the ventilation flow patterns to be less stable. It was agreed that the plan would allow the space to be divided into three by glazed screens supported by the crossover bridges. Another possibility was to divide the atrium into three courts, separated by accommodation.

2.8 Other matters

At this meeting I obtained detailed climatic data giving average hourly values of solar radiation for months, orientations and inclinations, air temperatures humidities and wind roses. This data will be of use in subsequent analysis.

I also visited the EXPO site, and saw the bioclimatic landscape designs. I had a meeting with S Alvarez, a member of the Faculty of Engineering who had been involved with the monitoring of the passive landscape designs, and we discussed the performance. We also discussed the kind of analytical tools that were available at the faculty, for thermal analysis of the project building.

The most successful performer was the sunken rotunda, confirming that even outdoors, cool volumes of air could be held stable and separated from warmer moving air above.

I also visited the school at Guillena by Alberich and Lopez de Asiain. Here I saw the use of a freely ventilated atrium, and the lightshelf /louvre combination for south-facing classrooms.

3.1 Meeting in Dublin

A brief meeting was held with Jaime Lopez. The purpose of this meeting was for him to show me modifications at some level of detail which were a response to our discussions in Seville.

The new section is shown in fig 5.

It was agreed that a partially opaque atrium roof would be appropriate. My opinion now is that if the atrium is to contribute significantly to lighting of the surrounding spaces, the glazed area is a little too small. I recommend that a physical model study is carried out to check the daylighting levels.

Planting in the atrium must give careful consideration to available light levels in winter.

It was agreed that if moveable internal shading, probably of cloth sails on mechanical rollers, were installed as indicated, the shading on the south facing glazing need not be movable, and there could be fixed overhangs giving seasonal shading control. However, these would have an influence on winter daylighting levels which must be allowed for. (See above.)

The nature of the ridge ventilation opening needed further consideration, both from its aerodynamic performance and its weather performance.

The division of the atrium had been attained by glazed screens.

There had been a change in the library layout; the bookstacks had been moved to the atrium wall. This lost the atrium-adjacent reading area, but was felt to be necessary for library planning reasons.

Cross ventilation would still be available (above the bookstacks to high level extracts) but a daylighting contribution to the reading area from the atrium now seems unlikely. However this daylight might now be available to illuminate the bookstacks, provided UV filtration is incorporated in the glazing.

It was agreed that the space to the north of the library should be developed as an outdoor study area, if security problems could be solved. This outdoor patio using pergolas and other landscape features, could alleviate the thermal problems in the library itself, at times of overheating risk.

4.1 FLOW MODELLING STUDIES

This technique has been developed by P. Linden, Department of Mathematics and Theoretical Physics, University of Cambridge. It involves the use of a scale model, usually about 1:100, made from transparent acrylic sheet. The model is immersed, inverted in a tank of water which represents ambient air. Warm air is modelled by salt solutions, the higher the temperature, the higher the concentration. The salt solution is dyed so that flow is and mixing can be visualised. Vents are modelled by openings in the model, usually capable of being closed or varied.

The dimensional and time scaling, together with the viscosity and density variation of the fluids, can be shown to result in the system accurately modelling both spatial flow patterns and quantitative air flow (1). Flow patterns can be detected visually, usually recorded on video, and temperature variations by the density measurements of fluid samples.

The technique can be used to investigate transient conditions or steady state. In a typical example of the former, a model atrium may be filled with "air" at a scale temperature increment of 10°C above ambient. Vents will then be opened and the evacuation process of the "hot air",by buoyancy flow, will be studied.

For steady conditions, a heat input is modelled by an array of nozzles introducing a concentrated salt solution into the model. Strictly this is a convective source such as a ducted warm air system, but it is claimed that it can be used to model a hot surface such as a sun patch without much error.

Radiative and conductive heat flows cannot be modelled. Thus the significant heat loss through the glazed roof of an atrium cannot be studied. Flow due to wind pressure has not yet been modelled, but this is a possibility which will be investigated.

Plate 1 shows the acrylic model made of a part of the proposed building representing one of three atria and typical adjacent spaces - lecture rooms and library on the north and offices on the south, as indicated if fig 6.

Only the openings from the offices on the south side and on the third floor north side, were modelled. But since the flow within the lecture rooms and library was to be investigated, these rooms were modelled as complete enclosures, including the raked floor which provides the underfloor ventilation.

The experimental arrangement is shown in fig 7.

The detailled results of the study are described in a separate report (2). The visualization is recorded on video and is available from .... . Plates 2 - 6 are from the video.

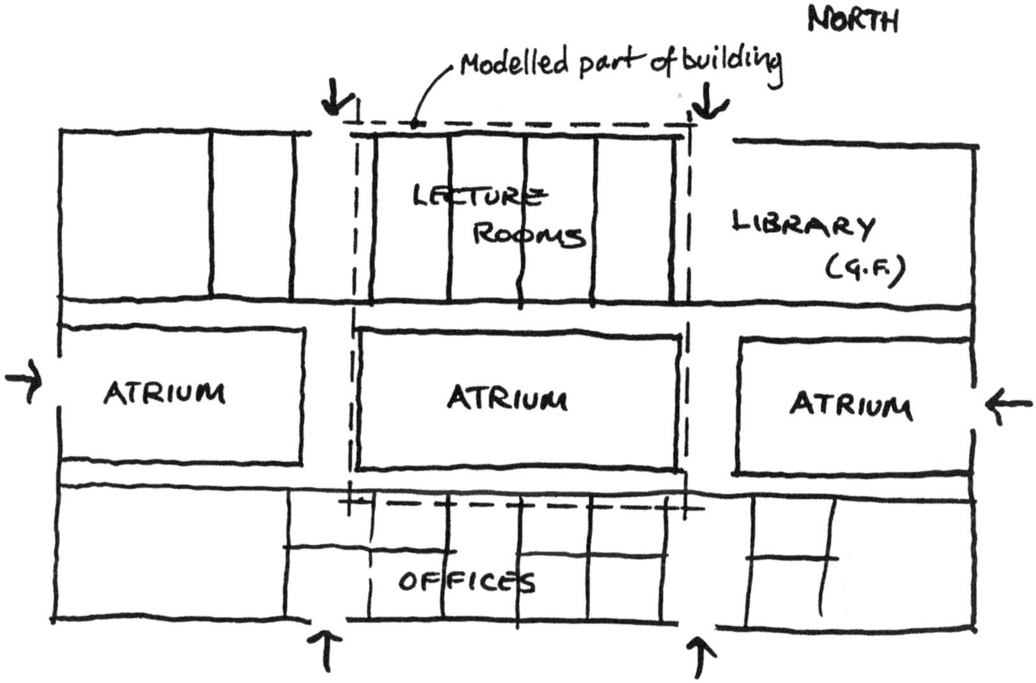

Fig 6 Plan showing part of building modelled in flow study

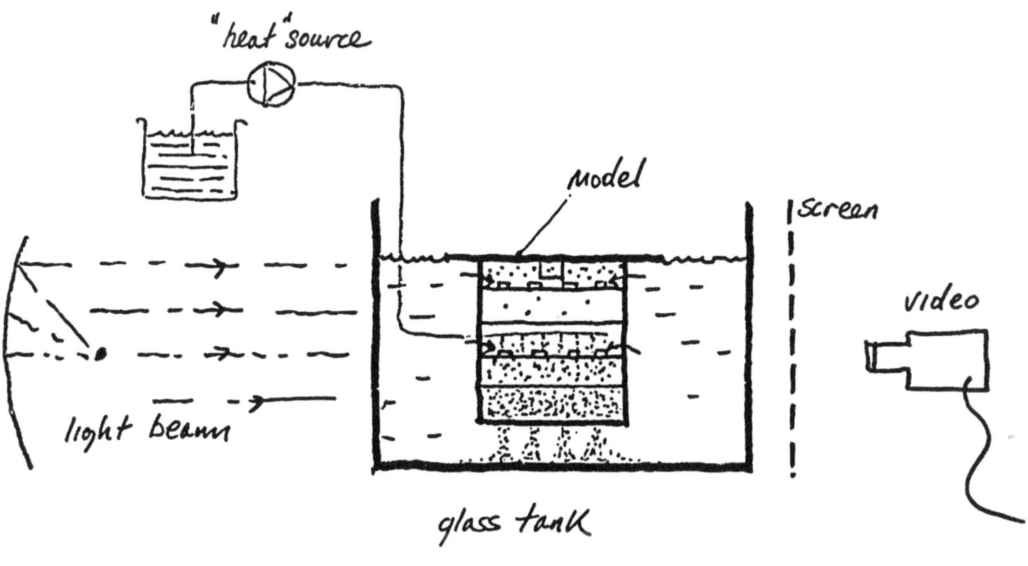

Fig 7 Experimental arrangement for flow modelling

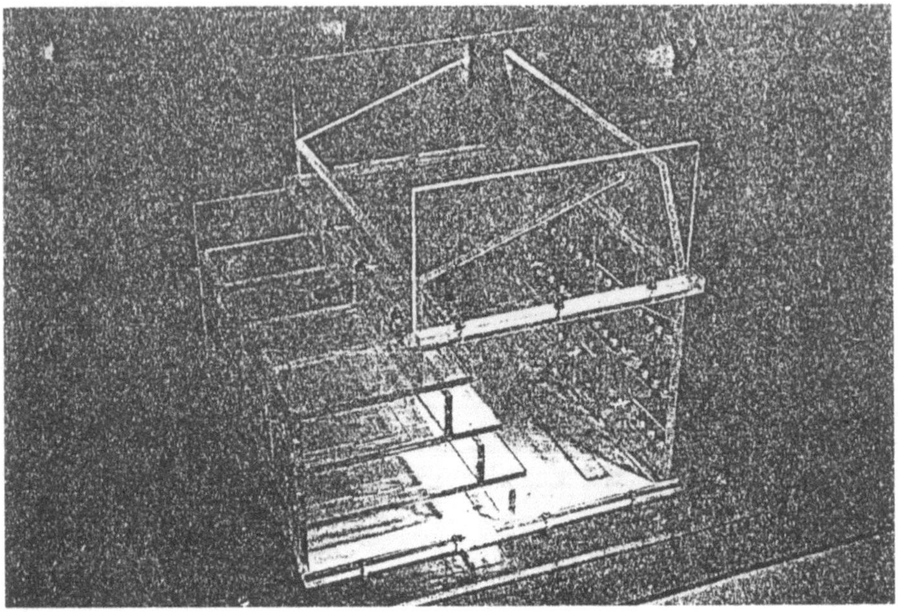

Fig 8 Acrylic model of part of proposed building

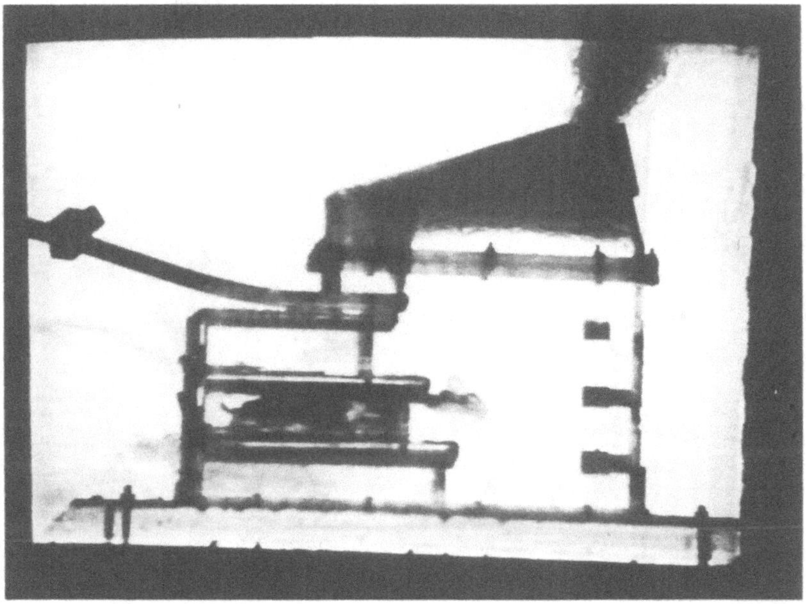

Plate 1 Ambient air drawn through lecture rooms by buoyancy of solar heated air in upper zone of atrium. Note neutral buoyancy air emerging into atrium.

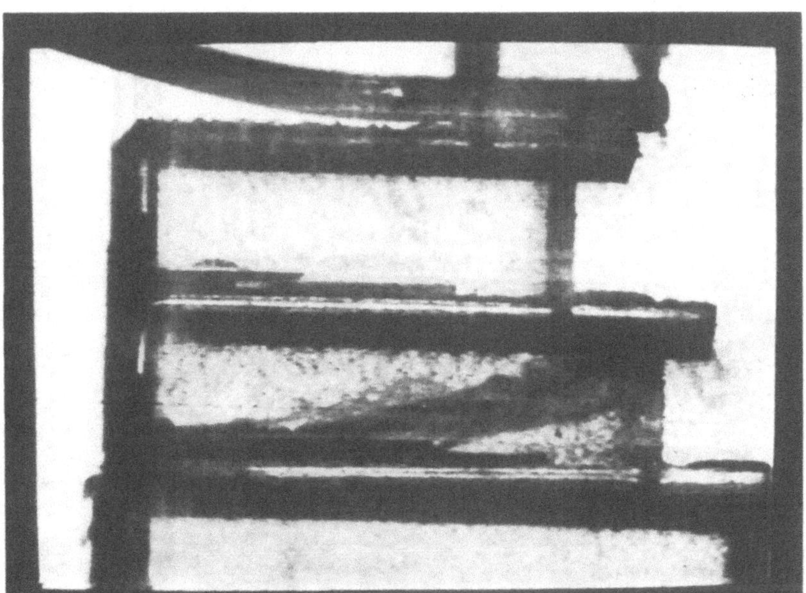

Plate 2 Close up of fresh air delivery via underfloor ducts.

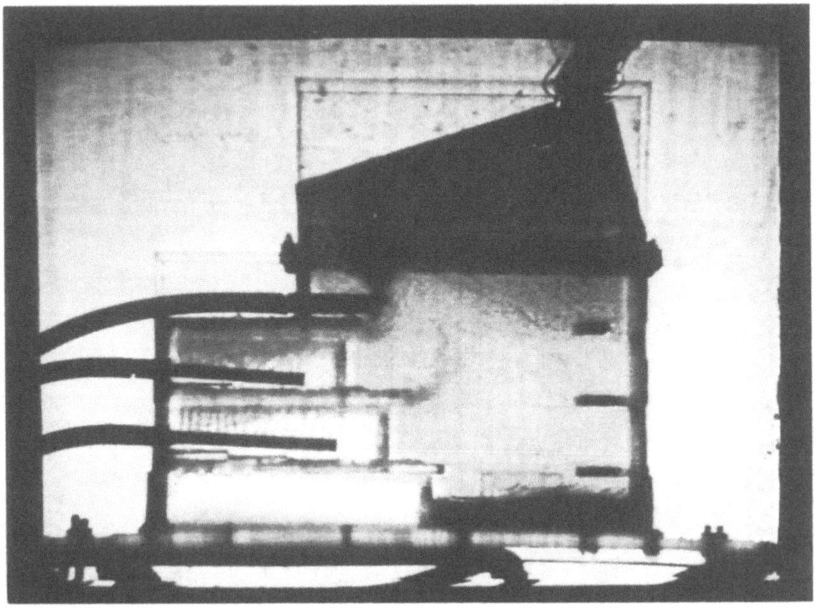

Plate 3 Heat gains from occupants in 1st and 2nd floor lecture rooms. Warm air plumes emerging rise immediately. Cool pool at base of atrium showing slight erosion due to ambient air being drawn through ground floor library.

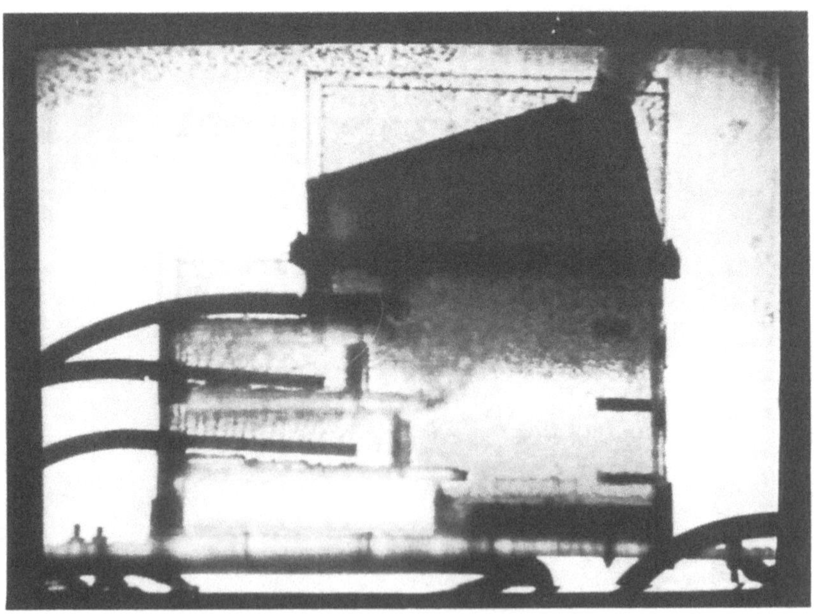

Plate 4 Reduced openings lower stratification layers. Cool pool stable due to zero vent through library.

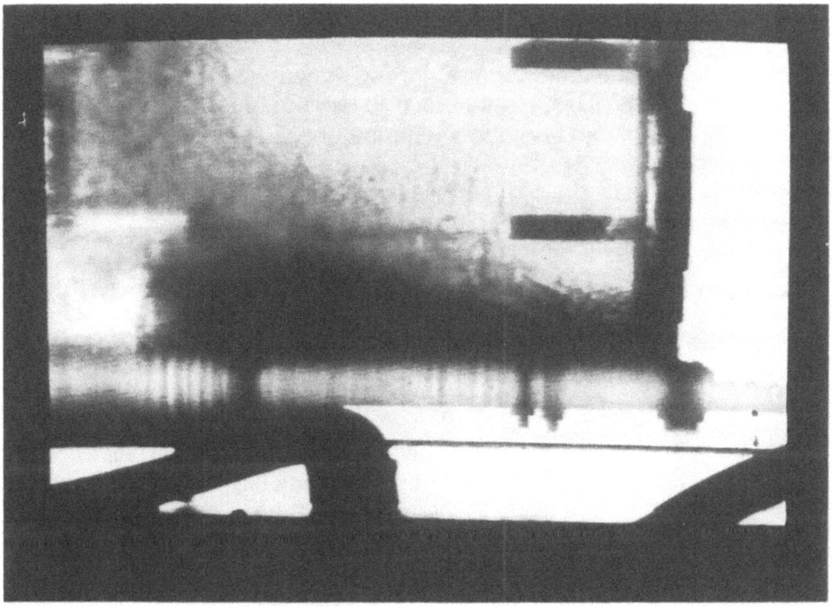

Plate 5 Erosion of cool pool due to ingress of ambient air from ground floor doorway

4.2 Summary and Conclusions of Flow Modelling Studies

1) Solar heated upper atrium will drive ventilation through the lecture rooms. This heated upper layer is established within about 20 mins (real time) and has a temperature of about 7°C above ambient.

2) An estimate of heat losses through glazing due to this temperature difference, but not modelled is as follows -

$$200m^2 \times 5W/m^{2o}C \times 7^oC = 7kW$$

which will have a negligible effect compared to the power of the solar gain of 150kW.

3) Position of separation layer between hot air and air at near ambient temperature, can be controlled vertically by relative sizes of top and bottom openings and magnitude of heat input. In the real case the latter could be controlled by the use of shading devices.

4) Generally the position of the separation layer is above the occupied zones of the atrium.

5) Flow rate through an unoccupied lecture room can be calculated from the total observed flow rate. This ranges between 3.5 and 4.6 m^3/s according to the number of lecture rooms being ventilated. This is close to the original simple calculations (fig 4) and the lowest value is sufficient to keep the effective comfort temperature (CET) in an occupied room to below 29°C with an outside design temperature of 31°C.

6) The underfloor ducts into the lecture rooms show good distribution of fresh air.

Heated (and humidified) air from the occupants appears to move vertically into the upper zone of the room creating a stratified layer close to the ceiling, which is at 4m. The hot air exhausting into the atrium is replaced by fresh air via the underfloor ducts. This leads to stratification within the lecture room height which will probably result in a lower comfort temperature at the occupant position, than predicted in (5), which assumes perfect mixing.

7) The additional gains from the occupants have a small effect on increasing the hot zone temperature and the flow rate (about 10%) However the building is modelled only about 1/4 occupied and this effect will be greater for full occupation. This will help to compensate the slight reduction in flow rate when all room vents are open as in (5)

8) The pool of cool air, initially 5.4°C below ambient, at the bottom of the atrium warms to about 3°C after about 2 hours due to mixing. However this is based upon the assumed initial condition, and also cannot show the effect of the continuing absorption of heat gains into the structure, cooled by night ventilation.

9) The flow visualization shows that the cool pool is only slightly disturbed by ventilation through the lecture rooms, provided they are occupied.If air is drawn through the lecture rooms which are unoccupied i.e. no heat input, the emerging jet of air at neutral temperature is much more destructive to stratification than the warm air from the occupied rooms.

   The visualization also showed that the ventilation openings from the library into the atrium should be modified to direct the plume upwards, rather than horizontal.

10) At first and second floor levels, the plumes emerging from the lecture rooms showed a tendency to curl back onto the walkways. This was probably because no balcony upstand was modelled, but the design of the balustrade should tend to deflect the plume away from the balcony and the rooms.

11) Suction from the heated upper part of the atrium can maintain the cool pool even when large openings exist at ground floor. However, this situation would be very unstable and easily disrupted by variable wind pressures. This would then cause major disturbance to the cool pool. It is recommended that the circulation routes connected directly to the outside occur at first floor level, and are protected by doors if possible. Open doorways, would reduce the ventilation through the lecture rooms.

5.1 CONCLUSIONS FROM DESIGN SUPPORT ACTIVITY

This scheme is notable in that it proposes a daylit, naturally ventilated building in a location with one of the hottest European climates. This is a bold proposition bearing in mind that many buildings of this size and use type would be air conditioned in locations in mid and northern Europe where the summer temperatures are far less extreme. The initial association of the building with the EXPO92 World Exhibition, means that, if built, it will have a very high profile and inevitably become an important exemplar for passive non-domestic building design in southern Europe.

From the consultants viewpoint, this project has been most interesting and enjoyable, partly due to the innovative nature of the building, but also because consultation has taken place steadily from the first proposal and there has been considerable design response to the technical input. It is interesting to briefly review the development of the design.

Initially, the use of the building had been designated only for the EXPO Commissariat. The original form was based around a sunken patio or courtyard, with much of the accommodation below ground level. This idea was not dissimilar from some of the design ideas for the external EXPO environment, where walkways and circular sunken courts were to be cooled by combinations of vegetation, fountains and shading. This landscape work had already proceeded to a pilot study and was being monitored during the design period of this proposal.

The issues discussed with the consultant was firstly the beneficial effects of the earth sheltering of the sunken accommodation and secondly whether the "cooling performance" of the patio could be improved by movable shading and if the patio could be used as a source of cool ventilation air.

It was pointed out that the effect of earth contact is dependent upon a high proportion of the accommodation being in contact with deep ground, and that the plan depth of the proposed building would not only preclude this, but also require mechanical ventilation and artificial lighting, both high energy users.

An investigation of the performance of diurnally operating shading for the patio to minimise radiant gain in the day and maximise radiant loss at night, was about to be undertaken. However, this was overtaken by events since a new end use of the building was established, as part of the University of Seville.This changed the accommodation use type, the most influential on the design being the provision of the lecture rooms and classrooms, and a new design proposal was made by the architect. The open patio developed into a narrow 3 - 4 storey atrium and the earth sheltering was abandoned. A section was proposed by the architect that owed something of its origin to a school already built at Guillena by Alberich and Lopez de Asiain.

At this stage the winter performance of passive solar utilization seemed to be receiving priority. However after discussions it was agreed that due to the small seasonal heating costs, most emphasis should be placed on the energy saving due to avoiding artificial lighting, and mechanical ventilation. Both of these are electricity users and in primary energy consumption almost certainly would account for the major energy demand.

Fortunately the proposed section, stepping back from the atrium, permitted double sided daylighting to the deep teaching rooms on the north side using the atrium as a daylight source. An interesting decision was to separate out the ventilation function from the daylighting of the windows. This, together with underfloor delivery, would permit blackout in the lecture rooms and acoustic and possibly smoke control via specially designed louvred ventilation openings. The idea of the atrium driving this ventilation by stack or wind-induced suction pressures was developed at this time.

Another important concept was the generation of temperature stratification in the atrium, maintaining by minimal daytime ventilation a pool of cool air, below ambient temperature, at the base of the atrium. Upper parts of the atrium would be ventilated at high rates to exhaust the gains from the densely occupied spaces on either side. This idea was prompted by looking at video of laboratory flow models of a UK law courts building. This was later tested and confirmed by a physical model test of the Seville building.

In the library, the deepest plan space of about 17m, the central section was to be the bookstack illuminated by special low energy sources and luminaires, with reading ares illuminated from the outside to the north and the atrium to the south.

A further change of section shifted the solar aperture of the atrium to a steeper inclination and included permanent geometric shading to reduce the solar gains in summer but improve winter gains. Other changes included bringing the walkways further into the atrium and moving the book stack to the atrium for reasons of library security. However a secure outdoor reading court was proposed for the north side of the building, to minimise the number of readers not provided with natural light.

This is how the building proposal remained at the last consultancy meeting in Dublin in March 1989. Subsequent fluid modelling has confirmed the ventilation proposal but not lead to any detailled recommendations or changes. However we believe that these results would be of use and influence when the detailled engineering design of the ventilation openings is carried out.

The building has undergone considerable development, partially under the influence of technical and environmental design considerations. From my point of view as consultant, this has been very rewarding and must be due to a large extent to the creative and positive way that Jaime Lopez, the architect has responded to the technical issues and demands.

I would also like to record my appreciation of the cooperation with Drs P Linden and G Lane-Serf during the flow modelling study, and to the B2000 Coordinator Cees den Ouden for the efficient and flexible way that these consultancies were commissioned.

Nick Baker 05.10.90

References

(1) Emtying filling boxes: the fluid mechanics of natural ventilation.
P F Linden, G Lane-Serff and D A Smeed
Journal of Fluid Mechanics (1990), Vol 212.

(2) Flow Modelling Studies.
P F Linden, G F Lane-Serff, and N V Baker
Cambridge Environmental Research Consultants Ltd
Sheraton House, Castle Park, Cambridge CB3 0AX

INCREASING DAYLIGHTING OF THE COURT BUILDING ORANIENSTRASSE 9 IN BERLIN

Gabriele Willbold-Lohr
Oldenburgerstraße 68, D-5000 Köln 60, Germany

1. INTRODUCTION

The court building of the typical dense urban structure in Berlin Kreuzberg need to be retrofitted for further use as ateliers for artists. The building facing a three sided court yard provides a quiet working environment well protected from the urban noise by a 5 storey block building. However the illumination of these spaces is very poor due to the dense urban structure. The goal for this study is therefore to improve the natural illumination of these working spaces for artists aiming low cost solutions. As the geometrical situation is very complex for analysis, the study is carried out by use of architectural scale models.

2. SITE ANALYSIS

The main building facing the street Oranienstraße is a 5-storey building which forms part of a block structure and thus creating a noise protected and quiet inner area. For reason of a higher exploitation of the urban area, the main buildings facing the streets are added by buildings perpendicular to the main buildings. For illumination purpose those buildings face to a three- or four sided light court. The illumination of these court buildings generally can be seen as poor daylit due to the relatively small distances between these buildings.

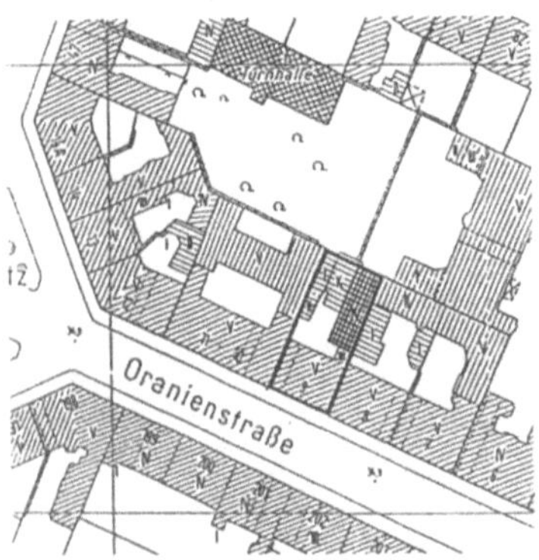

Fig. 1 Urban context

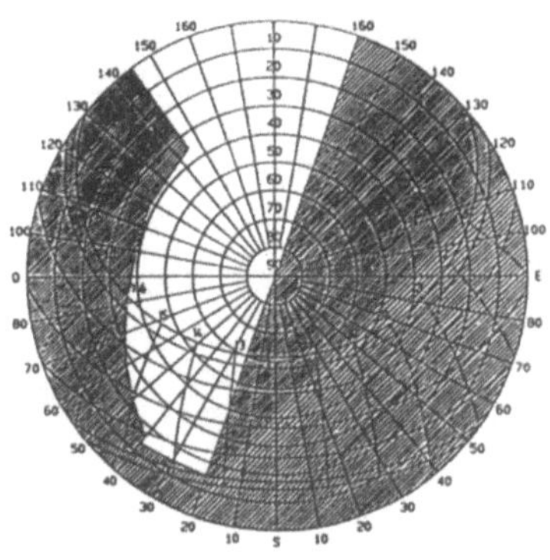

Fig. 2 Analysis of sunshine on the building facade

This three-sided court yard, sized 8.5 x 22 m, is open to a greater court yard and can be enterd from the street by an open hallway. The entrance of the court building is through this small court yard. For fire protection reasons the whole block structure is separated by so called fire walls which are situated perpendicular to the block structure and which are massive walls with no openings. Sometimes these fire walls have been used as a separating wall between two buildings, sometimes this fire wall is left unattached und thus creating a wider courtyard for better illumination of the court yard building. One of the main characteristics of these small court yards is the use of dark bricks as building material, both for the court yard buildings as well as for the fire walls.
In the case of the court yard of Oranienstraße 9 the fire wall remained unattached and the building material used for the fire wall as well as for the facade of the court yard building is dark brick. The facade of the court building is facing the 20 m high dark brick fire wall.
Photometric mesurements on site resulted in low reflectance values of these building materials: 0.15. Due to the fact that building measures already had been started photometric measurements inside the building did not provide reasonable results. However the visual appearance of the illumination of the spaces is very poor.

3. REQUIREMENTS FOR ILLUMINATION FOR ARTISTS USE

Depending on the type of work the artist is carrying out, the quantity of light differs - from low light levels for writing (200 - 300 lux) to very high light levels for designing jewelry (up to 1500 lux) and etching (2000 lux). As the type of work beeing carried out is not yet fixed at design stage, the quantity of light needed could not be specified.
Qualitative aspects of illumination needed to be considered providing a high quality of illumination - no direct sunlight entering the space, a uniform light distribution throughout the space and avoiding harsh shadows.

4. METHODOLOGY OF PREDICTION

As the urban situation is very complex (court yard) and the performance of daylight control elements hardly can be predicted with the use of graphical design tools or computer programs, the scale model as a design tool has been used for the study.
A scale model in scale 1:50 has reproduced the urban situation thus enabling the photometric evaluation of all spaces in the court building at simultaneous measurements. Therefore the scale model has been equipped with photometric sensors measuring illuminances and has been tested in the artificial sky of the RWTH Aachen.
Parametric studies so easily could be carried out taking into account the performance of different reflectances of the facing walls of the court yard as well as the performance of light directing elements.

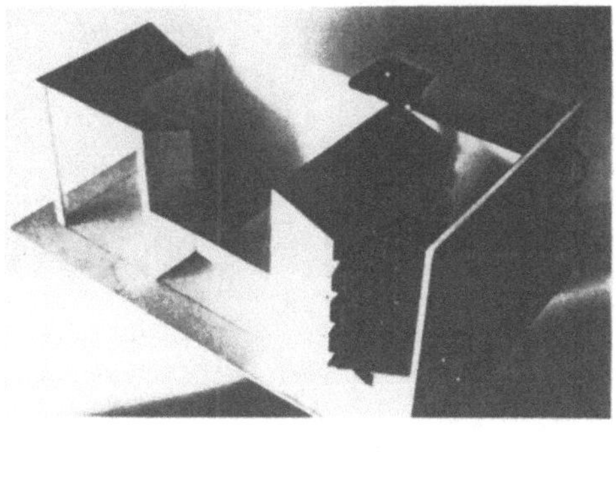

Fig. 4 Scale model setup

5. SCALE MODEL TESTING AND RESULTS

A parametric study has been carried out investigating the performance of the following paramters:

- altering reflectances of the facades facing the court yard.

  The present situation with the dark brick walls did present the reference case to which all the following studies will be compared. The reflectance of the brick walls has been defined as 0.15 and has been simulated using a dark cardboard providing a reflectance of 0.15. The reflectances of the facades has been changed to 0.70, which represented a white paint with a high reflectivity.

- double side lighting of the space

  adding a second window in the rear of the wall of the working space. This solution will create problems due to the fire regulations, as the rear wall of the working spaces represents the fire wall and no openings usually are allowed.

- using light control elements in the facade aperture of the court building.

  Reflectors above eye-level are designed to use zenithal light entering the court yard.

All results of the photometric testing have been converted to Daylight Factors and presented graphically as daylight distribution curves in the section of the building. Parallel to the photometric measurements simple scale model studies have been carried out to visualize the effect of various parameters.

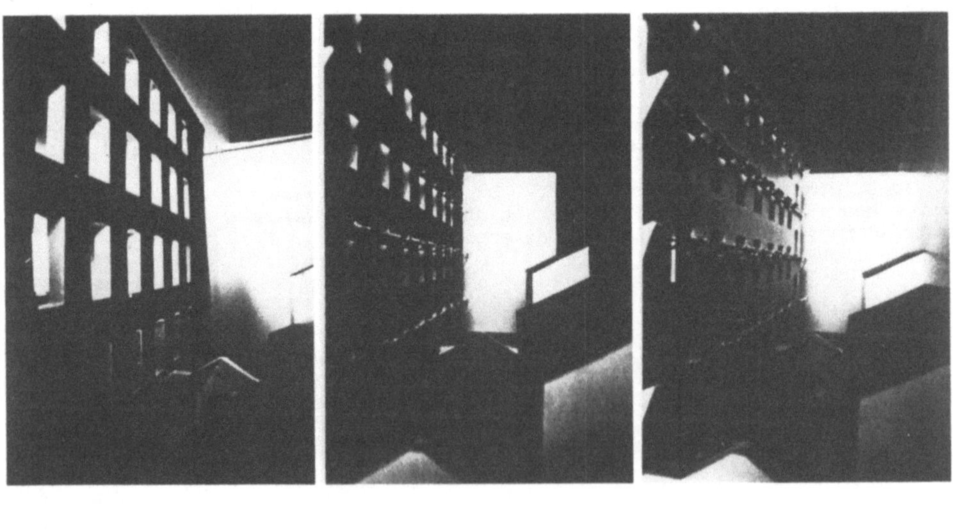

Fig. 5 Scale model facade aperture details
left: original situation, middle: adding light shelves to the original situation, right: white facade apertures with light shelves

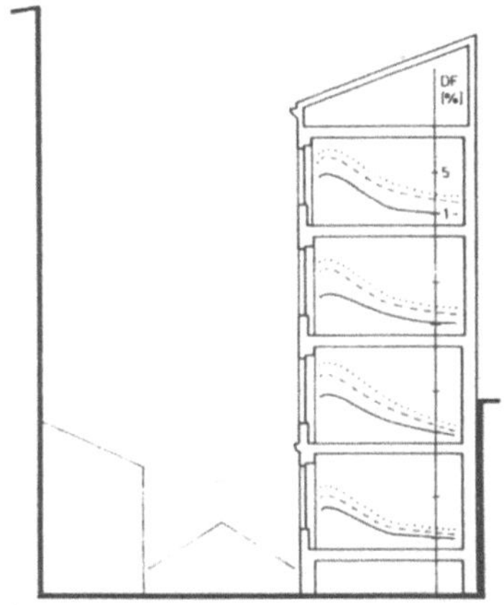

Fig. 6 Daylight distribution due to altered facade reflectances
solid line: reference
dotted line: white facades
dashed line: fire wall white, facade aperture dark

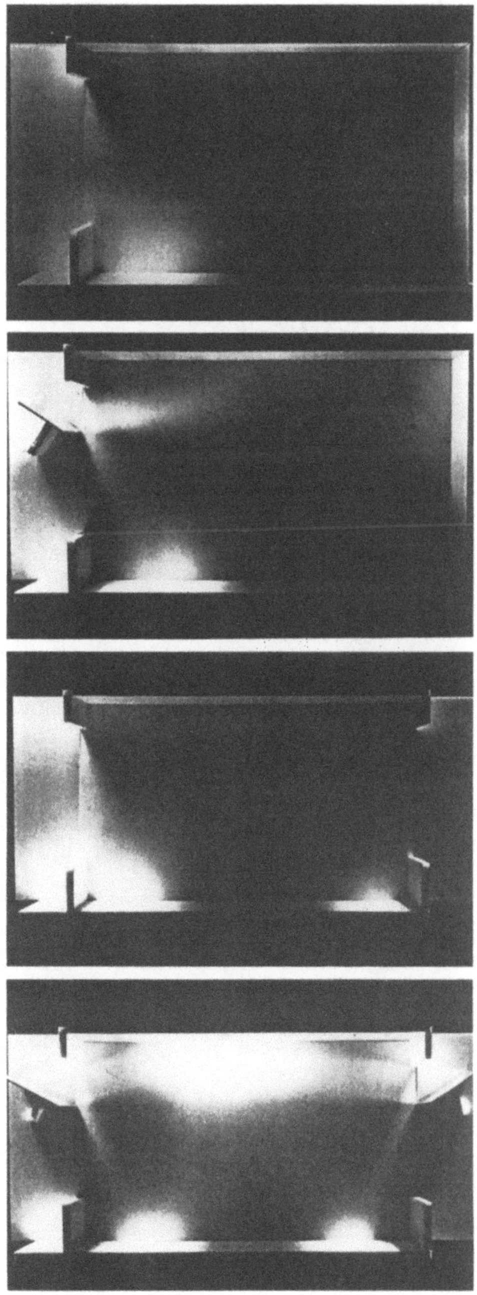

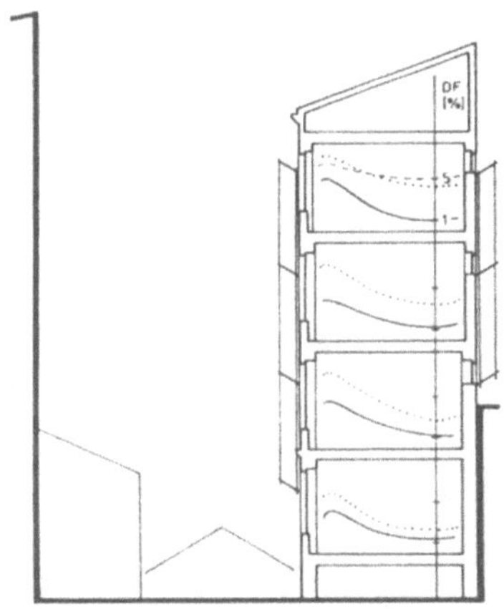

Fig. 8 Daylight distribution applying light shelves plus window in the rear wall of the space
solid line: reference
dotted line: white facades with control elements

Fig. 9 Twodimensional performance studies

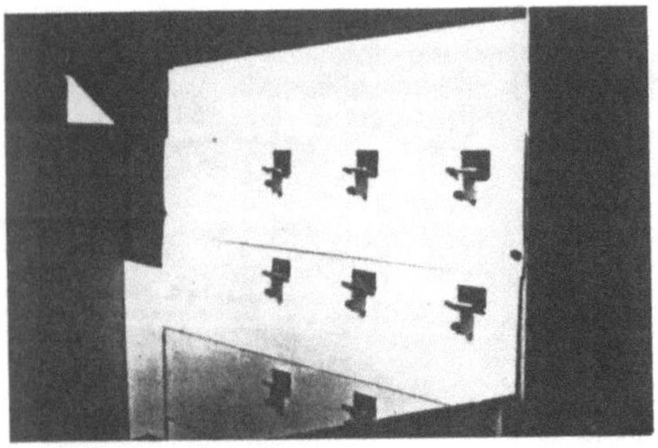

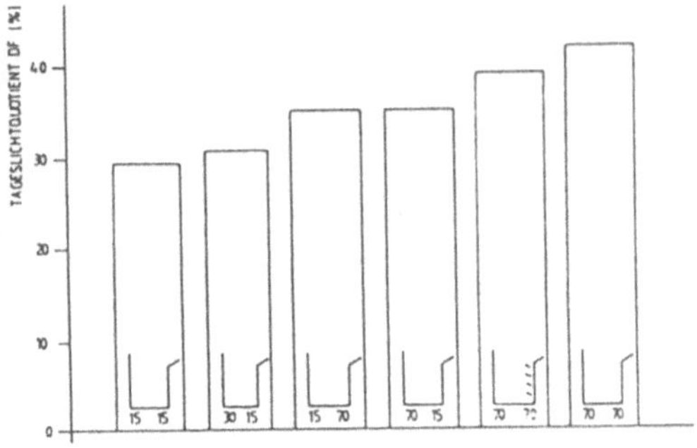

Fig. 10 Scale model rear wall (fire wall) with light shelves

Fig. 11 Horizontal Daylight Factor at centre of courtyard influenced by different strategies (the numbers given in the small schemes indicate the reflectances).

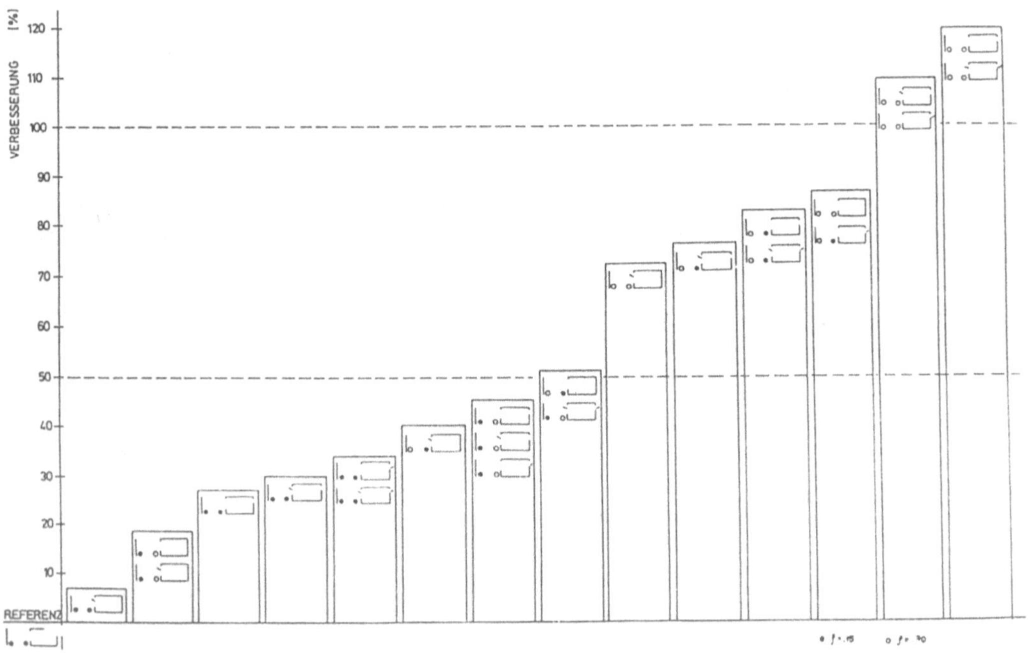

Fig. 12 Quantitative comparison of all tested strategies (the small schemes in the columns indicate the strategy, the dots in the schemes represent the reflectance of the facades, white dot: white facade, black dot: dark facade).

5.1 Results of scale model testing

23 different strategies have been tested with the scale model (see also Fig. 12). In this summary only general results will be discussed.

5.1.1 Quantitative improvement of natural illuminance levels:

Two strategies have proven best efficiency for increasing the amount of daylight in the spaces:

- painting the courtyard walls with white colours
- double side lighting in the spaces.

Compared to all tested strategies based on the present situation of the building structure (dark brick walls) the option to brighten the walls of the courtyard is the most successful concerning the increase of light levels in the spaces. Of special interest hereby is the fire wall facing the facade aperture of the artists building as this wall represents a major light source for the illumination of the spaces. Covering the courtyard walls with paint of a high reflectance and if possible of a glossy reflectivity, the amount of light in the spaces could be doubled compared to the present situation (reference case). For quantitative improvement of the illumination therefore this simple and low cost solution is the most effective.
Having a second window in the rear wall of the working spaces helps to improve the quantity of light for another 20%; but due to the fact that openings in fire walls usually are not permitted, these additional windows may cause delay in the permissing procedure for the building application. For quantitative aspects of the illumination the rear window is not required.

5.1.2 Qualitative improvement of the natural light distribution in the spaces:

Daylight control elements such as the tested reflectors turned out to serve both goals, the quantitative and qualitative improvement of the natural illumination. The efficiency of the quantitative improvements depends on the amount of light which can be quantified before applying daylight control elements: if the light levels are generally low (as given by the present situation, dark walls) daylight control elements contribute to an increase of the daylight level in the space. Is the light level in the space already high as given for white courtyard walls, then the use of daylight control elements effects a reduction of the light level. No tested strategy applying daylight control elements results in higher daylight levels than the simple measure of brighten the courtyard walls.

However daylight control elements provide two other advantages:

- qualitative improvement of the light distribution in the space
- specified direction of light to a defined working place (horizontal or vertical).

According the user requirements for uniformity of the natural illumination daylight control systems can be used to achieve this goal. The additional window in the rear of the wall is needed for the maximum of uniformity. Using light directing elements also involves special attention of the design of the ceiling of the space: either high reflecting and diffusing the reflected light or designed with additional reflectors to redirect the light to a specific working place within the room.

Greenery in the courtyard has to be designed very carefully as green leaves reduce the high reflectance. Partly greenery would not affect the illumination but covering the fire wall with greenery would reduce the light levels in the spaces nearly to those levels of the present situation. If greenery has to be there (which is not recommended) then only daylight control elements contribute both in quantitative as well as qualitative aspects to the illumination.

STUDY CENTRE, MACYNLLETH, WALES

Jon Walker, Mike Wilson, Yvonne Dean
Low Energy Architecture Research Unit,
Polytechnic of North London, Holloway Road, London N7 8DB, UK.

1.0 INTRODUCTION

The Low Energy Architecture Unit at the Polytechnic of North London have designed a study centre for the National Centre for Alternative Technology, with financial assistance from the CEC Building 2000 programme. Unusually for this programme PNL have taken the role of both of architects and energy consultants for the project. This has been successful in allowing an integrated design approach to be followed from the outset. The building includes:

* Planning arrangements and construction type is responsive to the occupancy mode of the building.
* Energy saving through daylight design to ensure minimum energy consumption.
* Passive solar ventilation pre-heat through six circulatory and sun spaces used as buffers.

Simulation has indicated that 47% of electric lighting is displaced by daylighting, despite evening and night time use. Ventilation pre-heat of 50% of ventilation air via buffer spaces gave annual savings of 7800kWh. Simulation for the final design has indicated annual heating season auxiliary energy use of 13600 kWh representing an energy saving of 87% over UK 1985 Building Regulations standards.

2.0 PROJECT DESCRIPTION

Accommodation will consist of 8 study bedrooms for 16 people with associated washing and eating facilities, one large teaching space, study areas, a library/reading room and a public demonstration and research room. Area heated : 400 m^2, volume heated 1325 m^3, total Area: 560 m^2, total volume 1703 m^3. The building orientation is offset 10 degrees to the west of south with a linear layout from east to west. The ground floor uses a 100mm screed for the installation of the underfloor heating system, separated from the ground floor slab by 100mm mineral fibre bats. The 150mm cavity in the timber frame walls is completely filled with mineral fibre bats. The roof has a total of 190mm of mineral fibre in the form of a cross lattice of 140mm and 50mm. All windows are double glazed. U values: Walls 0.26 W/m^2K, Roof 0.18 W/m^2K, Floor 0.32 W/m^2K

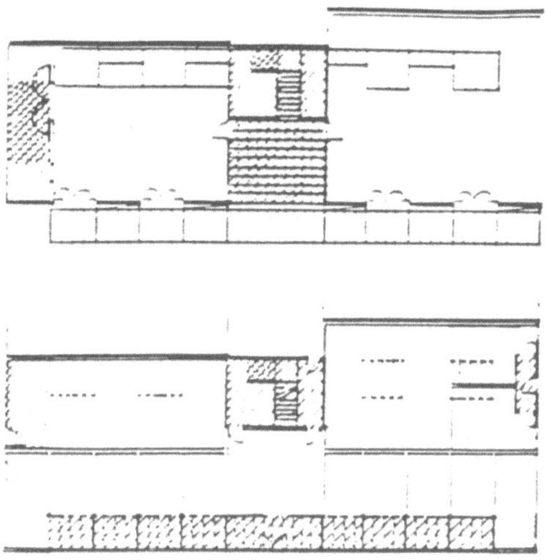

Plans of the Study Centre, North at Top, Buffer Zones Hatched. Top: First floor, showing library and teaching area surrounding stair landing/conservatory. Bottom: Ground floor showing skylights into northern rooms and study bedrooms adjoining the southern buffer space.

The upper floor is a fast response timber frame system rapidly benefiting from direct solar gain in the mornings, and pre-heated warm air entering from the south facing conservatory. The ground floor masonry construction, which accumulates the gain during the day and is warm for evening use when the study bedrooms will be occupied. As the study centre adjoins existing buildings, rooflights have been introduced along the whole length of the ground floor northern rooms, which simultaneously light these rooms and upper teaching spaces. A glazed stairwell penetrates the core of the building, introducing sidelighting for the study areas in the northern rooms on the ground floor, and leads to a south facing conservatory also providing circulation space. Together with a southern highly glazed corridor, which links all bedrooms, the kitchen, and bathrooms, all circulation areas are 'maximum daylit' spaces. The teaching area and library on the first floor receive daylighting from four sides and from rooflights located to provide an even distribution, with minimum glare. Passive Solar ventilation pre-heat has been introduced through six circulatory and sun spaces used as buffers, also reducing infiltration losses.

3.0 THERMAL AND DAYLIGHT MODELLING

The main aims of the design study was to question initial design decisions by means of a sensitivity study - changing just one factor and assessing the effect in terms of energy use and comfort. In the time frame available sensitivity studies were limited to glazed areas, glazing type, insulation thickness, ventilation pre-heat and summer comfort conditions. Daylight optimization took up the greater part of the study and was conducted in the following sequence:

1. Model studies to assess light distribution, quality and aesthetic considerations.
2. Daylight program to calculate average and minimum daylight factors of the base design for input to SERI-RES.
3. Stochastic analysis of lighting electricity use for range of switching point daylight factors from 0.5% to 16%. This generates an

understanding of electric lighting use as a function of daylight factor without altering the building details of the model.
4. Global changes of window areas/daylight factors to determine net lighting and heating energy consequences of fenestration changes for both minimum and average daylight factors. Glazed area increases of 20% and 40% and associated changes to daylight factors.
5. Selective changes to individual windows as required, based on results from steps 3. and 4.
6. As 2-5 for low-E glass.

In order to address both the daylighting and thermal modelling of the building a combination of complimentary tools were appropriate. For daylighting both physical scale models and the Essex IHE 'Daylight' computer model (ref.1) were employed. These provided the necessary daylighting input to the thermal simulation model SERI-RES Version 1.2 (ref.2,3,4). Using CEC Test Reference Year weather data (ref.5) annual simulations were performed for a range of variations upon the original design.

Thermal modelling of the study centre was carried out using the thermal simulation model SERI-RES Version 1.2, modelling the building in 17 zones. Details of effective shading coefficients and solar lost coefficients were based on results of UK test cell studies (ref.6). The weather data used for the annual simulation was a CEC test reference year for Aberporth which has a broadly similar climate and daylight availability. Aberporth is situated 70 km to the south west on the coast and is subject to 2342 degree days to an 18.3 deg C base.

Physical scale modelling (1:20) was carried out in the CIE Overcast Sky facility on four 1:20 models constructed to various designs and comparatively tested, resulting in the initial design. Much of the quantitative modelling was done by computer modelling. The chief advantage of computer daylight modelling over physical scale modelling is the ease with which design modifications may be made. Without recourse to scalpel or paintbrush, window sizes, transmission values and surface reflectance may easily be altered and new comparative daylight factors calculated.

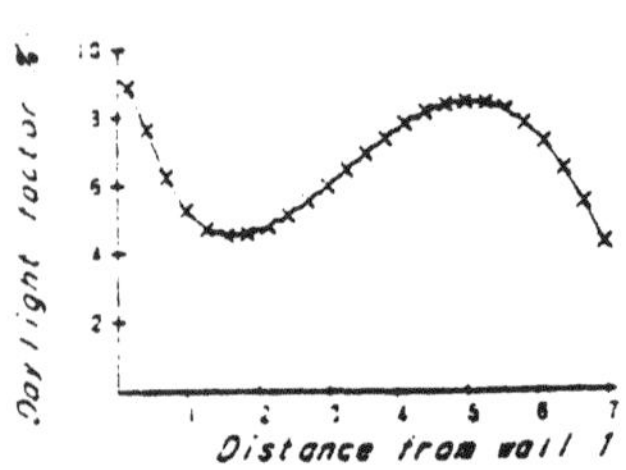

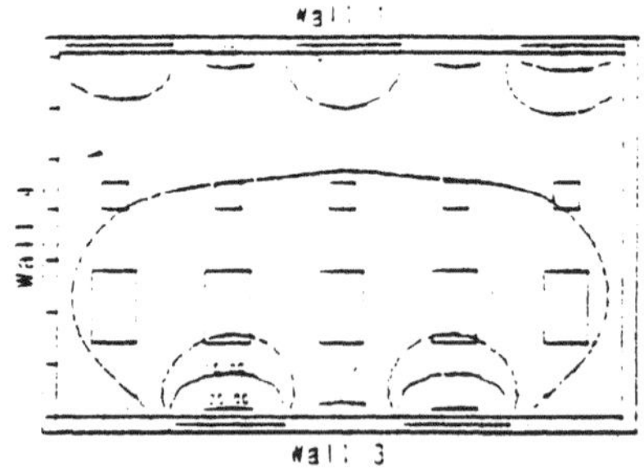

Example Contour Output from the Daylight Program for the Teaching Room in the Study Centre.

The lighting facility of SERI-RES can model both manual and automatic light switching systems, the latter either continuously variable or in steps. The manual switching facility of SERI-RES is based upon work by Hunt (ref.8) which includes an empirically derived probability for lights to be switched on at any given luminance. Following the daylight study, other aspects of the building were studied, most notable of these being the modelling of ventilation pre-heat from the buffer spaces. The difficulty with using SERI-RES to model the resulting airflows is that infiltration is the only airflow allowed but a simple solution to this has been suggested to overcome this drawback (ref.9). By adjusting infiltration rates, in combination with placing an appropriate resistance between the zones, it is possible to model re-circulatory airflow.

4.0 SIMULATION RESULTS

A series of general simulations were performed with switching point daylight factors varying from 0.5% to 16%, without altering fenestration details. This allowed a rapid overview of where potential lighting savings were most likely to occur. Sensitivity of lighting energy use to switching point daylight factor had a characteristic pattern for all zones, dropping rapidly as the daylight factor rose to 2%-4% before levelling off.

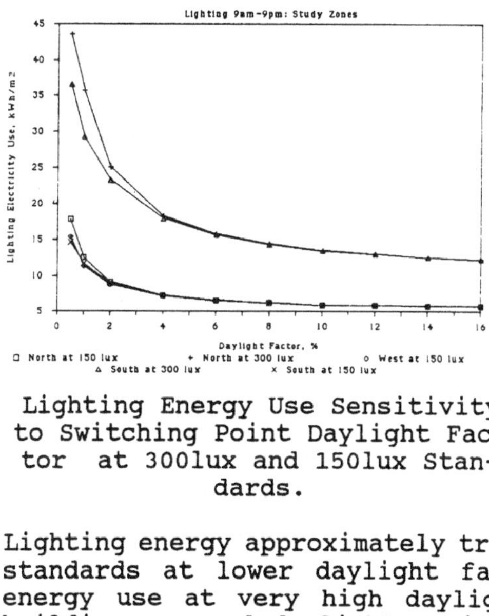

Lighting Energy Use Sensitivity to Switching Point Daylight Factor at 300lux and 150lux Standards.

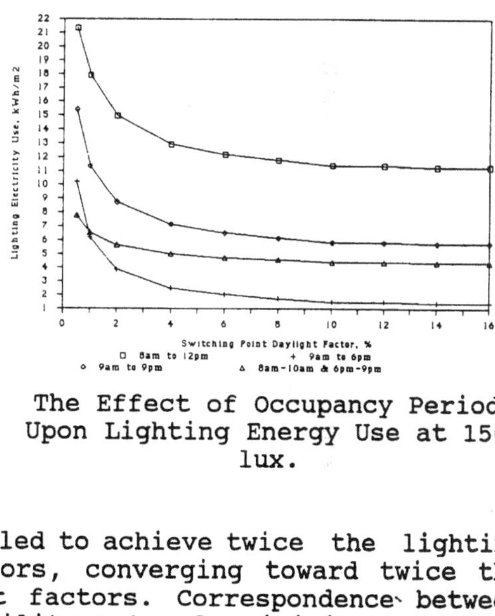

The Effect of Occupancy Period Upon Lighting Energy Use at 150 lux.

Lighting energy approximately tripled to achieve twice the lighting standards at lower daylight factors, converging toward twice the energy use at very high daylight factors. Correspondence between building use and daylight availability not only minimises electric lighting provision but also excludes those times when windows provide thermal losses but no daylight. For each run a zone by zone breakdown is necessary to measure the effects of new parameters on energy consumption, both total and for unit area. Total consumption is needed to predict the value of savings, while unit area consumption allows comparison on an equalised basis, showing up discrepancies. The figures below illustrate the two representations for the final design.

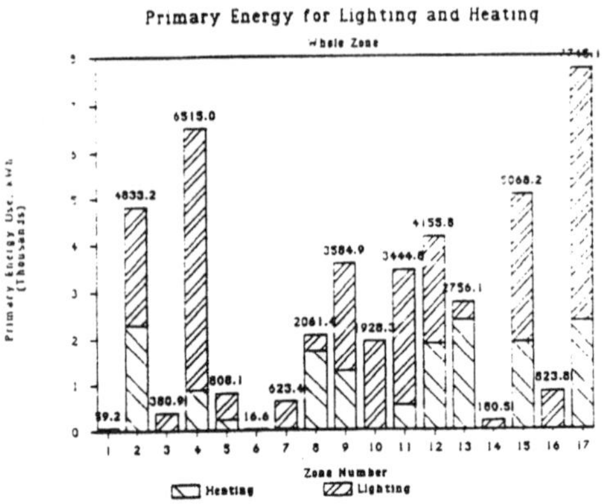

Zonal Breakdown of Lighting and Heating Primary Energy: Totals for Zones

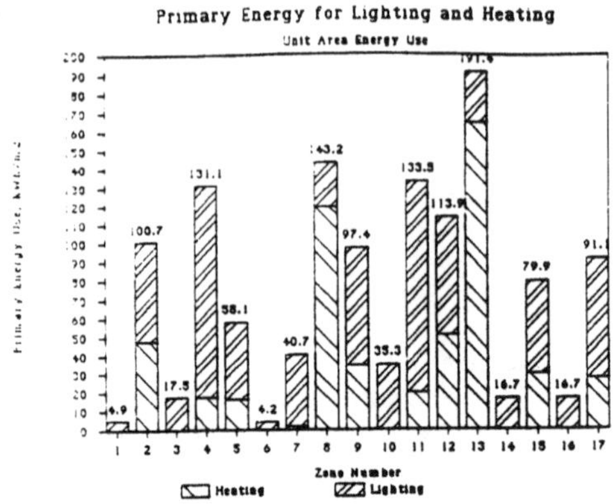

Zonal Breakdown of Lighting and Heating Primary Energy: For Unit Area

5.0 SUMMARY OF SIMULATION RESULTS

Based on working minimum daylight factor, an annual saving of 732kWh is predicted for a 20% increase in fenestration, all zones showing a reduction in total energy use. This trend continued for 40% fenestration increases, saving 1229kWh in the case of minimum daylight factor. As a proportion of the total lighting demand the net saving is 1.8%, or 1.1% of the total energy demand (worth 20 ecu per annum). Lights left on unnecessarily for a small proportion of the year will wipe out these savings, whereas the increased heating demand of 996kWh (2072kWh for +40%) is more or less guarantied, the lighting saving of 1645kWh predicted by automatic switching is conditional upon ideal manual switching behaviour by the occupants. A manual switching system will not realise the full potential lighting savings of greater fenestration predicted for an automatic system, assuming the same lighting standards are to be met. However the manual lighting facility of SERI-RES overrides the minimum illumination specified with an empirically derived probability for switching, shown below.

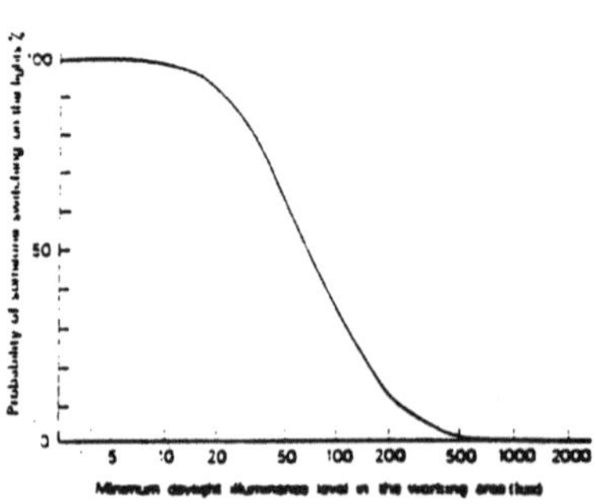

SERI-RES Empirically Derived Switching Probability - from Hunt

Using the manual switching facility of SERI-RES total energy use was predicted to be 4449kWh less than automatic switching, based on working minimum daylight factor. This surprising result suggests that, compared to Hunts data, lighting levels of 300lux and 150lux for manual switching may be generous and people may accept lower daylight levels.

The general sensitivity studies have shown that the base case design for the study centre is not overglazed. Automatic switching could realise some further small savings for a more highly glazed building but not significant enough to justify the cost of the automatic switching system, except possibly for the larger spaces. Taking into account the increased potential for overheating with larger glazed areas the base case window sizes, as originally developed in the model studies, have been retained.

5.1 Further Thermal Simulation Studies

The extensive use of thermal buffers in combination with well sealed windows ensures that a high proportion of ventilation air is pre-heated before entry into the heated zones. Assuming that it is possible to displace 50% of infiltration by pre-heated air, via vents to the buffer spaces, an annual space heating saving of 7800 kWh (200 ecu) is predicted.

Sensitivity to insulation thickness was also investigated for external walls and roof, from a base case of 150mm and 190mm mineral fibre respectively. On materials cost alone savings of 60kWh and 106kWh per 10mm increase yielded simple payback times of 33 and 57 years. This supports the initial design choice of fully filling existing cavities but not making expensive additional provision for insulation. The larger cavities occurring in timber frame buildings are therefore an advantage in this respect.

Summer overheating was investigated for the building assuming maximum ventilation rates of 3ach and 5ach for the occupied and buffer spaces respectively. Serious overheating occurred without the provision of blinds to the buffer spaces and south facing library and teaching area. With venetian blinds specified (50% effective shading) all occupied zones but the library maintained temperatures below 30°C. However with large opening stable doors and skylights as specified, stack flow ventilation greater than 3ach keeps the library temperature below 30°C on the hottest day of the test reference year.

Simulation of the final design has indicated that 47% of electric lighting is displaced by daylighting, despite evening and night time use. Ventilation pre-heat of 50% of ventilation air via buffer spaces gave annual savings of 7800kWh. Annual heating season auxiliary energy use of 13600 kWh representing an energy saving of 87% over UK 1985 Building Regulations standards. The following graphs summarise the main performance characteristics of the building.

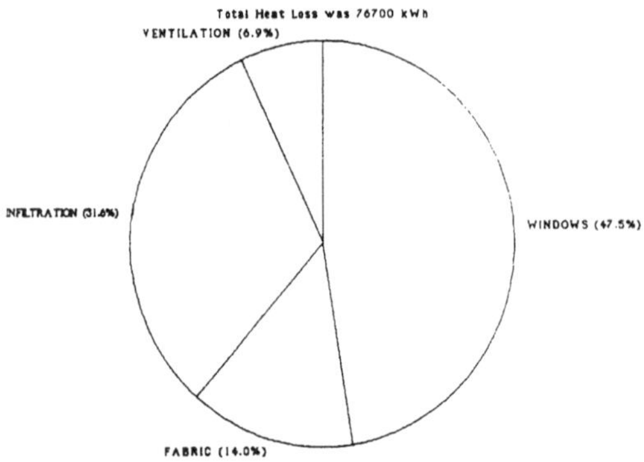

Heating Season Energy Loss Breakdown. Windows account for over 50% of heat loss in this well insulated building.

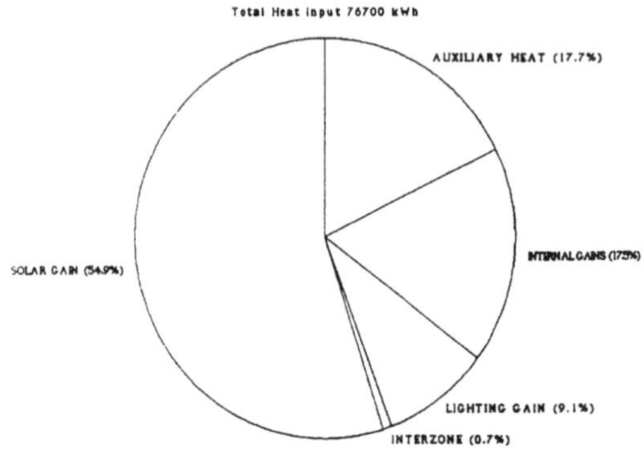

Contributions to Heating Season Energy Supply. Useful solar energy accounts for over 50% of total heat input. This is particularly high because of the extensive use of highly glazed buffer spaces.

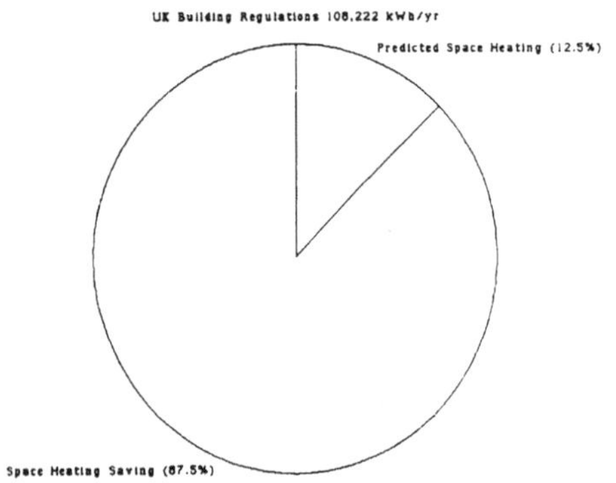

Comparison of Auxiliary Heat Demand for the Final Design with a Design to Current UK Building Regulations.

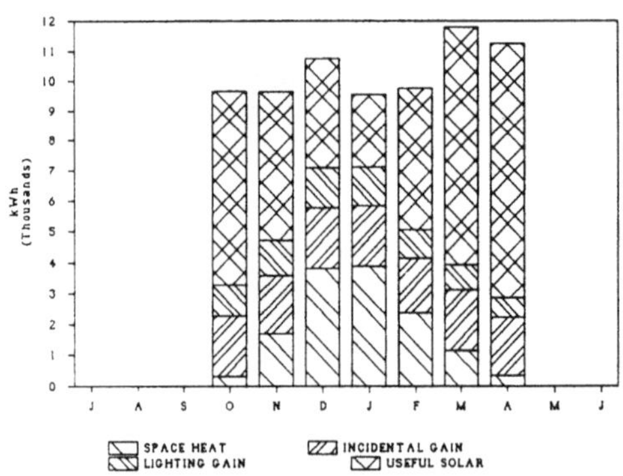

Monthly Contributions to Energy Supply Including Buffer Spaces.

6.0 CONCLUSIONS

From an early design stage the planning of the building aimed to minimise space heating and lighting energy use. Design features found to be useful, with potential for replication include:

* The use of a fast response timber frame structure on the first floor, but a slow response mass wall structure on the ground floor, relates directly to the timed used of the building.
* Corridors and the circulation has been moved to the perimeter of the building to avoid day time internal lighting.
* Multiple use of glazed spaces for solar gain circulation and ventilation pre-heat.

* Deep plan spaces have a mixture of daylighting sources (e.g. rooflights and atria) to achieve better light quality and distribution.

Physical models were very useful for planning purposes and lighting visualisation but cannot readily be altered as the design evolves. The computer daylight model provided rapid results. Dynamic thermal modelling is necessary if the multiplicity of factors contributing to overall energy use are to be taken into account. The main points arising from the modelling study were:

* Most daylighting benefit is obtained reaching daylight factors of 2%-4%.
* Large savings result from avoiding high lighting standards where not needed.
* Summer evening daylighting makes an important contribution in non-office accommodation.
* Ventilation pre-heat from buffer spaces provides large energy savings in one of the remaining areas of major heat loss for a well insulated building.
* Low-E coated double glazing was found to reduce space heating demand but taking into account the reduced light transmission was not economically viable overall.

A fuller account of the design, building performance analysis and daylighting studies is available elsewhere (refs. 10, 11).

References

1 I. Frame, S. Burch, Manual for 'Daylight' Program, Essex Institute of Higher Education UK, July 1988.

2 L. Palmiter, T. Wheeling, SERI-RES Solar Energy Research Institute Residential Energy Simulator Version 1.0, Solar Energy Research Institute, USA.

3 P. Haves, J. Littler, Refinements to SERI-RES, UK Energy Technology Support Unit Report S-1130, 1987.

4 P. Haves, P. J. Littlefair, Daylight in Dynamic Thermal Modelling Programs: A Lighting Faciltiy for SERI-RES, Polytechnic of Central London and Building Research Establishment UK, 1988.

5 Anon., Test Reference Years TRY - Weather Data Sets for Computer Simulations of Solar Energy Systems and Energy Consumption in Buildings, CEC DG XII, Research and Development, 1985.

6 Energy Monitoring Company, Test Cell Studies: 3-Solar Lost, 5-Window Coverings, Energy Technology Support Unit, 1987.

7 R. G. Hopkinson, P. Petherbridge, J. Longmore, Daylighting, Heinemann, London UK, 1966.

8 D. R. G. Hunt, The use of Artificial Lighting in Relation to Daylight Levels and Occupancy, Builing and Environment 14, 1979.

9 Principia Mechanica, Building 2000 Energy Performance Analysis - Swanley Downsview Primary Infants School, 1989.

10 J. Walker, M. Wilson, Y. Dean, Building Performance Analysis (CEC Building 2000): Study Centre, Machnlleth, Wales, CEC DGXII, 1989.

11 M. Wilson, J. Walker, Y. Dean, Daylight Evaluation in a Total Energy Context: Study Centre, Machnlleth, Wales, CIBS National Lighting Conference, 1990.

PERMACULTURE INSTITUTE, STYERBERG

Nick Baker, Cambridge Architectural Reserach and
The Martin Centre for Architectural and Urban Studies
6 Chaucer Road, Cambridge, UK

Two proposals had been made as illustrated in figure 1. The question was which of the two versions, the "wrap-around greenhouse" or the "atrium" would perform the best? Clearly, the wrap-around option would result in a smaller perimeter area to the heated building. On the other hand the atrium option benefits from a larger proportion of the greenhouse perimeter being in contact with the heated house.

Using the ATRIUM model, the two proposals were tested. ATRIUM is a monthly energy balance model which takes account of conductive and convective heat flows between the atrium and the parent building. Solar gains are accounted for using a utilization coeff. dependent upon the solar gain / load ratio. The two buildings were modelled in simplified form as indicated in figure 2.

The results, indicated below, show that for Hamburg climate data there is very little difference in annual heating energy between the two options. However the mean temperature of the atrium space is significantly higher than that of the wrap-around glasshouse.This was regarded as important since the use of the greenhouse space for human occupancy was anticipated.

However it seems likely that there is a cost penalty in the atrium option. Comparing buildings of different shape, which due to their shape incur different building costs, is difficult.

It was pointed out that to get the full benefits of the ventilation pre-heating from the atrium or wrap-around, passive airflow paths which promote a flow direction from the greenhouse to the heated space, must be provided. Both forms would need shading and ventilation in summer to prevent overheating in the greenhouses.

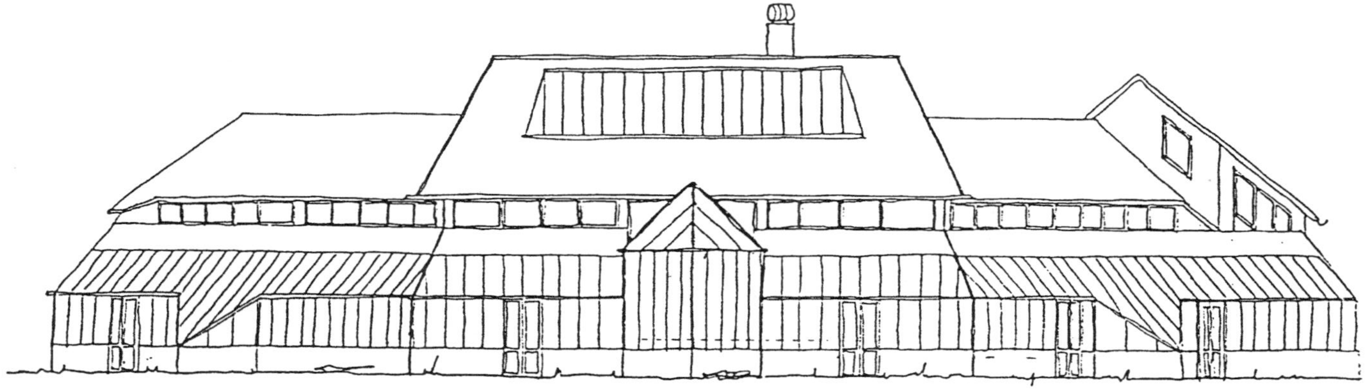

wrap-around

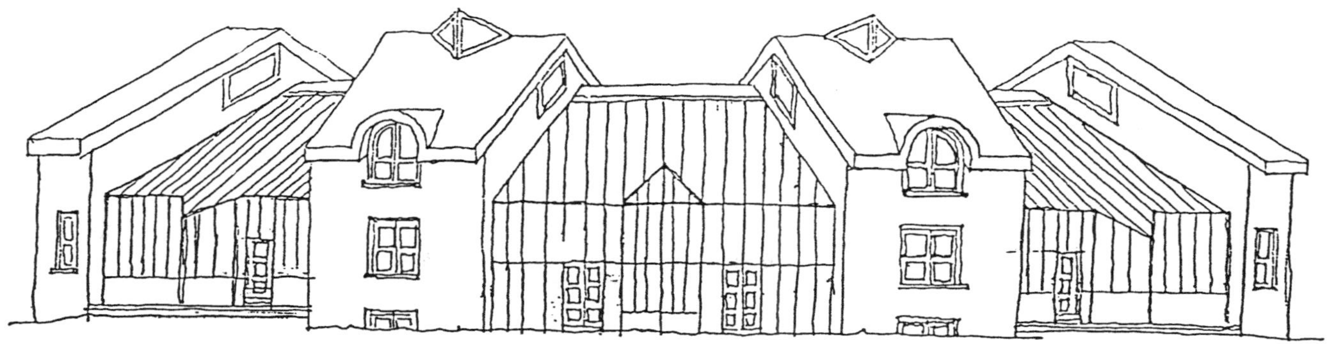

atrium

Fig 1 The Permaculture Institute - alternative proposals

	wrap-around	atrium
ave.temp.Jan	4°C	7.2°C
annual heating no vent pre-heat	25,052 kWh	24,903 kWh
with vent pre-heat	21,255 kWh	20,902 kWh

plan

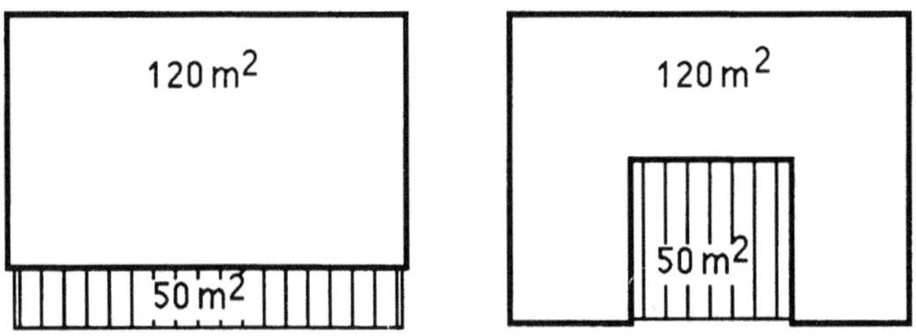

wrap-around greenhouse

atrium greenhouse

Figure 2: simplified input model for energy analysis.

BUILDING 2000

Herbert Schultz, Gärtnermeister
2903 Bad Zwischenahn, Bundesrepublik Deutschland

1. Projekt Ökologische Akademie Hosüne

1.1. Recommandation for plantation

From the plant physiology point of view following problems must be solved: the radiation area of this greenhouse faces south-west. This is disadvantageous. The heat radiation in winter is not sufficient to form a suitable room condition for plants, that means to heat accumulation masses who take effect in the night. Therefore the orientation to south is unavoidable. Because the solar house rises above the building the cooling area is not to underestimate, which causes problems especially in winter. Because this room will not be heated additinal he is to regard as a heat supplier only in restricted times. By selecting the plants this is to consider. Kiwi- and Wine plants are some examples of this group who do a time of rest in winter and who can tolerate frost periods.

1.2. The situation in winter

In these circumstances the stay for plants with a high heat requirement is restricted in winter. In long periods of frost it could come to cold snaps because the room is not heated. Between 10.00-16.00 o'clock there will be a suitable room condition for plants by solar radiation. Therefore only plant pots should be used, because they can be taken back to the inner region of the room if there is a danger of frost.

1.3. The situation in summer

Because of the high radiation in summer there is a problem with overheat, so there must be a sufficient ventilation in the ridge and a shading system on the outer wall. Otherwise the temperatures can come to 40-45°C in the noon by solar radiation. This is a stress situation for plants and also for people. The plants need because of the high temperature in summer a lot of water. The accumulation volume of the substratum is restricted by the size of the plant pots. For this reason an automatic irrigation is recommandable. Better will be to use hydro-cultivation pots. Clay or slate as substratum can provide the plants optimal over several weeks.

1.4. Example of plants who can tolerate short-term temperatures below zero

Aucuba japonica
Agapantus campanulatus
Rosmarinus officinalis
Pistacia lentiscus
Nerium oleander
Olea europaeus
Poncirus trifoliata
Punica granatum
Jasminum officinalis
Laurus nobilis

BUILDING 2000

Herbert Schultz, Gärtnermeister
2903 Bad Zwischenahn, Bundesrepublik Deutschland

1. Projekt Lebensgarten Steyerberg
Anlehngewächshaus am Zentralgebäude Westflügel

1.1. Recommandation for using and plantation

First of all the using of this greenhouse will contribute to meet the demand for vegetables and spices of the adjacent kitchen. From the vegetable view following conditions have to comply with: a subdivision in warm- and coldhouse is not possible, because of the missing additional heat the climate will be the same in both rooms. For the same reason the ceiling between the floors should be taken so far, that there is enough air circulation between the floors. Because of the small depth of the building (3.70 m respectively 3.10 m) there will be only a small loss of light in the rear region.

1.2. The situation in winter

Because the greenhouse will not be heated additional with primary energy the using of plants that keep a long time is only restricted possible. The application of warm water from the kitchen as heatsource is not advantageous. Recommandable is rather a laying of heating pipes into the ground beds on the outer wall to have a screening against seeping coldness from outside. Therefore PE-pipes (20x1,5 mm) have proved to be good and must layed 25 mm below the soil surface in distance of 30 cm.

1.3. The situation in summer

Considering the summerly, long hot weather condition there will be overheating problems in the greenhouse. The ventilation system must have such dimensions that temperatures in plant region do not go beyond 37°C to 42°C. If a shading is not necessary only plants with a high heat requirement, for example tomato, paprica, cucumber cold be cultivated. These cultivations should be planted in the front beds to form a comfortable climate in the interior for people.

2. Projekt Lebensgarten Steyerberg
Seminarraum, Vorraum, Zentralgebäude Westflügel

2.1. Recommandation for plantation

In these rooms a movable plantation in suitable plantpots is recommandable. Because of the weight the size of the plant pots should not be over 150x60x50 cm (length x width x height). According to the kind of substratum and its moisture content the volumetric weight will be between 0,4-0,8 kilogramme per litre. Following substratum mixture is recommandable to avoid high plantpot weights:

4 parts of compost	volumetric weight 0,8-1,2 kg/l
5 parts of white peat	volumetric weight 0,1-0,3 kg/l
3 parts of Lekaton 10-15 mm	volumetric weight 0,2-0,4 kg/l

Hence follows a volumetric weight of 0,35-0,65 kg/l, at which a moisture content of 50 per cent of the water capacity is considered. If plant pots are selfmade it is to take care that they are waterproofed. It is advantageous to cover the plant pots with high-grade steel.

1.2. The climatic situation

By using a shading system there is no problem with overheat in summer. In winter the minimum temperature must be at +6°C. At low night temperature in winter the shading installation should be closed to decrease radiation loss. Under these circumstances following plants can be used who are tolerant of variation in temperature:

Bougainvillea glabra
Camellia japonica
Campanula isophylla
Chlorophytum comosum
Citrus micrcarpa
Echevieria secunda
Fatsia japonica
Nerium oleander
Passiflora caerulea
Soleirolia soleirolii
Hedera helix
Cissus antarctica
Abutilon- Hybriden
Araucaria heterophylla

BUILDING 2000

Herbert Schultz, Gärtnermeister
2903 Bad Zwischenahn, Bundesrepublik Deutschland

1. Projekt Permacultur Steyerberg

1.1. Recommandation for using plant cultivations

From plant physiology point of view and for climatic reasons that construction is favourable where the buildings surround the solar houses. Because there will be also no additional heat in winter the heat requirement must meet from the radiating sun. In winter, even if the outside temperature goes down to -10°C the set in of frost may not happen, because of the surrounding buildings. However, during longer periods of coldness the room temperature can sink below the freezing point (-5°C). Therefore the selection of plants is of specially importance. It is ideal to integrate mediterranean plants who shade the interior in summer with theit leaves and who let the sun fall in when they have no leaves in winter. They can be plant in marketable pots, selfmade elements or in natural soil. Only such should be used who can endure long-term temperatures to -5°C. The precondition is, that the region of the roots does not freeze through. This is only possible, if the plants stand in natural soil, because there is more earth volume. This problem can be solved, if ground beds are layed out in front of the south-west inner wall of each greenhouse. Plants in tubs and throughts have to be taken away from the outer wall, if the outside temperature goes down. Additional, the shading system can be drawed. Because the area of the greenhouse does not rise above the building the soil is mostly protected against freeze. Nevertheless it is recommandable to isolate the foundation of the solar houses, to prevent that coldness oozes if there is a longer period of frost. In my opinion the shading of the adjoining buildings is compensated by more suitable room conditions and a better using. Also the walls contribute to a better air condition by radiating the heat in winter. Concerning the three greenhouses different climatic conditions will not to be formed. On grounds of the semi-circular arrangement of the buildings the warming will be temporal defered by the way of the sun. But this warming would be so unimportant, that there will be no plant physiological influences. For this reason a subdivision in different climatic conditions is not sensible.

1.2. The situation in winter

At the end of the vegetation period, from late autumn till winter, plants must be adapted to sinking room temperatures by reducing the watering. By this the ripeness and the resistance to coldness of the plants will be forced. This is necessary to prevent the plants from sprouting if there will be a short-term rise in temperature conditional on solar radiation.

1.3. The situation in summer

Conditional on the arrangement of the solarhouses there will be no problem with overheat temperatures for the plants. Because the ventilation system is arranged in the ridge the air change rate will be sufficient. In the morning- and evening hours a cast shadow will be formed by the surrounding buildings, who makes a shading superfluous. However, for the using by people a shading is sensible. Plants who need less light must stand in the inner east section. Because of the considerable height of the rooms also plants should be used who come up to 5 m.

1.4. List of suitable plants

Arbutus unedo
Eriotrya japonica
Escallonia ssp.
Eucalyptus gunnii
Jasminum officinalis
Laurus nobilis
Nandia domestica
Nerium oleander
Punica granatum
Rhynispermum jasminoides
Poncirus trifoliata
Aucuba japonica
Acacia dealba
Camellia japonica
Elaeagnus Maculata
Jacaranda mimosifolia
Hedera helix
Morus alba
Viburnum tinus
Pseudosasa japonica
Sasa pumila
Phyllostachus glaucescens
Yucca gloriosa
Chamerops humilis

TECHNOSITE DE REZE (FRANCE)

THERMAL BEHAVIOUR OF THE BUILDING

A. GRELAT C.E.B.T.P.
Groupe d'Etude et de Recherche multidisciplinaire
Domaine de Saint-Paul 78470 Saint-Rémy-les-Chevreuse

1. SCOPE OF THE STUDY

The subcontract was intended to cover particular studies of the thermal behaviour of REZE TECHNOSITE .

The project consists in a two storeys , triangular shaped office building organised around an Atrium.
Optimisation of this particular feature required the use of a convenient thermal simulation model.

Program BILGA of C.E.B.T.P. was used to cover the following topics :

- Optimization of solar gains in winter
- Ventilation pattern in the central Atrium
- Indoor thermal comfort in mid-season and in summer.

The expected results consisted in a set of design recommandations to the architect , and evaluation of heating needs to the client .

2. BASIC DESCRIPTION OF THE BUILDING

Basic drawings are shown in fig. 1 .

Ground floor is of slab-on-grade type with peripheric insulation , upper floor a 20 cm concrete slab .
Exterior walls are made of hollow concrete blocks with 8 cm insulation, cover of upper level of a metal sheet with insulation .

Openings are P.V.C.,single glass windows ,and wood and glass doors .

Principle of ventilation was to be defined .

Figure 1   REZE TECHNOSITE

Ground floor

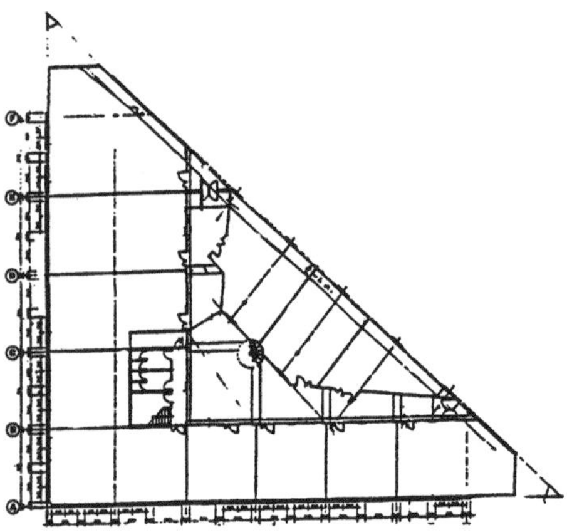

1st floor

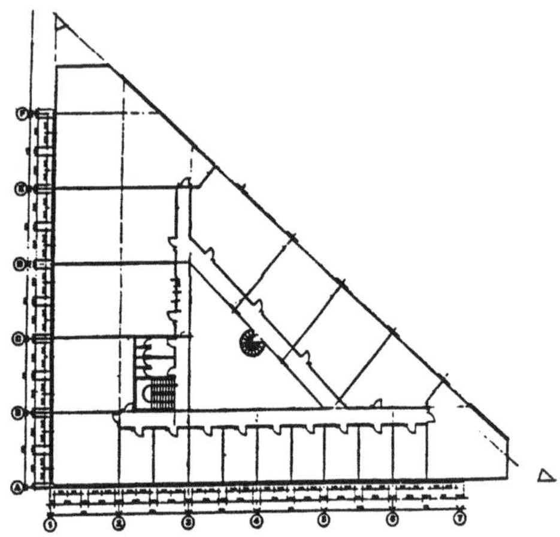

Elevations

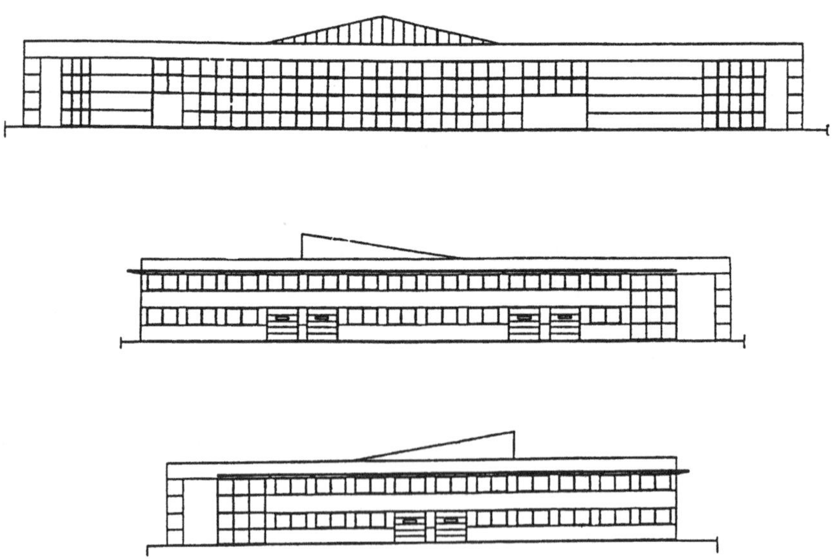

3. OPTIMIZATION OF THERMAL COMFORT IN SUMMER

3.1 Main parameters

Present study was intended to optimize following parameters :

- Ventilation sheme of offices and Atrium
- Inertia , color and insulation of roofing
- Protection of openings
- Shape and composition of glazed part of roofing
- Influence of floor coatings
- Influence of ventilation rates on thermal comfort .

3.2 Modelisation of the TECHNOSITE with program BILGA

BILGA is a multizone type thermal simulation model , developped by C.E.B.T.P. , using a detailed description of building envelope , indoor conditions , outdoors environment. Natural or mechanical ventilation is fully coupled with thermal simulation .A full comfort model is included .

The building was represented as 14 thermal zones , 6 for offices and 8 for parts of the Atrium .These are open on one another , allowing for air and radiative transfers.

Meteorological data was issued from Nantes , July 1983 considered as typical of hot summer conditions . (Temperatures : 16 to 32°C , high irradiation , low wind)

3.3 Study of the building in natural ventilation

A principle of solution for best summer thermal comfort using natural ventilation was first defined :

- High night ventilation rate / low day ventilation rate
- Adequate solar protection of openings
- Sufficient inertia .

Recommended dimensions of openings lead to ventilation rates of 3 to 6 vol/h at night , and 0.5 to 0.9 vol/h at daytime , owing principally to stack effect .

For different reasons - effraction protection , presence of polluting activities , the principle of natural ventilation was finally rejected . Also a variable rate mechanical ventilation with 2 vol/h night rate was studied , and found convenient ,but eliminated for technical reasons.

Even with basic design (case A1), acceptable summer conditions would have been otained using natural ventilation , particularly in ground floor offices , as shown on fig.2 .For upper level , lack of inertia results in a more degraded situation .

Fig 2 Basic design . Natural ventilation
Summer temperatures in offices

Ground floor

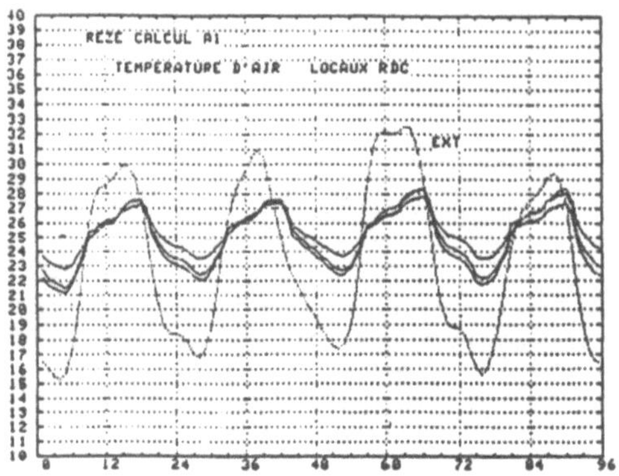

First floor

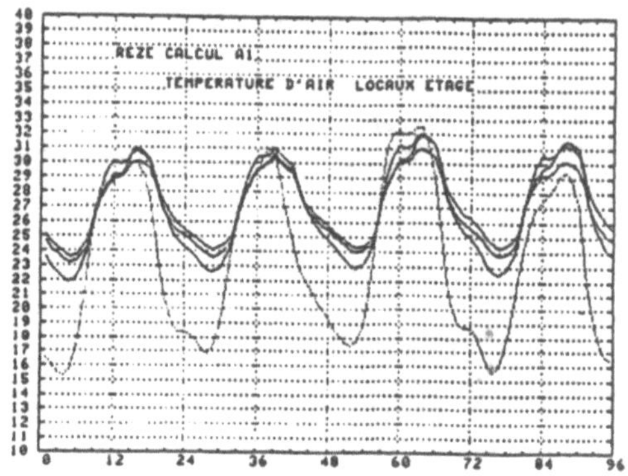

3.4 Influence of solar protections

Convenient solar protections might be exterior venetian blinds, with a solar factor of 0,14 (Case A2).
Substantial reduction of peak temperatures are obtained in offices (26°C max for ground floor , 28°C for upper level).

Fig. 4 Temperatures in offices with venetian blinds

Ground floor

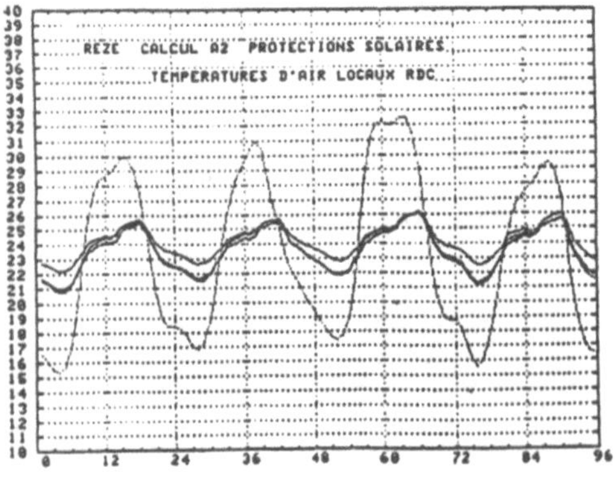

First floor

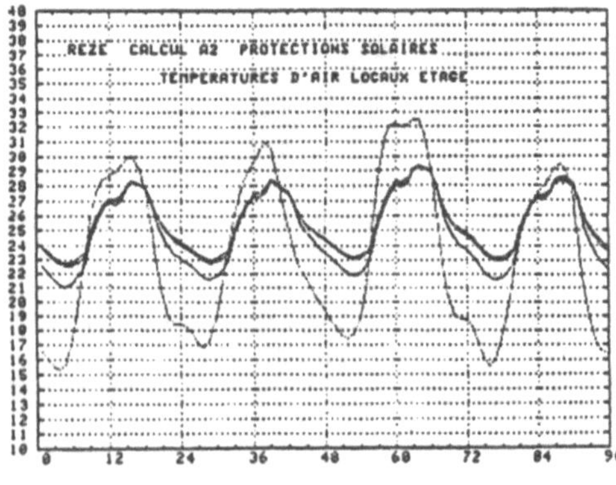

3.5 Influence of roof inertia an insulation

Next improvements may be obtained in two ways :

a) Increase inertia of 1st floor ceiling (Case A3) , or
b) Increase roofing insulation and solar reflectance

Solution a), consisting in replacing metal sheet roofing by 16 cm concrete slab , 4 cm polyurethane insulation ,watertight coating and 6 cm gravels , was found convenient :
peak temperatures were reduced to 25°C for ground floor , 26°C for 1st floor .
Solution b) was slightly less satisfying ,as peak temperatures for 1st floor remained 27°C .

Solution a) was finally accepted .

Fig 5 shows successive reduction of temperatures by solar protections (case A2) and roof inertia (case A3).

Figure 5 Comparison of cases A1 to A3

Ground floor

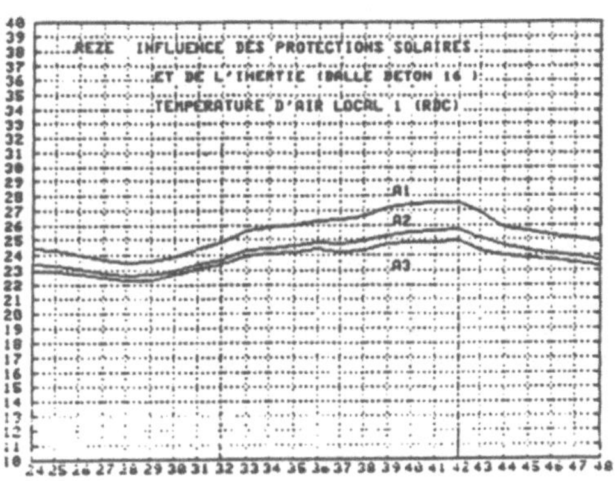

First floor

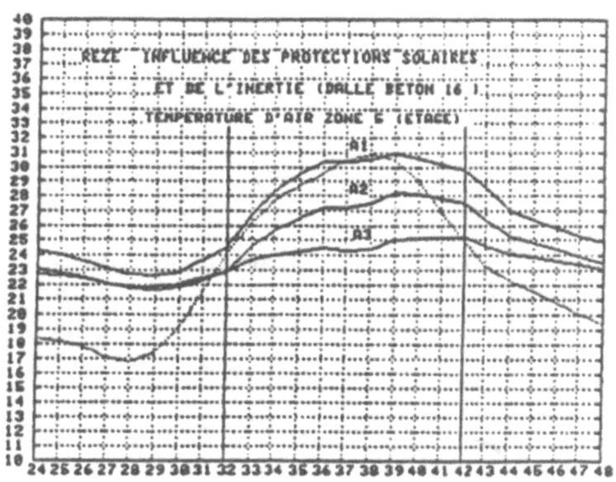

3.6 Other factors

Factors of less influence were also examinated :

- Influence of reflective roof glazing
- Influence of pile carpets

Both influences were found limited to 0.2°C for peak temperatures .

3.7 Influence of mechanical ventilation

As stated before , mechanical ventilation with constant rate was finally selected on safety and simplicity criteria for offices , while the Atrium was kept in natural ventilation.

Effect of this feature is rather detrimental to summer comfort.Figure 6 shows comparison between natural and mechanical ventilation using concrete roofing (cases A3 and B3).Peak temperatures are 2°C higher for case B3.

Figure 6 Influence of type of ventilation

Ground floor

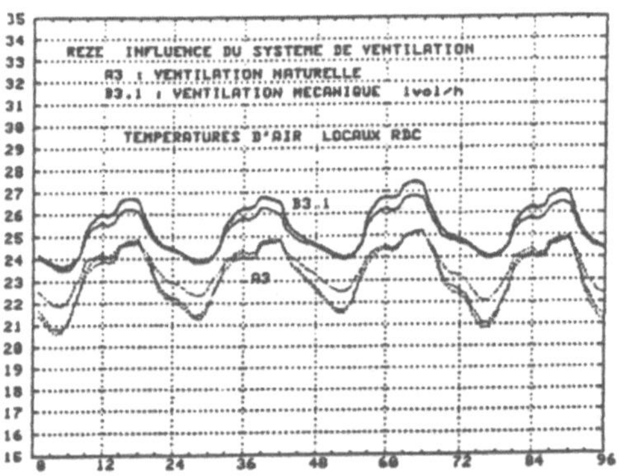

First floor

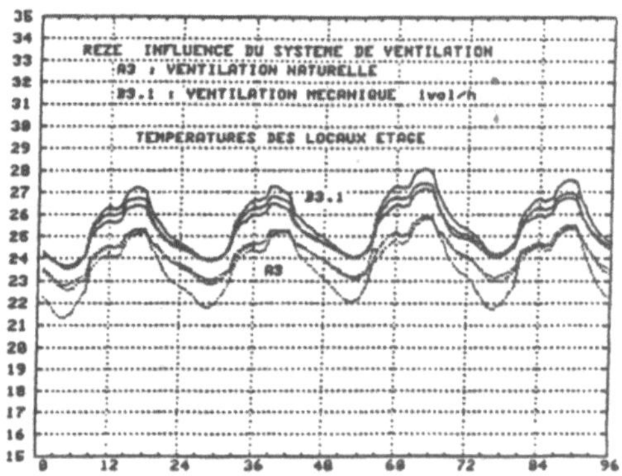

3.8 Conclusion for summer thermal comfort

The finally selected solution , including exterior venetian blinds on exposed walls , isolated concrete deck roofing and mechanical ventilation with maximum rate 1 vol/h for offices appears to lead to an acceptable solution for thermal comfort.

A further step to better comfort should have been an increase in night ventilation rates .

4. WINTER ENERGY CONSUMPTIONS

Energy consumption during heating season are calculated with program BILGA using meteorological data of NANTES 1970 .

Double glazing was selected for North façade only .

Main hypothesis for this calculation :

- Offices temperatures 19°C at open hours , 14°C min. at night
- Offices ventilation 1 vol/h at open hours, 0.5 vol/h at night
- Atrium ventilation 2 vol/h at open hours , 1 vol/h at night
- Occupancy (all together):
  8h-12h and 14h-18h 80 persons , 8000 W light and processes
  12-14h 30 persons , 3000 W light and processes,
  6 days per week .

Figure 7 shows Energy consumption and degrees-hours for the period .

Figure 7

Monthly energy consumption

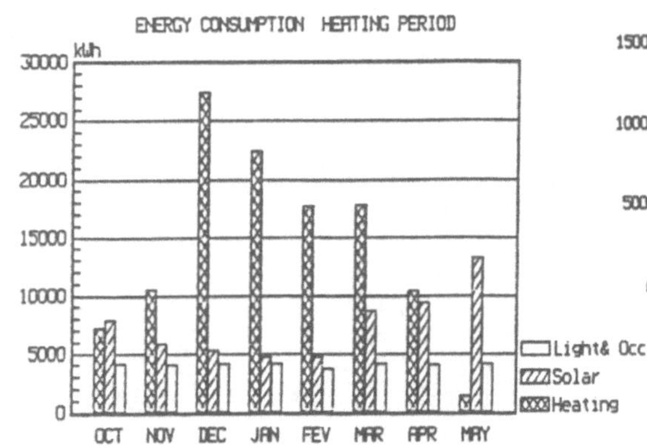

Degrees-hours
(basis 19°C)

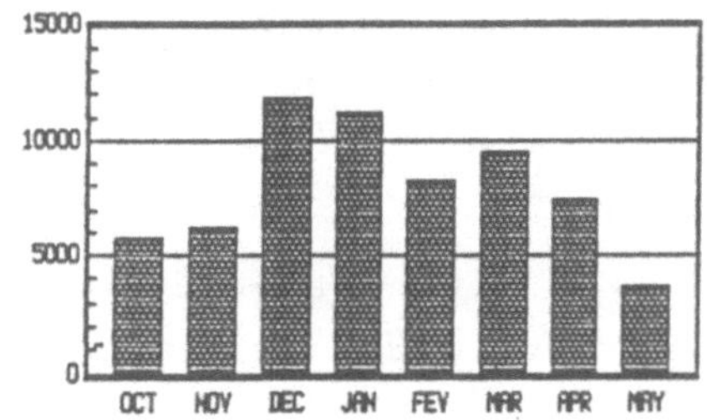

Refering to French Regulations , following coefficients were obtained from calculations :

Basic Heating needs : G1 = 0,56 $W/m^3.K$
G1V= 2708 W/K

Actual heating needs   B = 0,38 $W/m^3.K$
(taking into account occupancy , solar gains and regulation of ventilation and heating) .

BUILDING 2000.

NEWBURY CENTRAL LIBRARY, AN EXAMPLE OF A DAYLIGHT DESIGN STUDY.

C J Hancock* , J Littler. Research in Building
Polytechnic of Central London, 35 Marylebone Road, London NW1 5LS.

* present address Bickerdike Allen Ptnrs. Salusbury Rd. London.

SUMMARY

Public libraries require illumination to all spaces throughout long periods of the day. For deep plan buildings this results in a high electric lighting cost. Moreover heat gain from lighting displaces cheaper heating sources, and when occupiers allow high ventilation rates, significant useful energy is lost. Continuous use of electric lighting contributes to excessive peak cooling loads. Therefore, under these conditions, opportunities for solar energy lies in daylighting to reduce electrical consumption. This also provides a margin for useful wintertime solar gains, and for ventilation control exploiting solar pre-heat and stack venting. Design studies for this proposed library have predicted that upto 62% of the full lighting electrical load could be saved with daylight switched circuits. Modelling had demonstrated that satisfactory daylight distribution can be obtained in the design from a central lightwell combined with perimeter windows and clerestories.

1.0 CASE STUDY DESIGN.

The proposed building, designed by Berkshire County Council, is 1150m^2 of accommodation on two floors, comprising a lending and reference library, exhibition space, staff and meeting rooms, (Fig. 1). The upper floor will be open to the roof, which presents the possibility of a central roof-lit lightwell to supplement perimeter windows. The site offers unobstructed solar aspect between east and west, although existing buildings block daylighting from the north.

The design study has examined daylighting from the lightwell, three east facing gables serving the public zones, and west gables lighting staff areas and entrance.

2.0 PRELIMINARY ENERGY ESTIMATES.

Table 1 shows the preliminary estimates of the energy inputs for the proposed Newbury library, based on existing libraries without daylight/passive solar features. Measured in fuel cost the electric consumption far exceeds gas consumption and lighting is the major user of electricity. When considering energy loss, the current high standard of fabric insulation means that natural ventilation is the principle loss path.

TABLE 1 NEWBURY LIBRARY- ENERGY COSTS PER YEAR						
					Equivalent Energy	
	Total consumed kWh/yr	Fuel rate £/kWh	Total Cost £	percent of total %	Consumed kWh/yr per sq m	Installed power w/m2
ELECTRIC-lighting	43071	0.055	2347	44	37	19
ELECTRIC-power	26535	0.055	1446	27	23	12
GAS-Heating	125763	0.012	1547	29	109	55
Total	195368		5340	100	170	85

2.1 Fuel Cost Savings.

From existing examples of the use of daylight responsive electric switching systems electrical reductions in the order of 35-50% are possible.

For the Newbury library this would indicate savings above £1100 per year.

Passive solar designs to buffer the building and pre-heat the ventilation air plus restrictions on ventilation losses may typically save 55% of heating output. Thus further savings of the order of £800 may be achieved in the library design. Total saving may be in the order of £1900.

Assuming energy cost increases and discount interest rate at a standard 5% the total capital investment in extra passive solar measures to achieve savings within seven years would be £11000.

From the above preliminary analysis the priorities for further energy saving studies should be in the order :

1st- reduce electric lighting.
2nd- restrict ventilation and/or employ solar pre-heating of ventilation supply air and/or employ mechanical ventilation with heat exchanger.
3rd- Solar contribution to space heating (direct gain).
4th- Examine small power usage and electronic systems.
5th- Increase fabric thermal resistance.

5.0 PRELIMINARY ESTIMATES OF DAYLIGHT LEVELS,

With the simple Average Daylight Factor Formula [1],the effect of glass area and obstructions to windows may be quickly assessed, even before a building section has been developed.

$$DF_{av} = MTWO\ /\ [A(1-^{R2})]\%$$ For windows & rooflights

Subsequent work in this design study showed very close agreement between preliminary DF_{av} formulae values and more exhaustive computer and model predictions. The formula are robust when used with approximate values for the variables.

6.0 DAYLIGHT SWITCHING ZONES.

To maintain daylight responsive light switching, representative Daylight Factors are required in order to calculate switching illuminances as zone control values. Therefore the detailed analysis of daylight in the library was undertaken by defining ten switching zones. Each zone is differentiated by a particular daylight source and may be controlled under one switching strategy. Table 3 shows sample zones with calculated values of the Average Daylight Factor, and the control Daylight Factor (IRC+ SC) calculated by The BRE protractors.

TABLE 2 CALCULATION OF THE AVERAGE DAYLIGHT FACTOR						
ZONE NAME	ROOFLIGHT area sqM	WINDOW area sqM	AVERAGE IRC %	SKY COMPONENT %	IRC+ SC %	AVERAGE DAYLIGHT FACTOR %
L'well GF.	44	0	1.70	0.20	1.90	4.4
L'well base	44	0	3.80	16.90	20.70	9.7
staff 1st F		18	1.67	0.32	1.99	4.5

The positions chosen for switching control represent DFs close to the minimum. Commonly the Internally Reflected Component exceeds the Sky component for these DFs. Thus these DFs are particularly dependant on the reflectance values of interior finishes and furnishing. DFs, between 1.4% and 3.8%, indicate that electric supplementary electric lighting will be required. The lightwell provides a level of daylight deep into the ground floor public space that would ensure a more uniform distribution throughout this area. Calculations for the first floor public area shows daylight levels between 7% and 3.3%, and imply a reasonable distribution throughout the space.

7.0 SIMULATION OF LIGHTING CONSUMPTION

Using the building Energy simulation program SERI-RES, [2], the energy consequences of a number of lighting design options were examined. 300lux and 500lux target illuminance schedules were used, represent both the CIBSE guide recommendations [3] and Berkshire's practice. The target value for each zone is both the electric service illuminance and the daylight switching illuminance.

7.1 Switching Strategies

the effects on lighting energy of likely combinations of suitable switching for each zone were examined:

[A] Continuous dimming (variable control) to all lamps in all zones in response to daylight levels. High frequency dimmable fluorescent ballast under photo-sensor management will be required.

[B] Continuous dimming in the public areas of the library and manual switching in the staff and meeting rooms. Under manual switching the lights may be switched on by the staff in the morning if daylighting is inadequate. Lights are automatically switched off at 12.00 GMT and may be switched on again by staff if daylight is below target level. This offers the simplest switching strategy and may result in greater user satisfaction through user control.

[C] 4-step switching in the public areas and manual switching in the staff and meeting rooms. Stepped switching may be implemented with standard lighting gear, and is therefore likely to have the lowest installed cost.

7.2 Overall Savings Due To Daylighting

Figure 2 represents electrical energy for ambient lighting for ten zones, 1301 m^2 floor area, in use for 50 hours per week. The reduction percentages indicate the potential for daylight substitution for each switching strategy. From table 5, without daylight substitution 57,466 kWh (high) and 38,480 kWh (average) of electrical energy per year are required to light the ten zones. The corresponding values per square metre of floor are 44 kWh/sq m and 29.5 kW/sq m respectively. These are consistent with the preliminary estimate for the new library of 37 kWh/sq m calculated from the consumption of existing libraries.

TABLE 3 TOTAL LIGHTING POWER CONSUMPTION IN KWHR PER YEAR FOR AMBIENT LIGHTING (EXCLUDING TASK). From 9.00am to 7.00pm 5 days per week

[HIGH TARGET ILLUMINANCE- 500Lux] ELECTRICAL POWER PER YEAR	[1] kWh/yr	[2] kWh/yr	[3] %	[4] £
LIGHTING MODE				
BASE CASE-No daylight substitution	57466			
ROOFLIGHT+EAST WEST GLAZING*	25565	31902	56	1739
[AVERAGE TARGET ILLUMINANCE-300Lux] ELECTRICAL POWER PER YEAR	kWh/yr	kWh/yr	%	
LIGHTING MODE				
BASE CASE-No daylight substitution	38480			
ROOFLIGHT+EAST WEST GLAZING*	14531	23950	62	1305

[2]-SAVINGS IN ELECTRIC LIGHTING FROM BASE CASE
[3]-SAVINGS AS PROPORTION OF BASE CASE
[4]-VALUE OF SAVINGS @ 0.054/KwH
*Continuous dimming on electric lighting.

The maximum use of daylight would result from continuously variable electric lighting in response to internal daylight levels. Under this strategy 110 sq m of window, 14% of the building wall plus 44sq m of rooflight over the lightwell reduces electric lighting output by 56% (high) and 63% (average) of full consumption, worth £ 1739 and £1305 per year.
Strategy [C], manual + 4-step, switching reduces the lighting consumption to 53% of the full consumption for a 500lux target level. This value is 9% greater than the energy use under continuously variable control [A], worth £273 per year.

7.3 Seasonal Variation In Lighting

Figure 3 shows the seasonal variation in electrical consumption. The consumption under each switching strategy is expressed as a percentage of the monthly value with full use of electric lighting. Longer daylight hours and higher diffuse sky illumination during the summer period give significant reduced consumption , to 20-30% of full lighting. This clearly demonstrates that, as distinct from solar thermal gains, daylighting as a passive solar strategy is able to exploit the greater summer season gains to directly supplement building energy use.

8.0 -DISTRIBUTION OF DAYLIGHT WITHIN LIBRARY DESIGNS.

To assess the penetration of daylight within the proposed library design, a model of the design was constructed and tested within the Polytechnic's artificial sky. This work enabled preliminary calculated values for point daylight factors to be verified. In addition, through physical modelling, distribution patterns due to multiple or complex daylight openings could be plotted. The use of physical modelling for this work is more precise than manual calculation techniques or the simple computer calculation programs currently available. Distributions of DFs were measured normal to the glazing for both floors as shown in Fig 4.

8.1 Characteristic Daylight Gradient From Gables And Lightwell.

Figure 5 summarises the combined effect of daylight from the roof-lit lightwell and the glazed gables. It shows the Daylight Factors, measured on the horizontal plane, on the first floor along a line from the centre of gable 1 to the edge of the lightwell (Section C on floor plan, Fig 4). The glazed gable windows are 18% of the first floor external wall area, the rooflight is 5 % of the roof area. The ceiling line follows the roof pitch and therefore slopes from a height of 6.3m to 2.4m. Each gable features a main window and a clerestory which is set into the peak of the gable. The vertical windows have no obstructions. Surface reflectances are, walls= 0.45, floor= 0.3, ceiling=0.75.
The rapid diminution of daylight away from the gable window indicated that in spite of DFs exceeding 17% close to the glass, values fall below 4% within 5 metres. The Daylight gradient from this simple gable window as modelled is unsatisfactory. The Uniformity Ratio (UR), a measure of the daylight distribution, is 0.1. For this deep space, exceeding 14 metres to the centre, permanent supplementary electric lighting would be required for all but the 5-6m zone close to the gables to avoid internal gloom . Thus the introduction of secondary daylighting should aim for a shallower gradient. This could be represented by an increase in the UR value to at least the recommended minimum of 0.3.

The lightwell and reflective ceiling system shows a similar rapid diminution of daylight. A 4%DF would not be achievable beyond 3-4 m from the centre of the lightwell. The predominant vertical component of the light is apparent. However real improvement in light distribution is achieved when light well and gable wall glazing are used together for bilateral lighting. The illuminance gradient is less sharp and the horizontal DF is maintained above 3% along this line. The uniformity ratio of the DF in this line is now 0.42. There is scope to employ daylight reflectors at the gable glazing and beneath the rooflight to reduce the view of the sky from these positions. This design refinement will reduce the DF close to the glass and attempt to raise the level of reflected daylight deeper into the space. A 1/3 reduction in the DFs close to the gable window and rooflight (achieving an 11%DF within one metre of the window, and 8%DF beneath the rooflight) would result in a uniformity ratio of 0.65 with an average DF along this section of 6.3%DF. This could constitute an "ideal" distribution as a target for further design development.

8.2 Daylight Factor Distribution Through Computer Analysis.
Computer calculation of daylight distributions provides a rapid method of calculating DFs or illuminances for a grid of points within an interior, fig 7n. Programs, using established formula for calculating sky component and the reflection components [4], will produce reliable results within the limits of geometric complexity allowed in the program. Rapidity allows a number of design options to be examined at an early stage. In this study a program was used as a later supplement to, and a check against, earlier studies using physical modelling and manual calculations.
8.4 Comparison Between Predicted And Model Measurements.

The results from the design tools used in this study shown acceptable consistency in dealing with complex bilateral daylighting. Preliminary average DF values were well matched to values later predicted from a grid of points using computer analysis. Figure 6 shows DF values through sections of the design using all procedures. The results were encouraging, and increased the design team's confidence in these tools and the values assumed for the variables.
The conclusion of this comparison ;
the use of computer calculation provided a rapid and accurate assessment of daylighting in the general form of the design. . It would be the first choice by the designer. Modifying inputs, and splicing outputs to deal with more complex geometry is possible with experience. If further design times permits, detailed solutions on reflectors, lightshelves, may be refined through physical modelling or complex computing. However both are likely to demand considerable time from the team.

REFERENCES

[1] PJ Littlefair, VHC Crisp Average daylight factor prediction Proc. CIBS National lighting conf. 1984.
[2] P Haves P Littlefair DAylight in Dynamic thermal modelling programs: a lighting facility for SERI-RES Jan 1988.
[3] CIBSE,Lighting Guide- Libraries, & CIBS code for Interior Lighting,1984.
[4] Aylett & Coleman, Daylight program, Essex Institute of Higher Education.

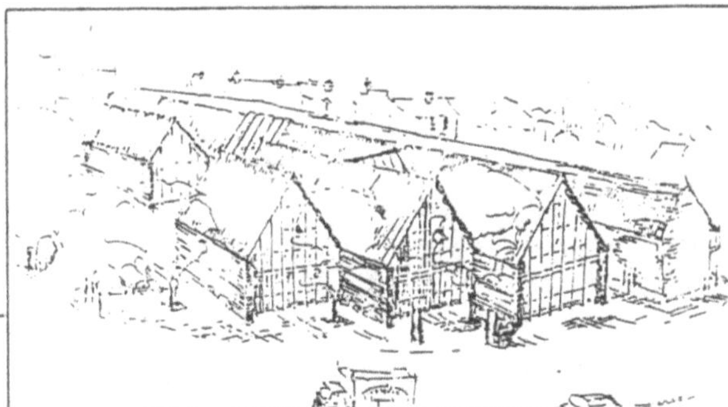

FIG 1a. PRELIMINARY SKETCH OF LIBRARY
VIEWED FROM THE WEST.

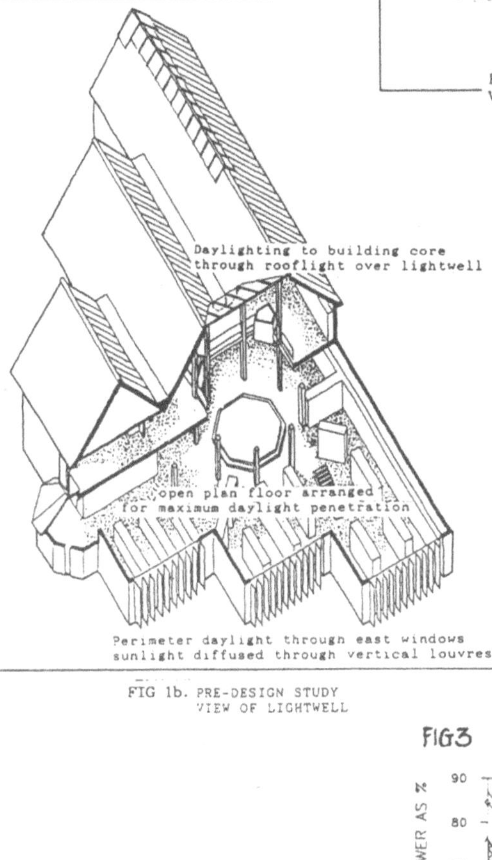

FIG 1b. PRE-DESIGN STUDY
VIEW OF LIGHTWELL

FIG 2. LIBRARY– ANNUAL ELECTRIC LIGHTING USE

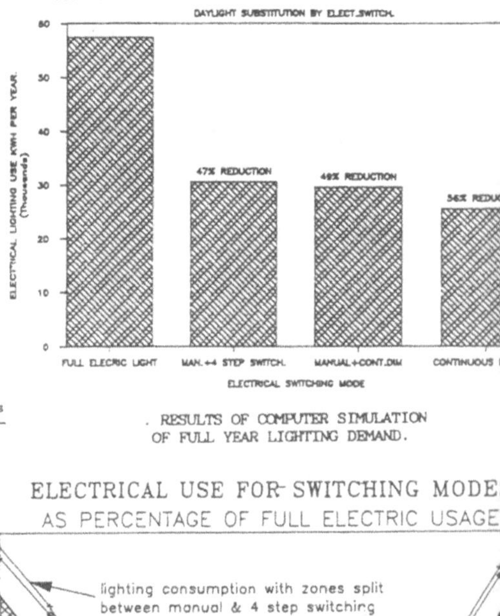

. RESULTS OF COMPUTER SIMULATION
OF FULL YEAR LIGHTING DEMAND.

FIG3 ELECTRICAL USE FOR SWITCHING MODES
AS PERCENTAGE OF FULL ELECTRIC USAGE

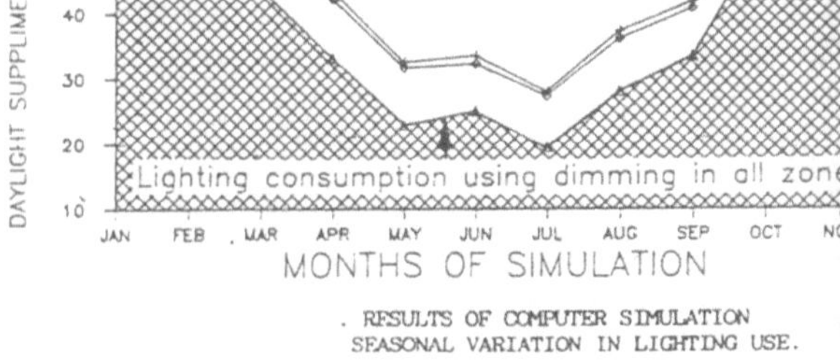

. RESULTS OF COMPUTER SIMULATION
SEASONAL VARIATION IN LIGHTING USE.

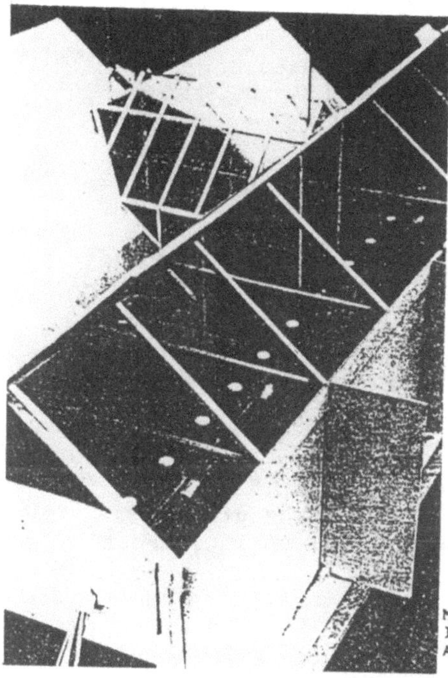

MODEL (WITH PART OF THE ROOF REMOVED) IN USE FOR MEASUREMENTS IN THE ARTIFICIAL SKY.

FIG 4 DAYLIGHT MODEL STUDIES

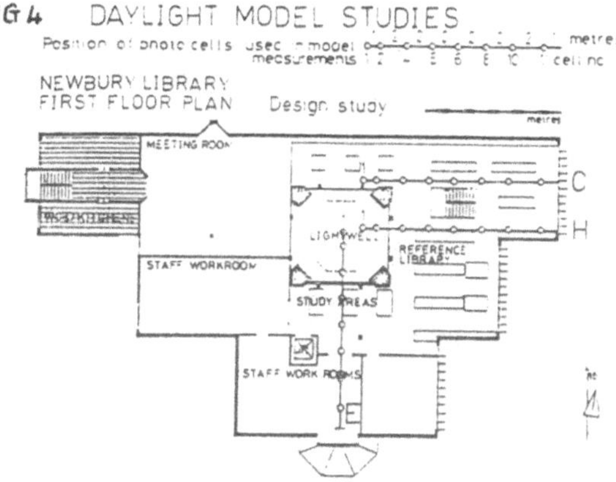

DAYLIGHT DISTRIBUTION ON THE FIRST FLOOR IN THE CORE OF THE LIBRARY. DAYLIGHTING FROM GABLE GLAZING, CLERESTOREYS AND LIGHTWELL. PLOTTING USING COMPUTER CALCULATION.

Plan showing daylight factors (%)

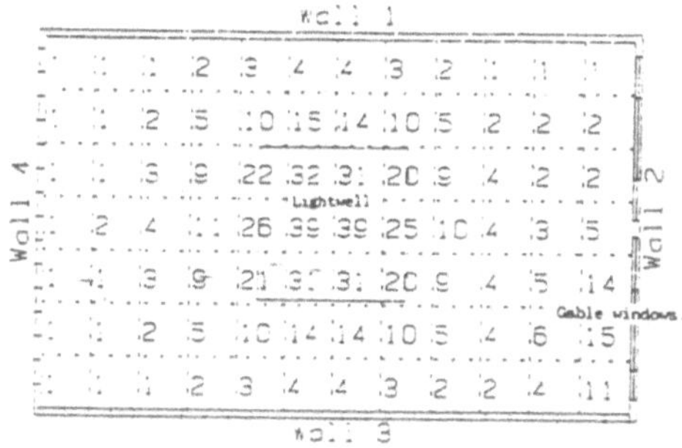

FIG 5 **DAYLIGHT PROFILE THROUGH FIRST FLOOR**

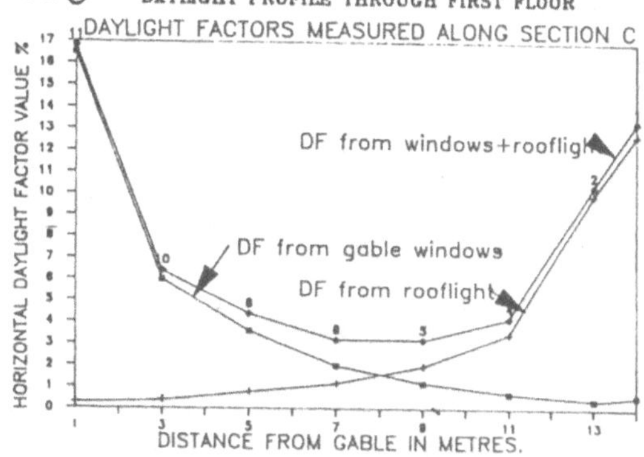

Plan showing daylight contours

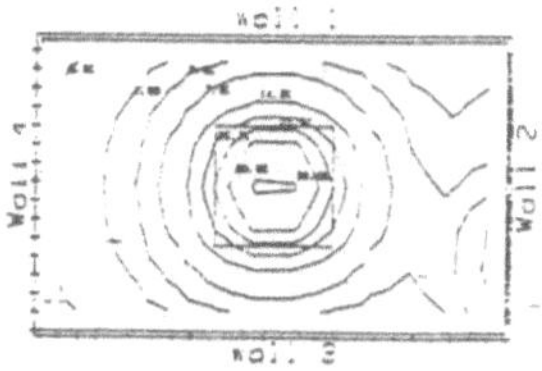

Min.	0.52
Max.	35.25
Average	8.04
Min/max ratio	0.01
Uniformity ratio (min/av.)	0.06

Graphs across centre of room

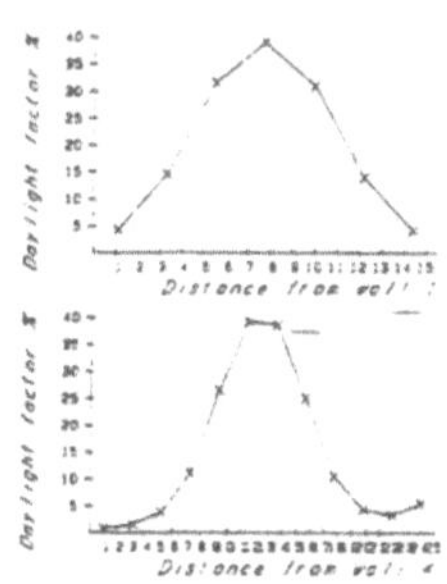

Calculations based on average internally reflected component.
Glass is assumed to be clean.
Room surfaces are assumed to be clean.

FIG 6 **COMPARISON OF COMPUTER CALCULATIONS WITH PHYSICAL MODEL MEASUREMENTS**

FIRST FLOOR NEWBURY LIBRARY STUDY.

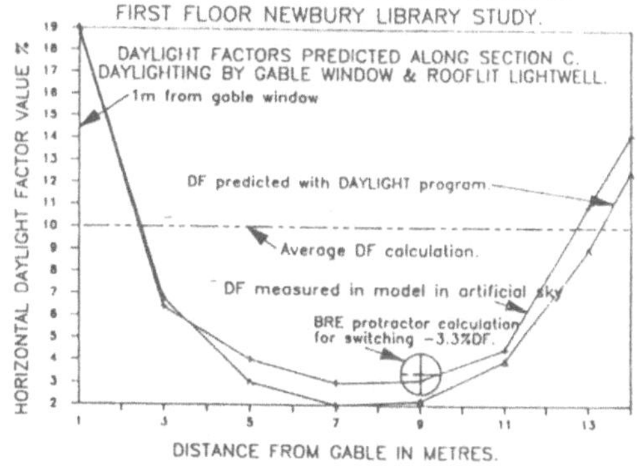

THE LEBENSGARTEN COMMUNITY CENTRE

Nick Baker, Cambridge Architectural Research and
The Martin Centre for Architectural and Urban Studies
6 Chaucer Rd, Cambridge, UK

There were two main areas of concern -

1) the lean-to glass house in relation to its use for horticulture and energy saving

2) the glazed attic space, shading, water-heating and ventilation.

1. THE GREENHOUSE

The importance of daylight availability to the growing areas was identified.Earlier proposed sections included too much obstruction.

A simple buoyancy model was used to calculate the openable area required for a tolerable temperature increment. This assumed "worst case" conditions - i.e. direct solar radiation on sloping surfaces of 780 w/m^2 and zero wind. This analysis suggests that with 20% transmittance shading and 2.8m^2 vents top and bottom, would be needed to prevent overheating. The shading serves the dual purpose of reducing solar thermal gain (i.e. air temperature), and also reducing radiant temperature and hence reducing the risk of leaf scorch. However, the shading must be movable in order to allow maximum transmission at times of low sky luminance.

The main energy saving function of the lean-to would be in the provision of pre-heated ventilation air to the house. One objection to this which was discussed, is the problem of humidity. However, for most of the heating season the temperature of the greenhouse would be well below that of the house, and this will ensure that the relative humidity of the pre-heated air when at room temperature, will be acceptable.In warmer weather this might be a problem, but then the green house could be more freely ventilated to reduce humidity.

For vent pre-heated air to be drawn into the house, there needs to be a passive extract in the house at high level, to ensure that the internal pressure remains negative with respect to the greenhouse. This would best extract from kitchen possibly via the stairwell, to a ridge terminal or high level Velux.

2. THE GLAZED ATTIC-SPACE (MEETING ROOM)

For the glazed attic space ventilation openings needed top and bottom - a total of 3.5m^2 each gives a 4°C temp rise assuming 25 people, plus projector, plus 20% of unshaded solar gains.This is rather large to accommodate conveniently, but is a result of the high internal gains and large solar gains. It was predicted using the buoyancy model. On days when there is wind, this will usually create a much larger ventilation rate for the same opening size, so the design conditions above may be considered to be a "worst case". Furthermore, temperature stratification in the room will result in the temperature of the air around seated occupants being lower by about 2°C than the average room temp.

In considering the benefit of the solar gains, it was observed that there is an absence of thermal mass, usually a requirement for the utilization of solar gains. Paving the floor close to the glazing, in order to increase thermal mass, could be considered if floor loadings permit.

Sunstrip absorber panels would be the best option for recovering heat from the shading devices, although these would not then be movable and at times when the solar gains to the room would be useful, they will reduce this. Further comments are indicated on the section shown in the figure.

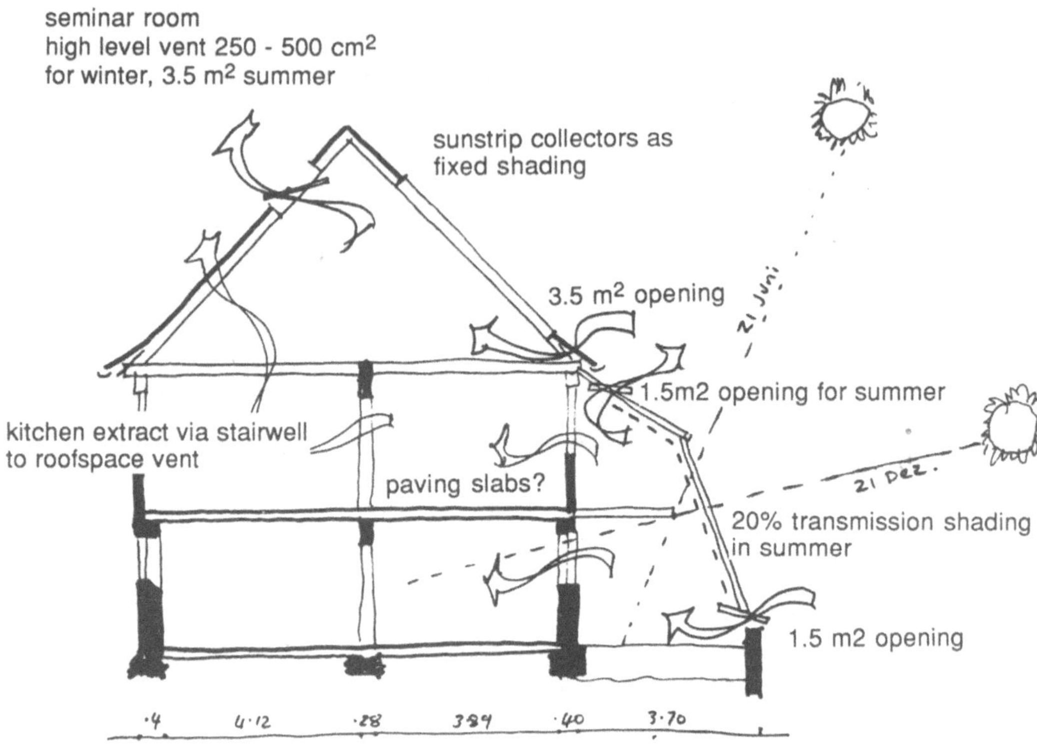

Figure 1 Section showing Lebensgarten greenhouse and seminar room

"HELLAS HOLIDAY HOTELS"

A.Argiriou, Protechna Ltd
Themistokleous 87, GR-10683 Athens,Greece

1. INTRODUCTION

"HELLAS HOLIDAY HOTELS" is a luxury hotel complex located in Agios Nikolaos, Crete (latitude 35, 1°N). The complex will operate mainly during summer. For this reason, the design of the complex was focused on passive cooling. Hybrid cooling techniques have also been used. Care was also taken in order to achive a satisfactory daylighting level in the buildings, without increasing the thermal gains.

The passive and hybrid cooling features of the complex are : shading, night vewtilation, evaporative cooling and air to ground heat exchangers. Rooms are day lit directly and inderectly and also by means of rooflights.

2. DESCRIPTION OF THE COMPLEX

The climate of Crete is characterized by high solar radiation throughout the year. The total annual solar radiation on a horizontal plane is close to 5.900 $MJ.m^{-2}$ [1]. The annual sunshine duration is of about 2.806 hours [2]. The mean monthly ambient temperature is 12,8^{O}C during January and 26,4^{O}C during July [2]. The maximum of maximum temperatures in July is close to 40°C and the mean maximum temperature during the same month is close to 30°C [3]. The annual cooling degree days (25°C base) are 200 [4], while the annual heating degree days (18°C base) are 700 [2]. The mean relative humidity during summer varies from 55 to 65 per cent. The corresponding mean maximum relative humidity varies between 70 to 85 per cent, while the mean minimum relative humidity varies from 35 to 40 per cent [5].

The complex covers about 180.000 m^2 and is adjacent to the bay. The whole site is inclined 20-30% to the north. The net floor area of the complex is 17.750 m^2 while the net volume is 84.370m^3. The complex consists of public facilities (restaurants, multipurpose hall, kindergarten, swimming pools and a health studio) together with 237 double bedrooms and 28 suites. The overall layout of the complex is illustrated in figure 1.

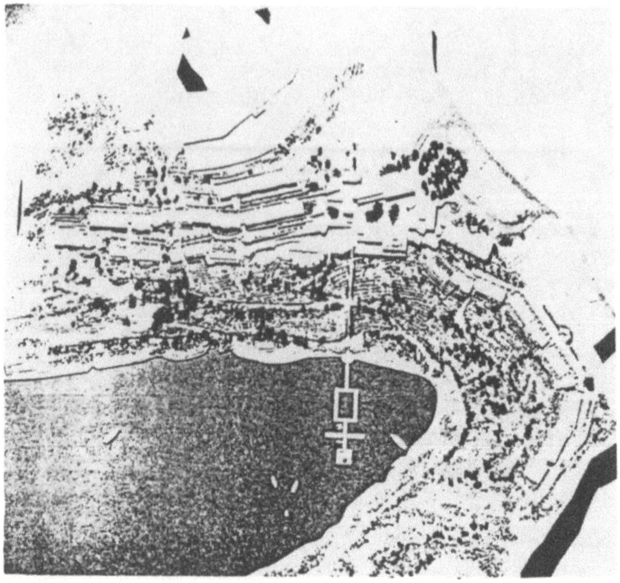

Figure 1. The maquette of the complex

3. PASSIVE FEATURES

Thermal gains from the environment are reduced by using an important insulation layer. The mean U-value for the roof is 0,32 $W.m^2.K^{-1}$ and for the floor 0,82 $W.m^{-2}.K^{-1}$. Various types of external walls are designed, their U-values ranging from 0,34 to 0,56 $W.m^{-2}.K^{-1}$. South opening are as small as possible for the same reason. Permanent or movable shading devices are used, in order to reduce the direct solar radiation impinging on the buildings during summer. Plantations are pergolas and deciduous trees further contribute to shading.

The microclimate is improved by means of evaporative cooling techniques. Evaporation is obtained through artificial fountains and water streams situated close to the buildings.Evaporation is also important because of the proximity of the sea and the existence of a swimming pool in the centre of the complex.

Internal heat gains due to artificial lighting are reduced. All rooms are mainly north daylit. Additional rooflights provide natural light into deep and otherwise unlit spaces.

Nocturnal cross or fan ventilation techniques are also used when the embient temperature is low during the night.

In spite of applying the passive cooling techniques mentioned above, an important cooling had still needs to be covered. This is achived by using air to ground heat exchangers.

4. AIR TO GROUND HEAT EXCHANGERS

The ground cooling system consists of 30 air to ground heat exchangers buried at a depth of 1,5m. Measurements of the ground temperature at various depths and simulations of the heat exchanger's performance have shown that their optimal operating conditions are obtained at this depth [6]. The area where are heat exchangers are buried is integrated in the complex; the surface is planted and used as an outdoor sitting area.

The lenght of each pipe is 30m and has a diameter of 20cm. The pipes are made of high resistant P.V.C. , in order to avoid corrosion and reduce the overall cost. The lower thermal conductivity of P.V.C. compared to metals does not affect significantly the overall heat transfer coefficient between earth and ground. This is due to the fact that conduction in the ground dominates the heat transfer process between air and ground, minimizing the influence of heat conduction through the surface of the pipe.

The pipes are placed in parallel, at a distance of 3m from each other. The pipes are slightly tilted in order to avoid accumulation of condensed water, which is evacuated through a small hole located at the lower elbow of the pipe. The pipes are first covered with a small layer of sand in order to obtain a better thermal contact between the pipe and a ground.

The 30 pipes are disposed in 6 groups of 5 six exchangers each. The pipes of each group have a common inlet and outlet opening. The air circulation is achived by a fan placed at the inlet of each group. The fan is chosen in order to have an air velocity of 5-8 $m.s^{-1}$ in the tubes. The outlet of each group is connected to the conventional heating and cooling system of the complex. Figure 2 illustrates the layout of the air to ground heat exchangers.

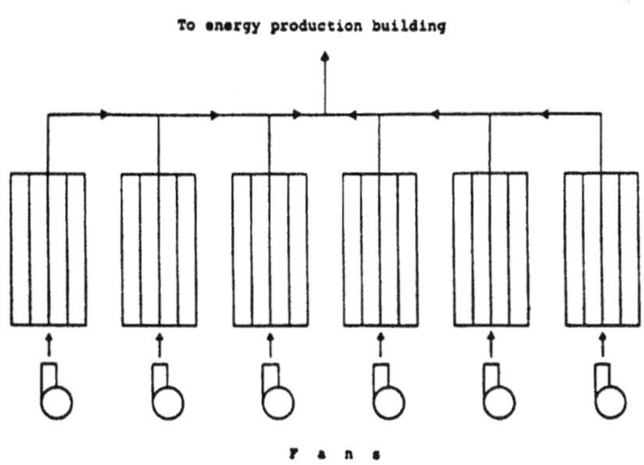

Figure 2. The layout of the air to ground heat exchangers

The outlet air temperature of the air to ground heat exchangers is calculated using the method proposed in [7]. The control and the coupling of the heat changers with the conventional heating and cooling system are described in [8].

5. THERMAL PERFORMANCE OF THE COMPLEX

Figure 3 illustrates the hourly inlet and outlet air temperature of an air to ground heat exchanger, for a typical day of June, July and August respectively. The temperature decrease

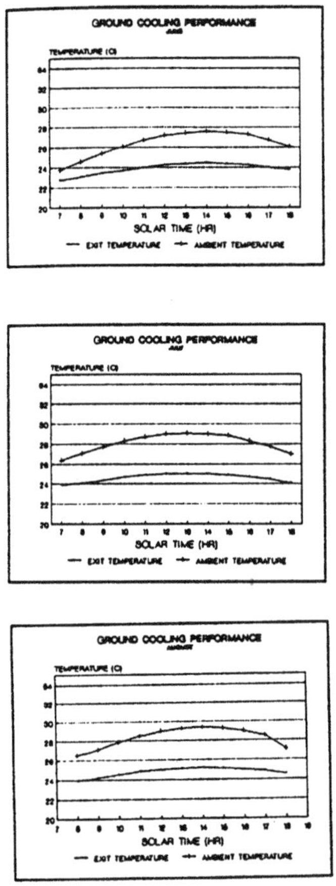

Figure 3. Inlet and outlet air temperature profiles

ranges from 2°C during the early morning hours to 4-5°C during noon. The thermal performance of the buildings of the complex has been evalutated, using a detailed simulation model in which all the passive components and techniques used are considered.

The contribution of the shading devices and of the ventilation techniques to the reduction of the monthly cooling load of the buildings is illustrated in figure 4. The main passive contribution

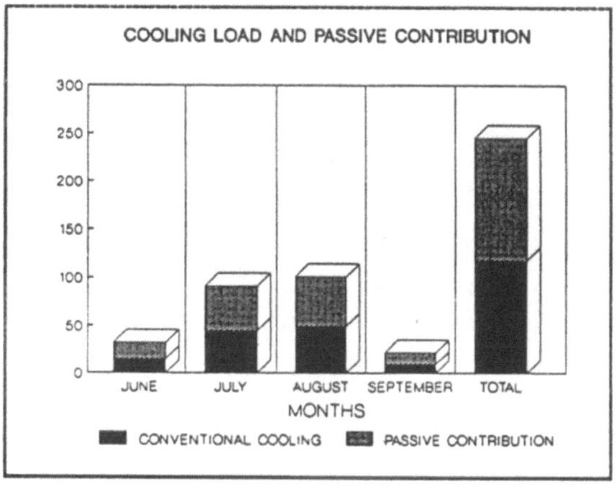

Figure 4. Monthly contribution of shading and ventilation

to the reduction of the cooling load is of about 52 per cent.

The contribution of the air to ground heat exchangers to the remaining cooling load is illustrated in figure 5. The mean annual contribution of the buried pipes is calculated to be of about 37 per cent of the remaining cooling load. The energy consumption of the fans used for the air circulation through the heat exchangers, is estimated to be around to 22 per cent of the cooling load provided by the pipes. The net energy gain from the ground cooling system is close to 29 per cent.

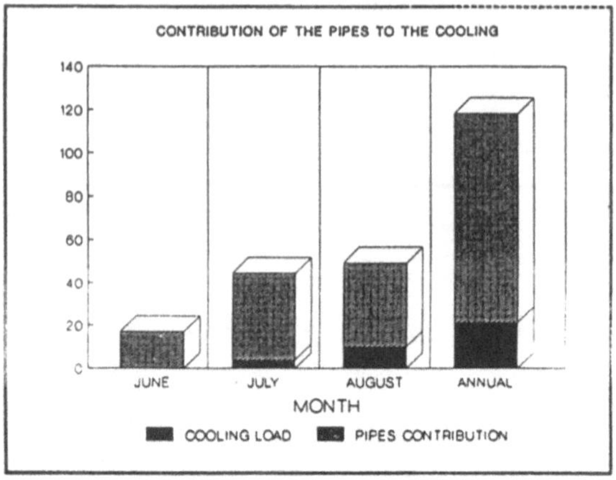

Figure 5. Contribution of the air to ground heat exchangers

Figure 6 illustrates the contribution of the passive systems to the heating load. The mean annual passive contribution to the heating load was calculated to be of about 82 per cent. Therefore the total annual contribution of the passive features to bothe heating and cooling is close to 80 per cent.

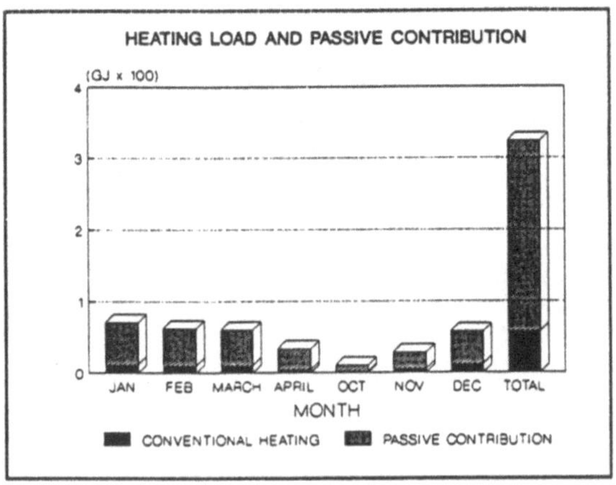

Figure 6. Contribution of the passive systems to the heating load

6. CONCLUSIONS

From the results briefly presented it is shown that the architectural design of the buildings of the "Hellas Holiday Hotels" tourist complex, toghether with the use of passive and hybrid cooling techniques lead to important energy savings. Especially the use of earth to air heat exchangers can contribute substantially to cover the necessary cooling load. However the use of such a system, requires a special control strategy in order to optimize the management of the available cooling sources.

References

1. M.Santamouris, Solar Radiation data of Greece, Athens 1988
2. A.Pelekanos and G.Papchristopoulos, Proc. of the 1st National Conference on New and Renewable Energy Sources, Vol. A, p. 41-75, Thessaloniki, 1982
3. D.A. Kouremenos and K.A. Papachristopoulos, Temperature characteristics of 35 Greek cities, Athens, 1984
4. National Meteorological Service, Unpublished work, 1989
5. D.A. Kouremenos and K.A. Antonopoulos, Relative Humidity in 40 Greek cities, Athens, 1985
6. M.Santamouris et al, Ibid 2, p. 107-114, 1982
7. H.J. Levit, R.Gaspar and R.D. Pieacentini, Agricultural and Forest Meteorology, 47, 31, 1989
8. A. Tombazis, A.Argiriou and M.Santamouris, Int. J. Solar Energy, 9, 1, 1990

DAYLIGHTING AND THERMAL REPORT FOR BARCELONA HOTEL

Orlando de Urrutia, architect.

Research team directed by Prof. Rafael SERRA
Escola T.S. d'Arquitectura de Barcelona (Univ. Polit.Cataluña)
Av. Diagonal, 649 08028 Barcelona (Spain)

1. INTRODUCTION

In this Building 2000 project a study of daylighting and thermal behaviour has been carried out with the collaboration of the author in order to facilitate him in the design of the project.

The study of the illumination conditions has been carried out in these two ways:

- Firstly the illumination level of the bedrooms has been tested in a model under exterior real sky conditions.
- Secondly the illumination conditions of the bathrooms and corridors achieved by sun-ducts has been calculated based on a graphic approximation and tested in a cross-sectional model of the ducts using simulated sun rays.

The thermal behaviour has been calculated based on a simplified algorithmic evaluation system inorderto calculate the interior thermal conditions of the rooms.

Throughout the design period the architect has been approised the global thermal behaviour of the building and particularly of the aspect of cooling by way of natural ventilation.

2. DAYLIGHTING REPORT

2.1 Design of sun-ducts

The sun-ducts have been designed in order to illuminate the bathrooms of the four floors of the building by means of natural light while, at the same time allowing the evacuation of air by these same ducts (natural ventilation).

The upper part of the duct is designed in a curved shape in order to allow maximum entrance of sunlight which is considered better than a non curved surface design so that in our design the height of the opening is three times greater than the width of the lower section. What is more, the fact that light is dispersed after the first reflexion guarantees a uniformed amount of light at the exit of the duct.

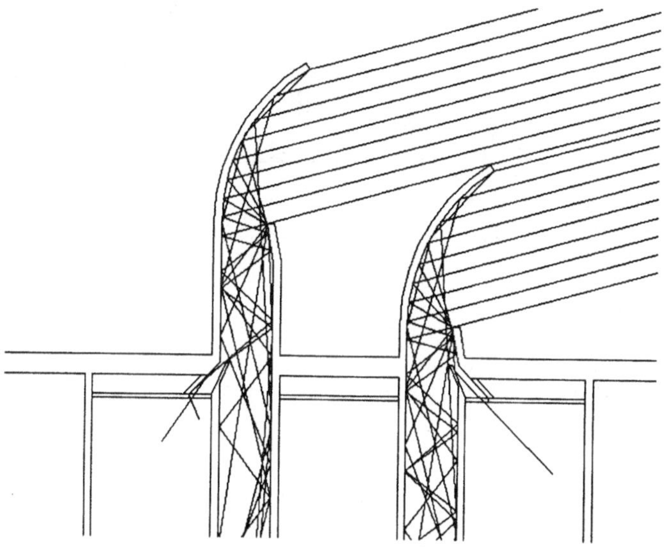

The section of the ducts is rectangular (1.9 m x 0.7 m). This section may be easily integrated into the layout of the hotel, and located between bathrooms and corridor. This situation permits maintenance to be carried out from these corridors.

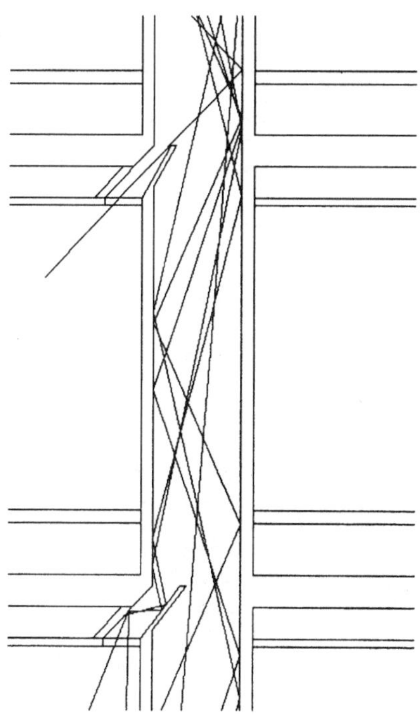

The light exits in each bathroom are of a lineal type and light passes through the ceiling into the bathroom with an aperture of 1.9 m x 0.3 m. From these apertures the connexion to the principal duct is made by a secondary duct of different dimensions.

The size of the secondary ducts has been calculated to guarantee the same quantity of light in each bathroom.

2.2 Sun-duct calculations

- Upper light collection duct: 1.9 m x 2.1 m = 4 m^2
- Duct section: 1.9 m x 0.7 m = 1.3 m^2

CALCULATION OF DAYLIGHTING THROUGH THE SUN-DUCT:

1) <u>Blue sky</u>:
Illuminance level of 50,000 lx on the vertical plane of the upper opening of the duct.
- Luminous flux entering the opening: 4 x 50,000 = 200,000 lm
- Luminous flux in the duct: 200,000 lm x 0,85 = 170,000 lm

<u>Evaluation of light in each exit</u>:
- <u>4th floor bathroom (with three reflexions on average)</u>:
  Exit dimensions: 1.9 m x 0.08 m = 0.15 m^2
  Flux entrance: F4 = 170,000 x 0.8 x 0.15/1.3 = 10,043 lm
  Flux remaining: 170,000 x 0.8 - 10,043 = 77,000 lm
- <u>3rd floor bathroom (with two additional reflexions on average)</u>:
  Exit dimensions: 1.9 m x 0.14 m = 0.27 m^2
  Flux entrance: F3 = 77,000 x $(0.8)^2$x 0.27/1.3 = 10,235 lm
  Flux remaining: 77,000 x $(0.8)^2$ - 10,235 = 39,045 lm
- <u>2nd floor bathroom (with two additional reflexions on average)</u>:
  Exit dimensions: 1.9 m x 0.27 m = 0.52 m^2
  Flux entrance: F2 = 39,045 x $(0.8)^2$ x 0.52/1.3 = 10,000 lm
  Flux remaining: 39,045 x $(0.8)^2$ - 10,000 = 15,000 lm
- <u>1st floor bathroom (with two additional reflexions on average)</u>:
  Exit dimensions = duct section
  Flux entrance: F1 = 15,000 x $(0.8)^2$ = 9,600 lm
- <u>Average value of illuminance in each bathroom (on an average of 10,000 lm)</u>:
  Utility factor with light wall surfaces = 0.52
  Useful surface area of bathroom = 3.75 m^2
  E = 10,000 lm x 0.52 / 3.75 m^2 = 1,390 lx

2) <u>Overcast sky</u>:
Illuminance level of 5,000 lx on the vertical plane of the upper opening of the duct, and the sequent decrease in the reflexion measurements obtained is proportionate and in ratio with the decrease in the illuminance level on the opening of the duct.
Results obtained: E = 139 lx

2.3 Sun-duct conclusions

1) <u>Illuminance Level with Direct Sunlight</u>:
In winter and summer a very high level of luminosity is produced, providing sufficient illumination for occupations and activities requiring a high degre of luminosity.
2) <u>Illuminance Level without Direct Sunlight</u>:
Even in winter and in the least favourable light conditions a sufficient level of luminosity is produced and the use of this light is further makimised by favouring the distribution of light in the area.

2.4 Sun-duct sections with different solar angles.

Scale: 1/200

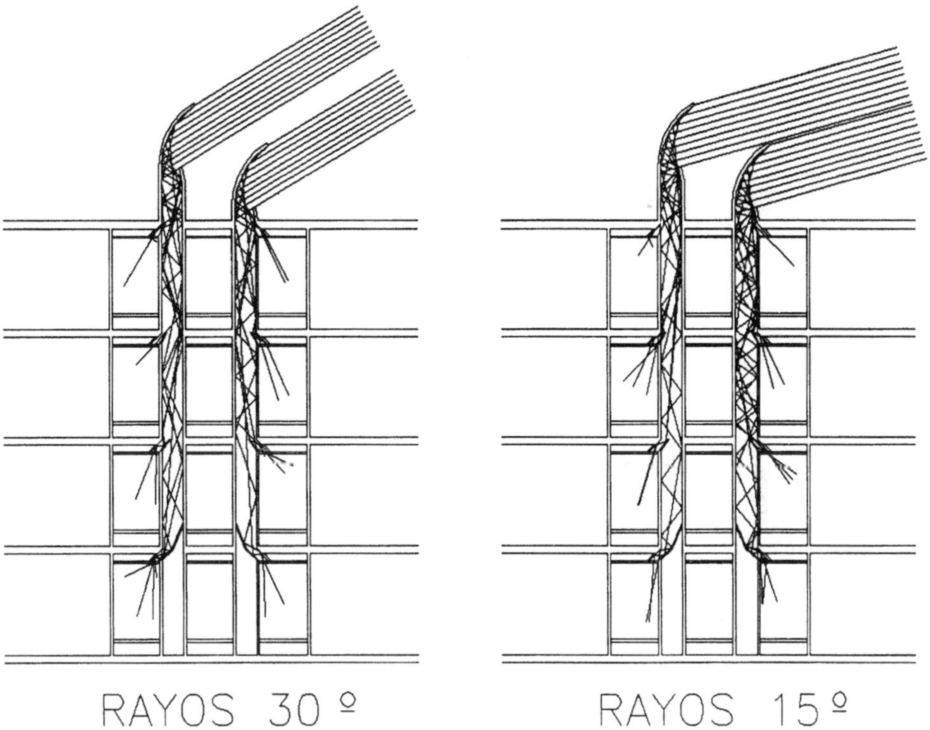

2.5 Measurement of daylighting levels in rooms.

These interior measurements have been carried out using a model which groups together a cell of three rooms.

- The measurements of this cell have been made with two distinct orientations:

  a) South façade (-20º) Av. Diagonal.
  b) West façade (-20º) secondary street.

- The measurements were taken with and without sun light simulating three different: winter and summer solstices and spring-autumn equinox at 12 o'clock solar time.

- For each orientation carried out on the cells:

  1) Measurements with transparent glass (simulating transparent windows).
  2) Measurements with translucent or light diffusing glass (simulating control elements on the windows).

2.6 Room lighting conclusions

1) <u>Measurements with blue sky</u>

a) The measurements with <u>transparent glass</u> demonstrate that for all kinds of conditions simulated the glass surfaces allow excessive light entry above all in the areas nearest to the façade. There is a large reduction of light level towards the interior which creates a sharp contrast.

b) The measurements with <u>translucent glass</u> demonstrate the necessity of a system of this type which permits the reduction of light level but which, at the same time will improve the quality of light that penetrates the interior. Also it should be taken into account that control elements can be applied (venetian blinds, etc...).
The measurements obtained display not only more suitable values in quantity but also a more homogeneous o distribution of light throughout the spaces.

2) <u>Measurements with overcast sky</u>

In an overcast sky situation the values are more homogeneous and sufficiently high for interior comfort. In any case, control elements continue to be recommendable.

2.7 Graphic representation of room measurements.

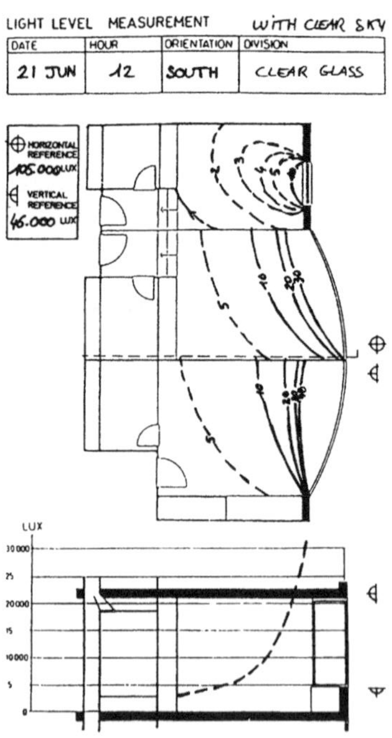

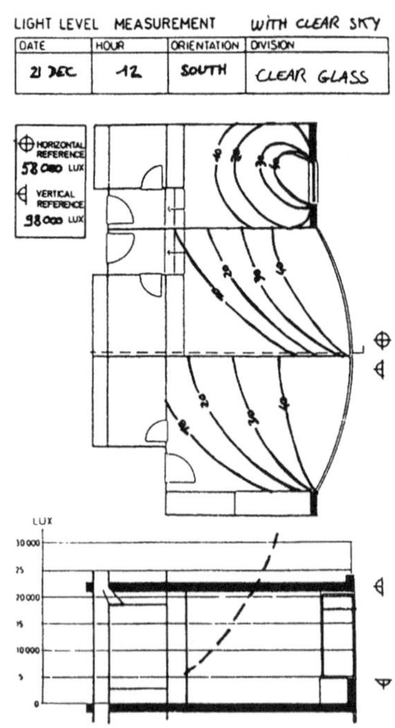

3. THERMAL REPORT

From the thermal point of view the critical elements of the building are to be found in the glass surfaces that separate the interior spaces from the exterior. This is the case of some of the intermediary rooms whose glass surfaces are greater than 50% of the floor surface.

The orientation of the apertures of the rooms south or to the north, exacerbates the problems of overheating in summer (in the case of southern orientation) and of excessive cold in winter (in the case of northern orientation).

To enable us to make a quantitative evaluation of the above mentioned factors we have calculated a thermal average for two central rooms situated on an intermediate floor on the south and north façades.

The characteristics of the exterior enclosures are as follow:
- wall with insulation in the interior
- water storage wall
- double glazing.

3.1 Thermal calculations.

G': Coefficient of thermal losses by unit of volume (without ventilation).

$$G' = 0.9 \text{ W/m3ºC}$$

- Winter : G ventilation = 0.34 W/m3 ºC
  G winter = 1.24 W/m3 ºC
- Summer : G ventilation = 2.72 W/m3 ºC
  G summer = 3.62 W/m3 ºC

We now apply the thermal balance equation to determine the interior temperature (Ti) obtained in the room (both North and South in winter and summer).

$$Ti = Te + (I + D) / G$$

Te = the average exterior temperature (ºC)
I = average radiation gain by unit of volume (W/m3)
D = average internal gain by unit of volume (W/m3)

We take the following values:

- Winter:	Te = 8 ºC	- Summer:	Te =	24 ºC
	I(South) = 17 W/m3		I(South) =	15.8 W/m3
			I(North) =	3.2 W/m3
	D = 2.5 W/m3		D =	2.5 W/m3

and we obtain:

- winter: Ti(South) = 23.7 ºC — Summer: Ti(South) = 29 ºC
- winter: Ti(North) = 10 ºC — Summer: Ti(North) = 25.6 ºC

Winter: We can observe the difference in temperature that exists between the bedroom of the North façade and that of the South. To homogenize the temperature of these two zones the surplus energy of the South façade is directed to the North.

The transmission of heat is carried out by air convection caused by the difference temperature inthe zones of the two façades. The convection circuit is effected by means of the upper and lower ducts that cross the corridor separating the two rows of rooms.

Summer: The rise in temperature in the South bedroom is critical.
This can be controlled only by way of solar protectors (venetian blinds, awnings, etc.) in order to reduce the transmission of heat and at the same time facilite air circulation.
In this way acceptable temperatures are obtained in both the North and South façade.

Up till now only daytime conditions have been taken into account. During the night the effect of thermic inertia must be considered in the opaque closings and the water storage wall.

After h hours the interior temperature will have suffered an increased calculated as follows:

$$Ti = (Ti - Te' - \frac{I' + D}{G}) \cdot (1 - e^{-hG/Md})$$

Where Te' : Average nightime temperature (ºC).
I' : Average solar gain by radiation during the night, by unit of volume due to thermic inertia (W/m3).
Md : Thermal mass by unit of volume that actuates during the nightime period (Wh/ºCm3).

$$Md = 30 \text{ Wh/ºCm3}$$

Winter :	Te'	=	4 ºC
	I'(south)	=	0
	h	=	16 hours
	Ti(south)	=	8.3 ºC
	Ti(north)	=	2.2 ºC

Summer :	Te'	=	21 ºC
	I'(south)	=	0
	h	=	8 hours
	Ti(south)	=	4.5 ºC
	Ti(north)	=	2.4 ºC

3.2 Thermal conclusions

The elevated temperature and oscillation of temperature value which is produced in winter in the South façade is due to the fact that the temperature rises because of the effect of direct solar gain throughout the day while during the night when there is no sun the temperature drops. A redistribution of solar energy will lessen the interior oscillation temperature value in the South façade, there by not only decreasing overheating but also allowing better energy storage which will in part compensate for the nightime drop and better the averag temperature.

In short, at the moment temperature differences arise between the North and South façade. We believe it is an opportune moment to provide systems which would help to decrease these differences.

Here are some possible solutions:

A) Winter:

- Produce air circulation through double glazing winter taking advantage of the air that we have preheated in the rooms of the North façade.
- We can take advantage of the heated air in the double-glazing of the South façade to maintain normal room temperature by way of passing air through the double-glazing to the North façade.

B) Summer:

- Protection from solar radiation (for example venetian blinds, awnings, etc.).
- Increasing the circulation of fresh air in the rooms (with the possibility of further increasing it during very hot periods).
- For South façade rooms if protection is installed in the space between the double glazing, sufficient ventilation can be obtained through this space.

ENERGY CONSERVATION STRATEGIES AND ASSESSMENT OF A TOURIST RESIDENCE COMPLEX

Koronakis P.
Pr. Mech.Engrng, Technological Education Institute of Piraeus
250 Thevon & P.Ralli st., 12244 Aegaleo, Athens, Greece

ABSTRACT

This work presents the results of a case study developed with the aid of a number of computer programs used for design purposes. Feeling strong about energy conservation and the creation of a attractive-comfortable habitat, a certain number of design strategies and techniques have been studied regarding the buildings concerned, an all-year-around tourist residence complex located in a rural area 4 km outside the port city of Hania in Crete. The right architecture and structural desing of the buildings, the appropriate for the local climate energy design strategies followed, and the convenient site morphology mark this tourist complex for its low to nil auxiliary energy demand throughout the year. Thus, winter season heating requires only 30% of what it would have be needed otherwise, while summer cooling could alltogether be abolished if users were satisfied with thermal comfort PMV index not higher than 0.6.

1. INTRODUCTION

Estimation of the various passive and hybrid components contribution, and system performance assessment at the room or building level, are the aspects been examined regarding,

- Summer cooling and winter heating
- Daylighting performance
- Thermal comfort status
- Further energy saving design improvements

In order to proceed with calculations the weather dara file of Hania had to be constructed, in a proper form, and as it was required by the software used. Solar radiation intensities on the main vertical orientations were computed from hourly global solar radiation measurements on the horizontal provided in [1] using the Orgill and Holland [2] model and assuming a uniform sky diffuse solar radiation. Resulting values as well as other useful weather data are shown in Table 1.

2. BUILDING AND SITE DESCRIPTION

The hotel complex consists of three heavy wall, partly set in earth multilevel buildings built around a central open space plaza where the swimming pool is located. Two of them buildings, A and B, are for residence housing all apartment units, while building C accomodates the main reception and all catering facilities. The building site is located on the south side of a small hill with the buildings following the slopping terrain totally unobstructed.

Building A is a three level construction housing seven double room apartments covering 300m^2. Building B is a four level construction housing twenty double and single room units covering a total of 674 m^2. Building C covers 187 m^2 distributed in three levels. All rooms are faced due south, each one with a south door+window combination providing 4.7 to 6.0 m^2 of double pane clear glass glazing. In addition, some rooms are added with a glazed door placed on the eastern or northern wall,besides some units been designed with a typical bathroom window. Light color venetian blinds are placed on the inside of all south facing glazed surfaces in order to reflect natural directing it to the white painted ceiling and walls in addition, all glazed surfaces are protected with curtains which could be used to provide privacy, inhibit sun rays, or reduce thermal loss,according to what is needed at the time. Horizontal awnings consisted of retractable overhangs, made of white tent canvas, extended down the whole lenght of the south facade during the summer, late spring, and early autumn months in a full or half drawn mode, shading the sun and inhibiting overheating.

3. SUMMER COOLING SCHEMES AND BUILDING PERFORMANCE ASSESSMENT

With hotel residents dressed in season, relative humidities (R.H.) up in the 50's and temperatures going over the 28 oC mark, cooling has to be provided in order to bring thermal comfort back to normal. The methods and design techniques adopted in this work providing cooling to all units follow a pattern presented in the chart below.

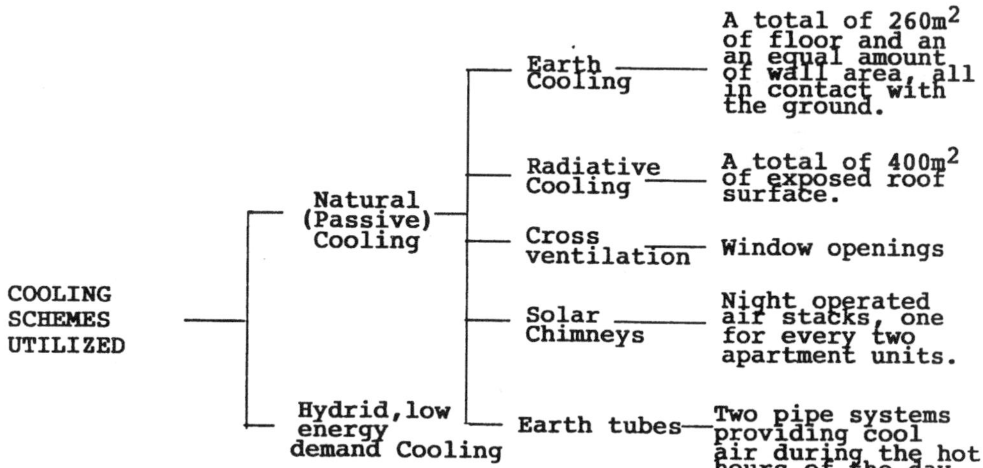

With floor and wall U-values (only surfaces in contact with the ground) of 1.94 W/m^{2}K and 0.65 W/m^{2}K, respectively, the average monthly amount of heat escaping to the earth for the months of June, July, August and September (J,J,A,S), for the builgings concerned, is estimated at 6.5, 7.5, 6.0, and 2.5 GJ, respectively. Figure 1 gives the rate of heat dissipating to the possibility of producing extremely humid air (R.H>85%) at the outlet. When the system operates normaly (convective cooling only), conditioned air's characteristics are found using the corresponding charts in [3]. For building A, the earth tube system consists of two parallel, 0.2 m diameter, 25 m long PVC tubes, while for B, the two parallel PVC tubes have a 0.4 m diameter, and a 60 m length. Both pipe systems are buried 4 m beneath the surface, sloping at 1%. Figure 4 shows the designed system's performance in terms of inlet-outlet air temperatures. Calculations of the average monthly amounts of heat both earth tube systems (buildings A and B together) could remove from fresh air, when they are operated for 6 hours every day (i.e. noon to 6 pm), give, in GJ, 11, 12, 12, 11 and 8 for the months J, J, A and S, respectively.

A study for the calculation of the average monthly cooling loads performed on A, B and C buildings, based on 25 oC temperature and 2 ACH, has revealed that for the months J, J, A and S, the corresponding cooling load values are 52.4, 70.5, 77.5 and 21.2 GJ, (total for A+B+C). Table 2 presents contributions to cooling by each component separately and together, as well as percentagewise, for all months concerned. Similary, Figure 5 shows the % contribution to cooling of each component but for the whole summer season. As it can be seen, passive and hybrid cooling could cover more than 52% of the cooling load on a monthly basis, and 64.5 % seasonaly, when thermostat is set at 25 oC. In case internal temperatures of up to 27 oC, or 28 oC are acceptable then no auxiliary cooling (i.e. air conditioning) is ever necessary.

4. HEATING SEASON THERMAL PERFORMANCE OF BUILDINGS

Estimation of building's thermal performance during each heating season month was based on results from the Programme TCM-heat, a simplified design tool. Heating of the units is considered continuously supplied by radiators and the room temperature controlled by thermostats set at 21oC. Computer output results of monthly values of solar gains, heat losses, (auxiliary) heat demad for all buildings are presented in Tables 3, 4 and 5. Figure 6 gives the annual % contribution to heating of the solar, casual and auxiliary energy utilized. A significant, almost 70% solar contribution shows the important role to energy saving that passive solar systems could play in the National, and World economy, as well as the environment, when they will be adopted by the building construction industry.

5. DAYLIGHTING ASSESSMENT OF APARTMENT UNITS

Checking a hotel room for daylighting conditions it might seem needless and unnecessary unless, of course, rooms are rented during off season months to students, teachers, etc on a regular basis, as it is in this case. Daylighting performance of each type of room space available in the hotel was investigated using the Programme MICROLITE version 1.0. Output results in terms of Daylight factor values for the three types of rooms considered are earth from all concerned surfaces of the buildings as a function of the room and ground temperatures.

The average radiative night specific heat loss from the roof, for the months J, J, A and S, is estimated at 28, 35, 30 and 25 W/m^2, respectively. Thus, the average radiative heat losses for each one of these months amount to 9, 12, 12 and 6 GJ, respectively. It should be pointed out that by adding the last two natural cooling effects to the hotel building cell structure considered having 140 Wh/m^2K thermal capacity, the resulting temperature drop is estimated at 2 K, at least.

Cross ventilation is enhanced by window openings located on opposite walls occupying 15% to 20% of the available window area. Such an arrangement could create an air stream able to provide up to 50 ACH (Air Changes per Hour), mainly depending upon the suction effect created by the wind on the low pressure side opening. Cross ventilation could greatly improve comfort during warm and humid nights. However, ventilation cooling is rather limited to short time intervals during a normal summer day period, or when favorable winds prevail, since the site is inside the wake of the natural hill barrier when the cool north summer winds blow (mainly during the day).

Rectangularly shaped solar chimneys with heavy concrete walls, the three thermally insulated and the fourth painted black and single glazed, extend vertically over the roof of each floor with the larger side facing west to south west. This is in accordance with the fact that during the month of July a vertical surface facing west would receive more solar energy than any other chosen orientation, throughout the Mediterranean zone latitudes. Figure 2 shows a cross section element of such a chimney, with thermal analysis and design calculations given in [3]. The particular solar chimneys are designed to fuction for 12 consecutive hours during each summer night moving air up their stacks at an average speed of 0.1 m/s. The total amount of heat removed by all chimneys, during each monthly period, for the months, J, J, A and S is estimated at 7.5, 9.5, 10.5 and 9.0 GJ, respectively.

The earth tube air cooling system has been designed to supply up to 3 ACH of conditioned air during the hot hours of the day (i.e. noon to 6 pm). The pipe system can supply air to the building according to a dual operational functioning mode: It can cool the flowing inside the tube air by convection only (regular earth tube) operational mode), where air loses sensible heat to the ground, or it can cool the moving air by convection and evaporation if air is forced to flow over the free surface of a (lower temperature) water body, the latter being the case when low R.H. (<55%) values prevail. Figure 3 shows the outlines of such a pipe system with air and water streams flowing inside the buried tube. The idea is to cool air by combined sensible and evaporative cooling, thus, reducing the required pipe length as much as possible, while producing air of thermal comfort value equal to that of air cooled convectively only inside an equal diameter but longer tube. When circumstances arise, water flow to the pipe can be stopped and the pipe be drained completely by gravity. This should occur automatically when R.H. of inlet air becomes higher than 55%, thus avoiding the shown in Figure 7. The 21st of December, noon time with average cloudy conditions was taken as the model day for calculations.

6. THERMAL COMFORT ASSESSMENT

Building's thermal comfort assessment for each month of the year was based on the Predicted Mean Vote (PMV) index value calculated using the Bruel & Kjaer Demonstration Program of Comfort Equation. Results of PMV index values were found after feeling the program with appropriate for each month values of air and radiant temperatures, R.H., air speed, thermal protection provided to tenants by the clothing they wear, and finally the human activity level, PMV index values thus obtained never surpassed the 0.40 and the -0.42 mark. Results have also been obtained by considering extreme cold and extreme warm conditions that tenants possibly could experience inside their rooms. In this case the upper and lower limits were 0.62 and -0.84, respectively. Thus PMV was always between -1.0<PMV<1.0. Figure 8 gives the average PMV index values expected inside the rooms, under normal conditions.

7. SENSITIVITY STUDIES, FURTHER ENERGY SAVINGS

So far, the shown thermal performance evalutation of the buildings was based on constructional details and parameter values according to original plans. It is possible, though, to improve energy saving by altering some key factors. So, by varying the south facing glazed surface area, the number of ACH and the overall U-value of floors in contact with the ground, one could observe significant changes towards higher energy savings. Table 6 shows the resulting new values as they change after each step.

REFERENCES

1. Measurements for the Development of Solar and Aeolic Potential of Greece for Energy Purposes.Scientific Publication of the Public Power Co. (PPC) of Greece. Vols I, II, III, and IV, (1976-1981).
2. Orgill, J.F. and Hollands, K.G.T., Correlation Equation for Hourly Diffuse Radiation on a Horizontal Surface, Solar Energy, vol. 19, p.357, 1977
3. Koronakis, P., Altrnative Cooling Schemes and Use of Earth Tubes and Solar Chimneys, 2nd European Symposium on Soft Energy Sources and Systems at the Local Level, Hania, Crete, Greece, October 1989.

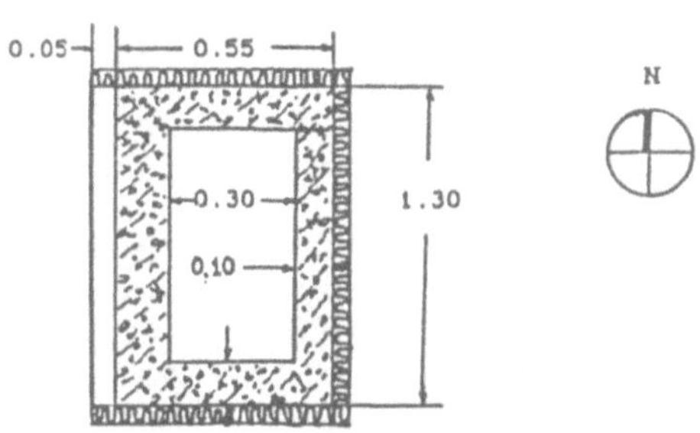

Figure 2. Horizontal cross section characteristics of a solar chimney. Dimensions in m.

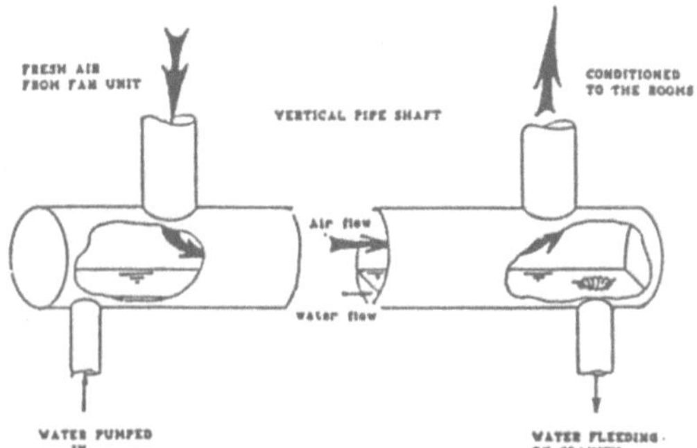

Figure 3. Outline of the earth tube system

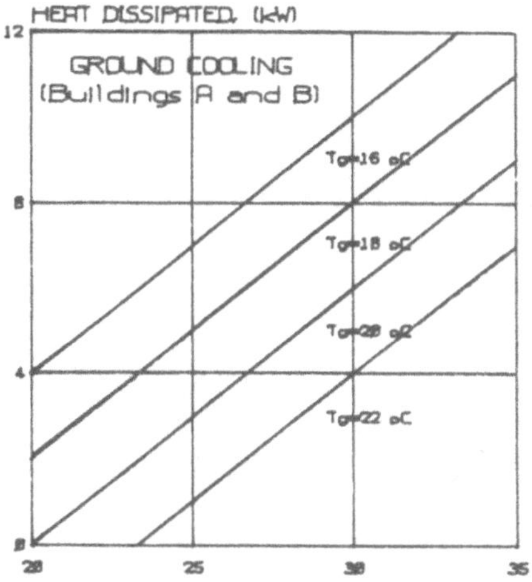

Figure 1. Rate of Heat Dissipated

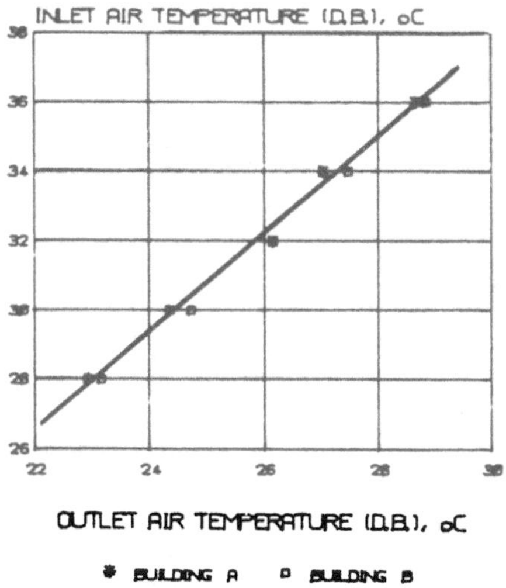

Figure 4. Earthtube System Performance

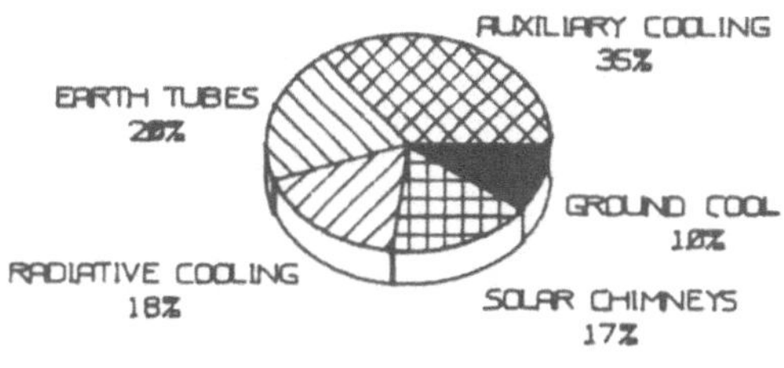

Figure 5. Annual Contribution to Cooling

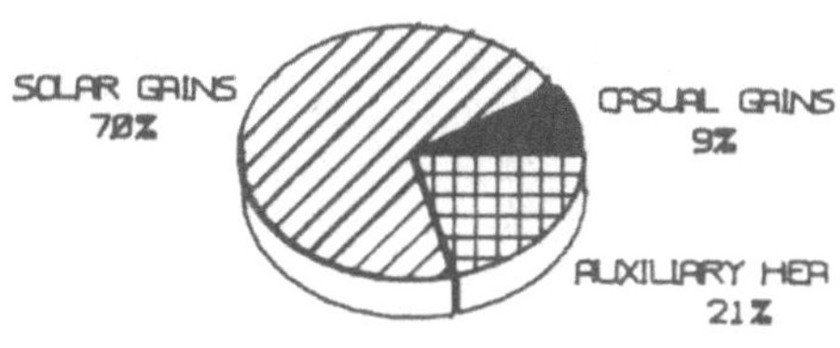

Figure 6. Annual Contribution to Heating

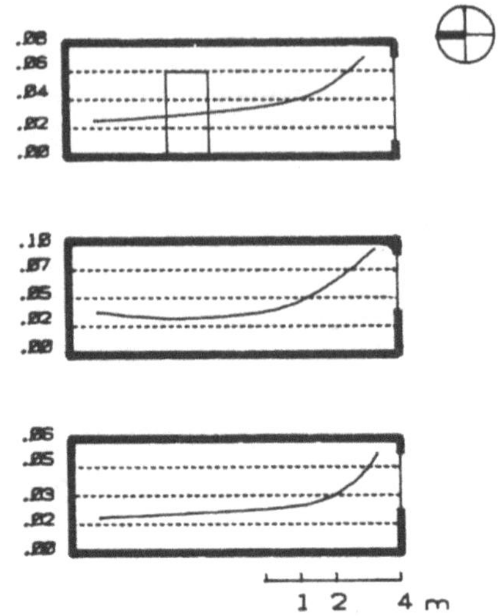

Figure 7. Daylight factor inside each type of room corresponding to the 21st of December, noon time, cloudy day

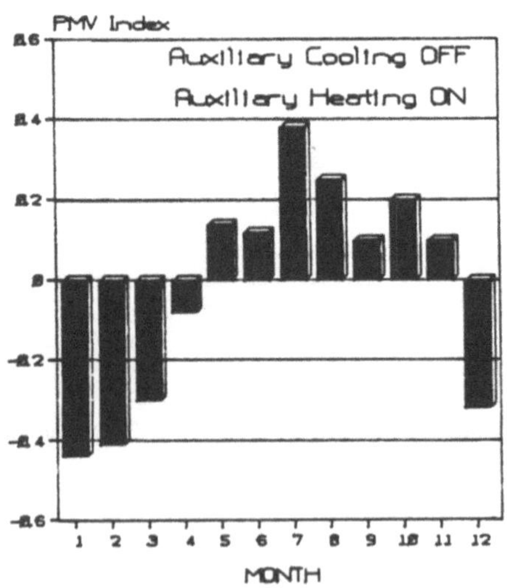

Figure 8. Estimated mean PMV values

Table 1. Monthly average climatic data for Hania Latitude 35°30' N, Longitude 24°02' E

MONTH	AIR TEMPER (oC)	R.H. (%)	V.P. (Pa)	D/LIGHT HOURS (hr)	S/SHINE HOURS (hr)	DEGREE DAYS (18oC)	FRACTION EV.+NIGHT DEG/HOURS
JAN	11.9	73	971	10.5	3.5	188	0.62
FEB	12.2	70	984	11.0	4.5	156	0.61
MAR	13.4	68	1104	12.0	5.8	143	0.64
APR	16.4	67	1224	13.1	7.4	104	0.86
MAY	20.4	64	1543	14.0	10.3	9 (7)	-
JUN	24.7	59	1809	14.4	11.7	- (26)	-
JUL	26.9	57	1969	14.0	12.2	- (41)	-
AUG	26.6	58	1955	13.2	11.4	- (41)	-
SEP	23.6	65	1822	12.0	9.0	- (17)	-
OCT	20.0	70	1610	11.0	5.8	9	0.87
NOV	16.8	73	1277	10.0	5.7	50	0.78
DEC	13.5	73	1117	9.6	3.6	123	0.63

( ) Values in parentheses denote COOLING deg/days,(25oC)

Table 2. Contribution to cooling by individual system components (passive, hybrid), in GJ/mo

COOLING SYSTEM	JUN	JUL	AUG	SEP	SEASON SUM
Ground Cooling	6.5	7.5	6.0	2.5	22.5
Radiative Cooling	9.0	12.0	12.0	6.0	39.0
Solar Chimneys	7.5	9.5	10.5	9.0	36.5
Earth Tubes	12.0	13.0	12.0	8.0	45.0
CUMULATIVE	35.0	42.0	40.5	25.5	143.0
COOLING LOAD	52.4	70.5	77.5	21.2	221.0
% CONTRIBUTION TO COOLING LOAD	66.8	59.6	52.3	100	64.5

Table 3. Heating season average monthly values of thermal parameters for buildings A and B

MONTH	INTERNAL TEMPERAT (STND DEV) (oC)	HEAT LOSSES (GJ)	CASUAL GAINS (GJ)	SOLAR GAINS (GJ)	AUX HEATING (GJ)	% CONTRIB TO HEATING: CASUAL GAINS	SOLAR GAINS	AUX HEAT
OCT	27.3 (1.6)	73	7.1	65.9	0	9.7	90.3	0
NOV	23.6 (1.3)	65	7.1	57.5	0.4	10.9	88.5	0.6
DEC	21.5 (0.3)	80	7.1	50.2	22.7	8.9	62.7	28.4
JAN	21.3 (0.2)	94	7.1	48.2	38.7	7.5	51.3	41.2
FEB	21.4 (0.3)	83	7.1	46.3	29.6	8.5	55.8	35.7
MAR	21.6 (0.4)	82	7.1	54.7	20.2	8.6	66.7	24.6
APR	22.8 (1.0)	61	7.1	52.0	1.9	11.6	85.2	3.1
SEASON'S TOTAL		538	50	374	114	9.3	69.5	21.2

Table 4. Thermal performance of buildings A and B after a string of consecutive cold and cloudy days (heating season)

MONTH	THERMOST SET TEMP (oC)	AMBIENT TEMPER (oC)	HEAT LOSSES (MJ/d)	CASUAL GAINS (MJ/d)	AUX HEAT (MJ/d)
OCT	21.0	16.5	1493	280	1213
NOV	21.0	12.8	2661	280	2381
DEC	21.0	9.5	3694	280	3414
JAN	21.0	7.9	4200	280	3920
FEB	21.0	8.2	4103	280	3823
MAR	21.0	10.0	3535	280	3255
APR	21.0	13.0	2579	280	2299

Table 5. Thermal performance of buildings A and B after a string of consecutive warm and sunny days (heating season)

MONTH	INTERNAL TEMPER (oC)	AMBIENT TEMPER (oC)	HEAT LOSSES (MJ/d)	CASUAL GAINS (MJ/d)	AUX HEAT (MJ/d)
OCT	27.8 (1.8)	22.0	2570	234	0
NOV	24.3 (1.6)	18.8	2471	234	0
DEC	22.3 (1.0)	15.5	2875	234	428
JAN	21.9 (0.7)	13.9	3289	234	785
FEB	22.0 (0.7)	14.2	3185	234	629
MAR	22.1 (0.8)	15.4	2791	234	400
APR	22.9 (1.1)	18.4	2004	234	0

Table 6. Tabulation of sensitivity analysis results after a series of modifications, on a seasonal basis

No.	CONSIDERED MODIFICATION	HEAT LOAD (GJ/pa)	CASUAL GAINS (GJ/pa)	SOLAR GAINS (GJ/pa)	AUX HEAT (GJ/pa)	NEW AUX HEAT VS ORIGINAL (%)
1.	Original design	538	50	374	114	100.0
2.	As (1) plus 20% more S. glazing	584	50	440	94	82.4
3.	As (2) + floors U=0.6 W/m^2K	562	50	440	72	63.1
4.	As (3) but 1 ACH to unoccupied, 1.5 to occupied	528	50	440	38	33.3

FSC
www.fsc.org
MIX
Papier | Fördert gute Waldnutzung
FSC® C083411

Kolibri 360 GmbH
Ferdinand-Jühlke-Straße 7
99095 Erfurt, Deutschland
produktsicherheit@kolibri360.de